Texture and Colour in Image Analysis

Texture and Colour in Image Analysis

Editors

Francesco Bianconi
Antonio Fernández
Raúl E. Sánchez-Yáñez

MDPI • Basel • Beijing • Wuhan • Barcelona • Belgrade • Manchester • Tokyo • Cluj • Tianjin

Editors

Francesco Bianconi
Department of Engineering
Università degli Studi di Perugia
Perugia
Italy

Antonio Fernández
School of Industrial Engineering
Universidade de Vigo
Vigo
Spain

Raúl E. Sánchez-Yáñez
Department of Electronic Engineering
Universidad de Guanajuato
Salamanca
Mexico

Editorial Office
MDPI
St. Alban-Anlage 66
4052 Basel, Switzerland

This is a reprint of articles from the Special Issue published online in the open access journal *Applied Sciences* (ISSN 2076-3417) (available at: www.mdpi.com/journal/applsci/special_issues/texture_colour_image_analysis).

For citation purposes, cite each article independently as indicated on the article page online and as indicated below:

LastName, A.A.; LastName, B.B.; LastName, C.C. Article Title. *Journal Name* **Year**, *Volume Number*, Page Range.

ISBN 978-3-0365-1378-2 (Hbk)
ISBN 978-3-0365-1377-5 (PDF)

© 2021 by the authors. Articles in this book are Open Access and distributed under the Creative Commons Attribution (CC BY) license, which allows users to download, copy and build upon published articles, as long as the author and publisher are properly credited, which ensures maximum dissemination and a wider impact of our publications.

The book as a whole is distributed by MDPI under the terms and conditions of the Creative Commons license CC BY-NC-ND.

Contents

About the Editors vii

Preface to "Texture and Colour in Image Analysis" ix

Francesco Bianconi, Antonio Fernández and Raúl E. Sánchez-Yáñez
Special Issue Texture and Color in Image Analysis
Reprinted from: *Applied Sciences* **2021**, *11*, 3801, doi:10.3390/app11093801 1

Carlos F. Navarro and Claudio A. Perez
Color–Texture Pattern Classification Using Global–Local Feature Extraction, an SVM Classifier, with Bagging Ensemble Post-Processing
Reprinted from: *Applied Sciences* **2019**, *9*, 3130, doi:10.3390/app9153130 5

Fabrizio Smeraldi, Francesco Bianconi, Antonio Fernández and Elena González
Partial Order Rank Features in Colour Space
Reprinted from: *Applied Sciences* **2020**, *10*, 499, doi:10.3390/app10020499 25

Dengyong Zhang, Shanshan Wang, Jin Wang, Arun Kumar Sangaiah, Feng Li and Victor S. Sheng
Detection of Tampering by Image Resizing Using Local Tchebichef Moments
Reprinted from: *Applied Sciences* **2019**, *9*, 3007, doi:10.3390/app9153007 39

Mingxing Tang, Zhen Huang, Yuan Yuan, Changjian Wang and Yuxing Peng
A Bounded Scheduling Method for Adaptive Gradient Methods
Reprinted from: *Applied Sciences* **2019**, *9*, 3569, doi:10.3390/app9173569 49

David González-Patiño, Yenny Villuendas-Rey, Amadeo-José Argüelles-Cruz and Fakhri Karray
A Novel Bio-Inspired Method for Early Diagnosis of Breast Cancer through Mammographic Image Analysis
Reprinted from: *Applied Sciences* **2019**, *9*, 4492, doi:10.3390/app9214492 65

Subrata Bhattacharjee, Hyeon-Gyun Park, Cho-Hee Kim, Deekshitha Prakash, Nuwan Madusanka, Jae-Hong So, Nam-Hoon Cho and Heung-Kook Choi
Quantitative Analysis of Benign and Malignant Tumors in Histopathology: Predicting Prostate Cancer Grading Using SVM
Reprinted from: *Applied Sciences* **2019**, *9*, 2969, doi:10.3390/app9152969 81

Christos Bontozoglou and Perry Xiao
Applications of Capacitive Imaging in Human Skin Texture and Hair Analysis
Reprinted from: *Applied Sciences* **2019**, *10*, 256, doi:10.3390/app10010256 97

Rafał Obuchowicz, Karolina Nurzynska, Barbara Obuchowicz, Andrzej Urbanik and Adam Piórkowski
Use of Texture Feature Maps for the Refinement of Information Derived from Digital Intraoral Radiographs of Lytic and Sclerotic Lesions
Reprinted from: *Applied Sciences* **2019**, *9*, 2968, doi:10.3390/app9152968 111

Rocco Furferi, Lapo Governi, Luca Puggelli, Michaela Servi and Yary Volpe
Machine Vision System for Counting Small Metal Parts in Electro-Deposition Industry
Reprinted from: *Applied Sciences* **2019**, *9*, 2418, doi:10.3390/app9122418 125

Yang Liu, Ke Xu and Jinwu Xu
An Improved MB-LBP Defect Recognition Approach for the Surface of Steel Plates
Reprinted from: *Applied Sciences* **2019**, *9*, 4222, doi:10.3390/app9204222 . . . . **139**

Lei Geng, Qinglei Meng, Zhitao Xiao and Yanbei Liu
Measurement of Period Length and Skew Angle Patterns of Textile Cutting Pieces Based on Faster R-CNN
Reprinted from: *Applied Sciences* **2019**, *9*, 3026, doi:10.3390/app9153026 . . . . **153**

Hongdong Wang, Meng Lei, Yilin Chen, Ming Li and Liang Zou
Intelligent Identification of Maceral Components of Coal Based on Image Segmentation and Classification
Reprinted from: *Applied Sciences* **2019**, *9*, 3245, doi:10.3390/app9163245 . . . . **169**

Jie Yu, Youxin Lin, Yanni Zhu, Wenxin Xu, Dibo Hou, Pingjie Huang and Guangxin Zhang
Segmentation of River Scenes Based on Water Surface Reflection Mechanism
Reprinted from: *Applied Sciences* **2020**, *10*, 2471, doi:10.3390/app10072471 . . . . **185**

Loris Nanni, Andrea Rigo, Alessandra Lumini and Sheryl Brahnam
Spectrogram Classification Using Dissimilarity Space
Reprinted from: *Applied Sciences* **2020**, *10*, 4176, doi:10.3390/app10124176 . . . . **203**

Cefa Karabağ, Jo Verhoeven, Naomi Rachel Miller and Constantino Carlos Reyes-Aldasoro
Texture Segmentation: An Objective Comparison between Five Traditional Algorithms and a Deep-Learning U-Net Architecture
Reprinted from: *Applied Sciences* **2019**, *9*, 3900, doi:10.3390/app9183900 . . . . **221**

Marco Buzzelli
Recent Advances in Saliency Estimation for Omnidirectional Images, Image Groups, and Video Sequences
Reprinted from: *Applied Sciences* **2020**, *10*, 5143, doi:10.3390/app10155143 . . . . **235**

About the Editors

Francesco Bianconi

Francesco Bianconi received his MEng from the University of Perugia, Italy, and his PhD in computer-aided design from a consortium of Italian universities. He has been a visiting researcher with the University of Vigo, Spain; the University of East Anglia, U.K.; Queen Mary University of London, U.K.; and City, University of London, U.K. He is currently an associate professor with the Department of Engineering, University of Perugia, where he conducts research on computer vision, image processing, and pattern recognition, with special focus on texture and colour analysis for industrial and biomedical applications. Prof. Bianconi is an IEEE senior member, chartered engineer, and court-appointed expert; has been a TPC/IPC member of more than 30 international conferences and symposia; and is currently a member of the editorial board of two scholarly journals.

Antonio Fernández

Antonio Fernández graduated as an industrial engineer (equivalent to an MEng degree) in 1993 and obtained his PhD in 1998 from the University of Vigo, Spain, with a thesis entitled "Development of pulsed TV-holography techniques for the analysis of transient wave propagation in mechanical parts". He joined the Applied Physics Department at the University of Vigo in 1993 and then the Department of Engineering Design as an associate lecturer. He is now a senior lecturer in the same department. Prof. Fernández teaches Engineering Graphics, Computer Programming using Python, and Image Processing. His research interests include computer vision, image processing, and pattern recognition, with a special focus on texture analysis.

Raúl E. Sánchez-Yáñez

Raúl E. Sánchez-Yáñez is a doctor of science (optics), concluding his studies at the Centro de Investigaciones en Óptica (Optical Research Center, CIO) in León, Mexico, in 2002. He is also a master of electrical engineering and has a BEng in electronics, with both degrees received from the University of Guanajuato at Salamanca (Mexico), where he has been a full time professor since 2003. His research interests include colour and texture analysis for computer vision tasks, and computational intelligence applications in feature extraction and decision making.

Preface to "Texture and Colour in Image Analysis"

Texture and colour are optical stimuli that determine, to a great extent, the visual perception of objects, materials and scenes. It is no surprise, then, that texture and colour have received a great deal of research attention for at least forty years. The aptitude to process these stimuli in an effective way indeed plays a major role in the interaction between intelligent beings and the environment in which they live. Consequently, the ability to reproduce this behaviour within intelligent machines is fundamental in a wide range of applications: product inspection, object recognition, materials classification, computer-assisted medical image analysis, content-based image retrieval and remote sensing are just some examples.

In recent times, research in this topic has experienced significant changes. While the hand-crafted approach was the leading strategy up until not long ago, nowadays, Deep Learning has become the major focus. This book collects 16 technical contributions to the field (plus one editorial) from highly reputable researchers from around the world. The papers are grouped by subject and presented in the following order: Theory (1–4), Applications (5–14), Benchmarks and Comparative Evaluations (15), and Reviews (16).

Francesco Bianconi, Antonio Fernández, Raúl E. Sánchez-Yáñez
Editors

Editorial

Special Issue Texture and Color in Image Analysis

Francesco Bianconi [1,*], Antonio Fernández [2] and Raúl E. Sánchez-Yáñez [3]

1. Department of Engineering, Università degli Studi di Perugia, Via Goffredo Duranti 93, 06135 Perugia, Italy
2. School of Industrial Engineering, Universidade de Vigo, Campus Universitario, 36310 Vigo, Spain; antfdez@uvigo.es
3. Department of Electronic Engineering, Universidad de Guanajuato, Comunidad de Palo Blanco, Salamanca 36885, Mexico; sanchezy@ugtomx.onmicrosoft.com
* Correspondence: bianco@ieee.org; Tel.: +39-075-5853706

1. Introduction and Background

Texture and color are two types of visual stimuli that determine, to a great extent, the appearance of objects, materials, and scenes. The ability to process these stimuli enables humans and animals to interact with the environment they live in. As a consequence, texture and color have attracted a lot of research interest since early on. Color and texture analyses are also central to a wide range of applications including materials classification, surface inspection and grading, object recognition, biometric identification, content-based multimedia retrieval, remote sensing, and medical image analysis. In recent years, the appearance of new methodologies (notably deep learning) has elicited renewed interest toward the field. In this context, the objective of this Special Issue is to provide a forum for scientists and practitioners to discuss strategies, challenges, and perspectives in the discipline. The response of the community was substantial, which again confirms the interest in the topics; altogether, we received 26 contributions, of which 16 were deemed suitable for publication after peer review.

2. Theory

Four papers treated theoretical aspects of image processing, and two of them [1,2] were focused on color texture analysis. Navarro and Perez [1] introduced a method for pattern classification through color and texture features based on image partition. Their approach computes global and local features from different areas of the input image. Each pixel is represented as a quaternion and the color features are collected in a histogram obtained using Binary Quaternion Moment Preserving (BQMP). Textural information is extracted via Haralick's features from each partition to conform a feature vector, and a joint color–texture representation is obtained by merging the color code and a normalized texture descriptor. Smeraldi et al. [2] proposed a novel framework for color texture analysis based on partial orderings. Partial orders (PO) make it possible to compare multivariate data that, like colors, lack a natural order. In their work, the authors defined a general approach to extract rank features in color spaces via PO. They also extended a classical descriptor (the Texture Spectrum) to work with partial orders and showed that the partial-order version in color space outperformed the original grayscale descriptor.

Zhang et al. [3] presented a method for the identification of tampered images. In their solution, the input image is first filtered with Local Tchebichef Moments (LTM); then, the result is subtracted from the original image to obtain the 'residuals'. An error-correcting output code based on ensemble learning eventually classifies the images as tampered or not tampered.

In [4] Tang et al. addressed the problem of training Convolutional Neural Networks (CNN) and introduced a novel bounded scheduling procedure called Bsadam. The method first searches the upper and lower bound for Adam, then splits the training process into three steps: (1) the minimization step, (2) the convergence step, and (3) the uniform

scaling step. The proposed solution was effectively tested with simple neural networks, deep convolution networks, and recurrent networks for image classification and language modeling tasks.

3. Applications

Biomedical image analysis received much attention in this Special Issue with a total of four papers accepted. In [5], González-Patiño et al. addressed segmentation of mammograms as an optimization problem and considered three metaheuristic approaches: simulated annealing, genetic algorithms, and a bat algorithm. They used Dunn index as the fitness function to evaluate segmented regions, which were characterized by clinical data, intensity, texture, and shape descriptors. Then, for the diagnosis of breast cancer lesions, they proposed a new artificial immune system (AIS). The performance of the metaheuristic algorithms was compared to intensity-based segmentations obtained using the Otsu method, and the outcomes of the AIS were evaluated on six datasets. Bhattacharjee et al. [6] investigated automated grading of prostate cancer from histology images. Their method is based on four steps: (1) segmentation of the input images via k-means; (2) separation of the touching cells through watershed transform; (3) extraction of morphological features; (4) SVM-based classification into four Gleason grade groups—grade 3, grade 4, grade 5, and benign.

Bontozoglou and Xiao [7] explored assessing a person's condition from capacitive images of their hair and skin. Concretely, they attempted to determine whether a capacitive imaging sensor in combination with image processing algorithms such as gradient-based segmentation, gray level co-occurrence matrix, and normalized cross-correlation could be used in different hair and skin analysis tasks that are of great interest to the cosmetic and pharmaceutical industries, namely, the detection of skin polygons, the estimation of the bounding wrinkles length, and the observation of hair water sorption capabilities. The experimental results indicate that the proposed approach can be successful for detecting and tracking skin artifacts (e.g., wrinkles, moles, or scars) as well as skin age classification. Evidence indicates that capacitive imaging can also be applied to hair water loss studies.

In [8], Obuchowicz et al. examined whether additional digital intraoral radiography (DIR) image preprocessing based on texture analysis improves the recognition and differentiation of periapical lesions. They applied several texture models such as co-occurrences, first-order features, run-length matrices, gray-tone difference matrices, and local binary patterns to transform DIR images into feature maps. To improve the recognition of osteolytic and sclerotic lesions, the feature maps were further processed through k-means clustering. The ability of the proposed approach to yield information about the shape of a structure, its pattern, and adequate contrast was validated by two radiologists independently. The experimental results showed that the application of feature mapping to radiographic dental images constitutes a promising tool for the refinement and possible differentiation of periapical lesions.

Three papers investigated industrial applications. Furferi et al. [9] presented a computer vision system for counting small metal parts produced by electrodeposition. This manufacturing procedure is common in the fashion field and, since the raw materials are usually gold and silver, it is of paramount importance to reduce the amount of waste. The devised method employs a combination of image thresholding and morphological operations. Liu et al. [10] investigated online defect detection in the production of steel plates. This is a fairly common problem in the industry, and requires both speed and high recognition accuracy. The proposed solution relies on Multiblock Local Binary Patterns (MB-LBP), which the authors found to be superior to other methods such as the Gray-Level Co-occurrence Matrix (GLCM), the Scale-Invariant Feature Transform (SIFT), and the speeded up robust feature (SURF). Geng et al.'s work [11] is concerned with the problem of measuring the period length and the skew angle patterns of textile cutting pieces. This kind of semifinished product has been widely used in car seats and garment production.

Experimenting on a dataset of 5000 images, the authors demonstrated the suitability of a regional convolutional neural network (R-CNN) for the task.

Two papers addressed remote sensing problems. Wang et al. [12] described a technique to accurately identify maceral components in the fields of mining and geology. The correct identification of such components is central to a number of industrial processes such as hydrogenation, combustion, carbonization, and gasification. The proposed method employs a two-level coarse-to-fine clustering procedure to divide microscopic images into a sequence of regions with similar attributes (i.e., binder, vitrinite, liptinite, and inertinite). Yu et al. [13] addressed the problem of image segmentation of river scenes. To this end, they proposed a novel approach based on a reflection mechanism of the water surface. Their method employs a Multiblock Local Binary Patterns texture and hue variance in the HSI color space to detect the shadow area of the water's surface. A morphological operation with multiple dilation was employed to reduce false positives due to pseudo-water-patches.

The work by Nanni et al. [14] considered quite an original case study, that is, the automated classification of animal audio. For this task, the authors proposed the use a combination of Siamese neural network and different clustering techniques to train a support vector classifier.

4. Benchmarks and Comparative Evaluations

Using handcrafted features as visual descriptors has been the dominant paradigm in computer vision for many years. In the last decade however, consequently with the extraordinary advances in the field of deep learning, focus has been shifting from the model-based ('a priori') approach to 'a posteriori' strategies, where the features are learned from the data. Both methods have pros and cons; which one should be used in any specific application however, is far from clear. In this context, Karabağ et al. [15] comparatively evaluated traditional and deep learning methods for texture segmentation. In their work, they considered five well-known hand-designed methods (co-occurrence, filtering, local binary patterns, watershed, and multiresolution sub-band filtering) and a deep learning approach based on the U-Net architecture. The methods were evaluated on six classic mosaics of textured images. The main conclusion is that U-Net is effective for texture segmentation and provides equal or better than achieved with traditional texture algorithms. The authors also concluded that determining the correct configuration of the network is not a trivial task, and that variations of some parameters can easily lead to suboptimal results.

5. Reviews

Buzzelli [16] presented a valuable review of different approaches for automatic estimation of visual saliency—i.e., the perceptual property that makes specific elements in a scene stand out and attract the attention of the viewer. The work mainly investigates those domains where research attention is currently high, such as omnidirectional images, image groups for cosaliency, and video sequences. The paper also introduces domain-specific evaluation measures and provides quantitative comparisons among the different methods.

Author Contributions: All the authors have contributed equally. All authors have read and agreed to the published version of the manuscript

Funding: Partially supported by the Department of Engineering, Università degli Studi di Perugia, Italy, within the project *Artificial intelligence for Earth observation* (Fundamental Research Grant Scheme 2020).

Institutional Review Board Statement: Not applicable.

Informed Consent Statement: Not applicable.

Data Availability Statement: Not applicable.

Acknowledgments: This Special Issue would not have been possible without the valuable contribution of the authors, reviewers, copy editors and other members of the editorial team. We particularly wish to thank Tamia Qing, Section Managing Editor, for her continuous support throughout all the phases of the process.

Conflicts of Interest: The authors declare no conflict of interest.

References

1. Navarro, C.; Perez, C. Color-texture pattern classification using global-local feature extraction, an SVM classifier, with bagging ensemble post-processing. *Appl. Sci.* **2019**, *9*, 3130. [CrossRef]
2. Smeraldi, F.; Bianconi, F.; Fernández, A.; González, E. Partial order rank features in colour space. *Appl. Sci.* **2020**, *10*. 499. [CrossRef]
3. Zhang, D.; Wang, S.; Wang, J.; Sangaiah, A.; Li, F.; Sheng, V. Detection of tampering by image resizing using local Tchebichef moments. *Appl. Sci.* **2019**, *9*, 3007. [CrossRef]
4. Tang, M.; Huang, Z.; Yuan, Y.; Wang, C.; Peng, Y. A bounded scheduling method for adaptive gradient methods. *Appl. Sci.* **2019**, *9*, 3569. [CrossRef]
5. González-Patiño, D.; Villuendas-Rey, Y.; Argüelles-Cruz, A.J.; Karray, F. A Novel bio-inspired method for early diagnosis of breast cancer through mammographic image analysis. *Appl. Sci.* **2019**, *9*, 4492. [CrossRef]
6. Bhattacharjee, S.; Park, H.G.; Kim, C.H.; Prakash, D.; Madusanka, N.; So, J.H.; Cho, N.H.; Choi, H.K. Quantitative analysis of benign and malignant tumors in histopathology: Predicting prostate cancer grading using SVM. *Appl. Sci.* **2019**, *9*, 2969. [CrossRef]
7. Bontozoglou, C.; Xiao, P. Applications of capacitive imaging in human skin texture and hair analysis. *Appl. Sci.* **2020**, *10*, 256. [CrossRef]
8. Obuchowicz, R.; Nurzynska, K.; Obuchowicz, B.; Urbanik, A.; Piórkowski, A. Use of texture feature maps for the refinement of information derived from digital intraoral radiographs of lytic and sclerotic lesions. *Appl. Sci.* **2019**, *9*, 2968. [CrossRef]
9. Furferi, R.; Governi, L.; Puggelli, L.; Servi, M.; Volpe, Y. Machine vision system for counting small metal parts in electro-deposition industry. *Appl. Sci.* **2020**, *10*, 418. [CrossRef]
10. Liu, Y.; Xu, K.; Xu, J. An improved MB-LBP defect recognition approach for the surface of steel plates. *Appl. Sci.* **2020**, *9*, 4222. [CrossRef]
11. Geng, L.; Meng, Q.; Xiao, Z.; Liu, Y. Measurement of period length and skew angle patterns of textile cutting pieces based on faster R-CNN. *Appl. Sci.* **2020**, *9*, 3026. [CrossRef]
12. Wang, H.; Lei, M.; Chen, Y.; Li, M.; Zou, L. Intelligent identification of maceral components of coal based on image segmentation and classification. *Appl. Sci.* **2020**, *9*, 3245. [CrossRef]
13. Yu, J.; Lin, Y.; Zhu, Y.; Xu, W.; Hou, D.; Huang, P.; Zhang, G. Segmentation of river scenes based on water surface reflection mechanism. *Appl. Sci.* **2020**, *10*, 2471. [CrossRef]
14. Nanni, L.; Rigo, A.; Lumini, A.; Brahnam, S. Spectrogram classification using dissimilarity space. *Appl. Sci.* **2020**, *10*, 4176. [CrossRef]
15. Karabağ, C.; Verhoeven, J.; Miller, N.R.; Reyes-Aldasoro, C.C. Texture segmentation: an objective comparison between five traditional algorithms and a deep-learning U-Net architecture. *Appl. Sci.* **2019**, *9*, 3900. [CrossRef]
16. Buzzelli, M. Recent advances in saliency estimation for omnidirectional images, image groups, and video sequences. *Appl. Sci.* **2020**, *10*, 5143. [CrossRef]

Article

Color–Texture Pattern Classification Using Global–Local Feature Extraction, an SVM Classifier, with Bagging Ensemble Post-Processing

Carlos F. Navarro and Claudio A. Perez *

Image Processing Laboratory, Electrical Engineering Department and Advanced Mining Technology Center, Universidad de Chile, Santiago 8370451, Chile
* Correspondence: clperez@ing.uchile.cl; Tel.: +0-562-2978-4207; Fax: +0-562-2672-0162

Received: 12 June 2019; Accepted: 30 July 2019; Published: 1 August 2019

Featured Application: The proposed method is a new tool to characterize colored textures and may be applied in various applications such as content image retrieval, characterization of rock samples, biometrics, classification of fabrics, and in non-destructive inspection in wood, steel, ceramic, fruit, and aircraft surfaces.

Abstract: Many applications in image analysis require the accurate classification of complex patterns including both color and texture, e.g., in content image retrieval, biometrics, and the inspection of fabrics, wood, steel, ceramics, and fruits, among others. A new method for pattern classification using both color and texture information is proposed in this paper. The proposed method includes the following steps: division of each image into global and local samples, texture and color feature extraction from samples using a Haralick statistics and binary quaternion-moment-preserving method, a classification stage using support vector machine, and a final stage of post-processing employing a bagging ensemble. One of the main contributions of this method is the image partition, allowing image representation into global and local features. This partition captures most of the information present in the image for colored texture classification allowing improved results. The proposed method was tested on four databases extensively used in color–texture classification: the Brodatz, VisTex, Outex, and KTH-TIPS2b databases, yielding correct classification rates of 97.63%, 97.13%, 90.78%, and 92.90%, respectively. The use of the post-processing stage improved those results to 99.88%, 100%, 98.97%, and 95.75%, respectively. We compared our results to the best previously published results on the same databases finding significant improvements in all cases.

Keywords: colored texture pattern classification; global–local texture classification; color–texture features; color–texture feature extraction; bagging post-processing; BQMP and Haralick global–local feature integration

1. Introduction

Texture pattern classification was considered an important problem in computer vision for many years because of the great variety of possible applications, including non-destructive inspection of abnormalities on wood, steel, ceramics, fruit, and aircraft surfaces [1–6]. Texture discrimination remains a challenge since the texture of objects varies significantly according to the viewing angle, illumination conditions, scale change, and rotation [1,4,7,8]. There is also the special problem of color image retrieval related to appearance-based object recognition, which is a major field of development for several industrial vision applications [1,4,7,8].

Feature extraction of color, texture, and shape from images was used successfully to classify patterns by reducing the dimensionality and the computational complexity of the problem [3,9–16].

Determining the appropriate features for each problem is a recurring challenge which is yet to be fully met by the computer vision community [1,3,16]. Feature extraction and selection enable representation of the information present in the image, and limit the number of features, thus allowing further analysis within a reasonable time. Feature extraction was used in a wide range of applications, such as biometrics [12,14,15], classification of cloth, surfaces, landscapes, wood, and rock minerals [16,17], saliency detection [18], and background subtraction [19], among others. During the past 40 years, while a substantial number of methods for grayscale texture classification were developed [3,5], there was also a growing interest in colored textures [1,2,9,10,13,20,21]. The adaptive integration of color and texture attributes into the development of complex image descriptors is an area of intensive research in computer vision [21]. Most of these investigations focused on the integration process in applications for digital image segmentation [20,22] or the aggregation of multiple preexisting image descriptors [23]. Deep learning was applied successfully to object or scene recognition [24] and scene classification [25], and the use of deep neural networks for the classification of image datasets where texture features are important for generating class-conditional discriminative representations was investigated [26].

Current approaches to color texture analysis can be classified into three groups: the parallel approach, the sequential approach, and the integrative approach [11,27]. The parallel approach considers texture and color as separate phenomena. Color analysis is based on the color distribution in an image, without regard to the spatial relationship between the intensities of the pixels. Texture analysis is based on the relative variation of the intensity of the neighbors, regardless of the color of the pixels. In Reference [2], the authors first converted the original RGB images into other color spaces: HSI (Hue, Saturation, Lightness), CIE XYZ (Comission Internationale de l'Éclairage Tristimulus values), YIQ (Luminance In phase Quadrature), and CIELAB (Comission Internationale de l'Éclairage Lightness-Green-Blue), and then extracted the texture features and color separately. In a similar manner, as reported in Reference [13], the images were transformed to the color spaces HSV and YCbCr, obtaining wavelet intensity channel features of first-order statistics on each channel. The choice of the best performing color space was an open question in recent years, since using one space instead of another can bring considerable improvements in certain applications [28]. The quaternion representation of color was shown to be effective in segmentation in Reference [29] and the feature extraction method, binary quaternion moment preserving (BQMP), is a method of image binarization using quaternions, with the potential for being a powerful tool in color image analysis [6].

In the sequential approach to color texture analysis, the first step is to apply a method of indexing color images. As a result, indexed images are processed obtaining grayscale textures. The co-occurrence matrix was used extensively since it represents the probability of occurrence of the same color pixel pair at a given distance [9]. Another example of this approach is based on texture descriptors using three different color indexing methods and three different texture features [11]. This results in nine independent classifiers that are combined through various schemes.

Integration models are based on the interdependence of texture and color. These models can be divided into single bands, if the data of each channel is considered separately, or a multiband, if two or more channels are considered together. The advantage of the single-band approach is the easy adaptation of classical models based on a grayscale domain, such as Gabor filters [15,30–32], local binary patterns (LBP) or variants [5,7,8,33–38], Galois fields [39], or Haralick statistics [3]. In Reference [2], the main objective was to determine the contribution of color information for the overall performance of classification using Gabor filters and co-occurrence measures, yielding results almost 10% better than those obtained with only grayscale images. In Reference [4], the results reported using co-occurrence matrices reached 94.41% and 99.07% on the Outex and Vistex databases, respectively. Different classifiers, such as k-nearest neighbors, neural networks, and support vector machines (SVM) [40], are used to combine features. The latter was proven to be more efficient when feature selection is performed [41,42] or clustering is used [22].

There are other methods that reached the best results in color texture analysis on databases that are publicly available. In Reference [1], a multi-scale architecture (Multi-Scale Supervised self-Organizing

Orientational-invariant Neural or multi-scale SOON) was presented that reached 91.03% accuracy on the Brodatz database, which contains 111 different colored textures. These results were compared with those of two previous studies on the same database, reported in References [43,44], which reached classification rates of 89.71% and 88.15%, respectively. Another approach used all the information from sensors that created the image [4]. This method improved the results to 98.61% and 99.07%, on the same database, but required a non-trivial change in the architecture of the data collection. A texture descriptor based on a local pattern encoding scheme using local maximum sum and difference histograms was proposed in Reference [8]. The experimental results were tested on the Outex and KTH-TIPS2b databases reaching 95.8% and 91.3%, respectively. In References [32,33,36,39], the methods were not tested in the complete databases. In References [26,45,46], the methods used a different metric to calculate the classification.

A new method for the classification of complex colored textures is proposed in this paper. The images are divided into global and local samples where features are extracted to provide a new representation into global and local features. The feature extraction from different samples of the image using quadrants is described. The extraction of the features in each of the image quadrants, obtaining color and texture features in different spatial scales, is presented: the global scales using the whole image, and the local scales using quadrants. This representation seems to capture most of the information present in the image to improve colored texture classification. Then, a support vector machine classification that was performed is reported and, finally, the post-processing stage using the bagging that was implemented is presented. The proposed method was tested on four different databases: Colored Brodatz Texture [20], VisTex [4], Outex [47], and KTH-TIPS2b [48] databases. The subdivision of the training partition of the database into sub-images while extracting information from each sub-image of different sizes is also new. The results were compared to state-of-the art methods whose results were already published on the same databases.

2. Materials and Methods

The objective was to create a method for colored texture classification that would improve the classification of different complex color patterns. The BQMP method was used previously in color data compression, color edge detection, and multiclass clustering of color data, but not in classification [6,47]. The BQMP reduces an image to representative colors and creates a histogram that indicates the parts of the image that are represented by these colors. Therefore, the color features of the image are represented in this histogram. Haralick's features [3] are often extracted to characterize textures that measure the grayscale distribution, as well as considering the spatial interactions between pixels [3,9,23,38]. Creating a training set is part of the strategy for obtaining local and global features that contain all the information, local and global, for achieving correct classification. Different classifiers, such as k-nearest neighbors, neural networks, and support vector machines (SVM) [41], are used to combine features. In Reference [44], an SVM showed good performance compared to 16 different classification methods.

2.1. Feature Extraction Quadrants

The proposed method divides the images to obtain local and global features from them. The method consists of four stages: firstly, the images in the database are divided into images to be used in training and those to be used in testing. In the second stage, color and texture features are extracted from the training images on both global and local scales. In the third stage, color and texture features are fused to become the inputs to the SVM classifier, and the last stage is a post-processing stage that uses bagging with the test images for classification. These stages are summarized in the block diagram of Figure 1.

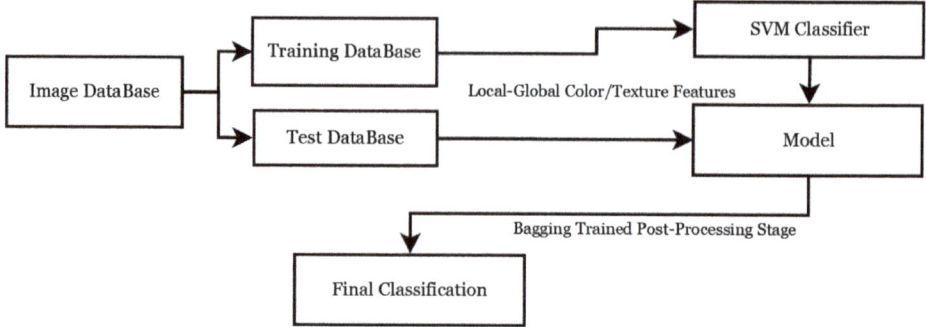

Figure 1. Block diagram of the proposed method.

To be able to compare the performance of our method with previously published results, we used the same partition, into training and testing sets, in each database. In the case of the Brodatz database, as in Reference [1], each colored texture image was partitioned into nine sub-images, using eight for training, and one for testing. An example of this partition is shown in Figure 2a–c. The Brodatz Colored Texture (CBT) database has 111 images of 640 × 640 pixels. Each image in the database has a different texture.

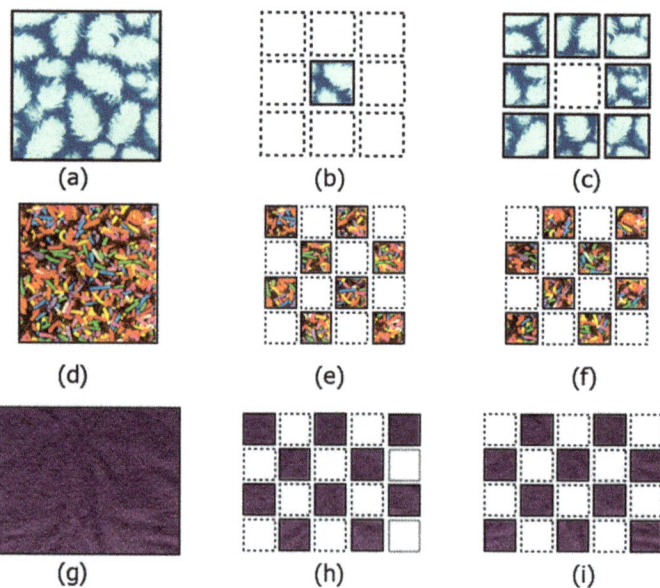

Figure 2. (**a**) The Brodatz image (D88) is used to create the training image (**b**) and test images (**c**). (**d**) The Vistex image (Food0007) is used to create the training images (**e**) and test images (**f**). (**g**) The Outex image (Canvas002) is used to create the training images (**h**) and test images (**i**).

For the Vistex database, the number of training and testing images was eight as in Reference [4]. In the case of the Outex database, the number of training and testing sub-images was 10 as in References [4,9,47]. The KTH-TIPS2b database was already partitioned into four samples, and we performed a cross-validation as in References [7,48].

In each case, we subdivided the training database and the test database using two parameters: n is the number of images divided by side, and r is the times we take n^2 local images from each sample. We can take r × n² local images to extract features from all the samples in each database. Figure 3 shows the image subdivision scheme for the training images. It can be observed that the partitioning scheme allows obtaining features from different parts of the image, at a global and local level. The method was designed to follow this approach so that no relevant information would be lost from the image.

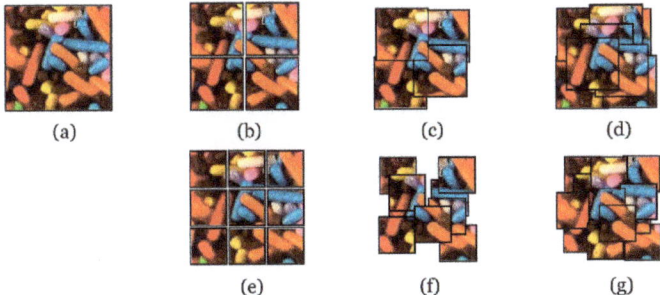

Figure 3. (a) A sample image from Vistex (Food0007) and the subdivision process. (b,e) Image is divided into n² images. (c,f) Image is divided into n² random blocks. (d,g) Image is divided into r × n² random blocks. In this example, (b–d) n = 2, (e–g) n = 3, and r = 2 (d,g).

BQMP and Haralick (without using co-occurrence matrices) are invariant to translation and rotation; the same features are obtained if an exchange of the position of two pixels is made in the image [3,6,47]. This suggests that there is spatial information present in the image that is not extracted by these features. In our proposed method, we use the two-scale scheme, local and global, to add spatial information to the extracted features. This is shown in Figure 4, and explained in detail in Section 2.4. The BQMP and Haralick features are extracted in each quadrant. The test images can be subdivided into local images from which the features are extracted. This allows the creation of a post-processing stage in which a bagging process can be performed. Our method is invariant to translation because of the randomness of the local image positions, but it is partially invariant to rotation because the features are concatenated in an established order. However, color features, as well as Haralick texture features, are invariant to rotation. There are problems where orientation dependency is desirable [49].

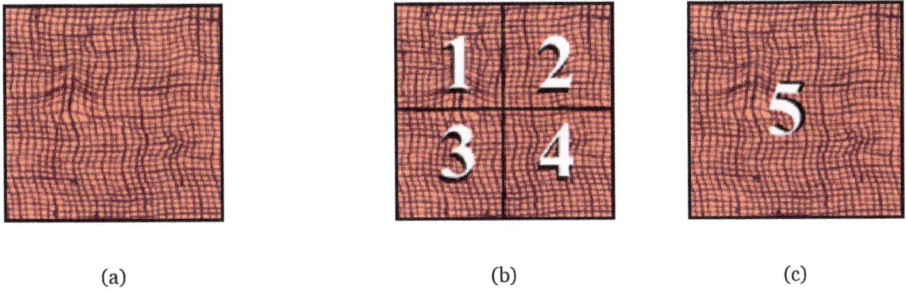

Figure 4. Example of (a) the original image. (b) The four local partitions and (c) the global partition. These five images generate two spatial scales: four local (b) and one global, adding spatial information to the extracted features.

2.2. BQMP Color Feature Extraction

After image subdivision, the BQMP was applied to each one of the local and global sub-images. This method is used as a tool for extracting color features using quaternions. Each pixel in the RGB space is represented as a quaternion. In Reference [6], the authors showed that it is possible to obtain two quaternions that represent a different part of the image, obtaining a binarization of the image in the case of two colors. For more colors, this method can be performed recursively n times, yielding 2^n representatives of an image, and the part of the image that each representative represents in a histogram. The process is repeated, obtaining a result with a binary tree structure. Figure 5 shows the BQMP method for the case of four different colors. The numbers show the color code and the number of pixels represented by each. Figure 6 shows the second iteration for the same case.

The feature vector is formed by concatenating the color code and histograms, through normalization. This vector is concatenated with the ones in other scales and the ones made using Haralick statistics, which are also normalized.

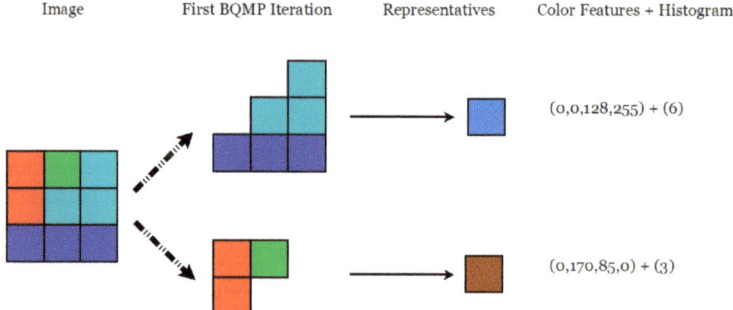

Figure 5. First binary quaternion-moment-preserving (BQMP) iteration example for the case of four different colors. Numbers show the color coded in (q_0, q_1, q_2, q_3) and the number of pixels for each color representative (histogram).

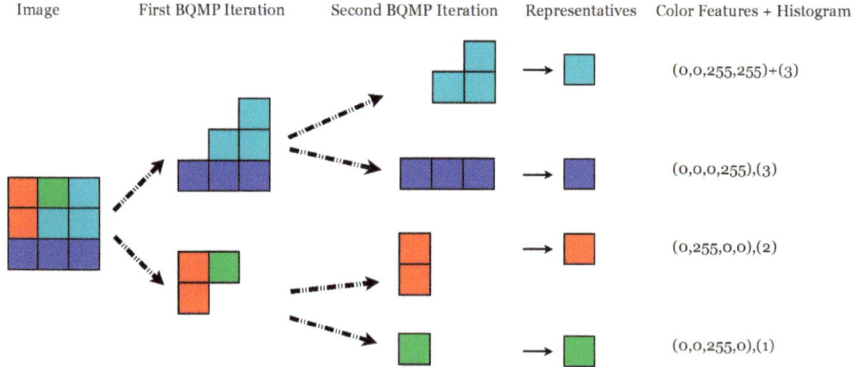

Figure 6. Second BQMP iteration example for the case of four different colors. Numbers show the color coded in (q_0, q_1, q_2, q_3) and the number of pixels for each color representative (histogram).

2.3. Haralick Texture Features

The Haralick texture features are angular second moment, contrast, correlation, sum of squares, inverse difference moment, sum average, sum variance, sum entropy, entropy, difference variance, difference entropy, and information measures of correlation [3]. Other measures characterize the complexity and nature of tone transitions in each channel of the image. The usual practice is to use the first 13 Haralick features with a co-occurrence matrix [4], but in this work, we extracted the 13 Haralick features directly from the images because they provide spatial information from different regions within each image.

The Haralick features used to extract the texture features were Equations (1)–(13), and Equations (14)–(21) explain the notation employed.

Angular second moment:
$$f_1 = \sum_i \sum_j p(i,j)^2. \tag{1}$$

Contrast:
$$f_2 = \sum_{n=0}^{N-1} n^2 \sum_{i=1}^{N} \sum_{j=1}^{N} p(i,j), \tag{2}$$

where $|i - j| = n$.

Correlation:
$$f_3 = \frac{\sum_i \sum_j (ij) p(i,j) - \mu_x \mu_y}{\sigma_x \sigma_y}, \tag{3}$$

where μ_x, μ_y, σ_x, and σ_y are the means and standard deviations of p_x and p_y.

Sum of squares:
$$f_4 = \sum_i \sum_j (i - \mu)^2 p(i,j). \tag{4}$$

Inverse difference moment:
$$f_5 = \sum_i \sum_j \frac{p(i,j)}{1 + (i-j)^2}. \tag{5}$$

Sum average:
$$f_6 = \sum_{i=2}^{2N} i p_{x+y}(i). \tag{6}$$

Sum variance:
$$f_7 = \sum_{i=2}^{2N} (i - f_8)^2 p_{x+y}(i). \tag{7}$$

Sum entropy:
$$f_8 = -\sum_{i=2}^{2N} p_{x+y}(i) \log(p_{x+y}(i)). \tag{8}$$

Entropy:
$$f_9 = -\sum_i \sum_j p(i,j) \log(p(i,j)). \tag{9}$$

Difference variance:
$$f_{10} = Var(p_{x-y}). \tag{10}$$

Difference entropy:
$$f_{11} = -\sum_{i=0}^{N-1} p_{x-y}(i) \log(p_{x-y}(i)). \tag{11}$$

Information measures of correlation:

$$f_{12} = \frac{f_9 - HXY1}{max(HX, HY)}, \tag{12}$$

$$f_{13} = 1 - \exp(-2(HXY2 - f_9))^{\frac{1}{2}}, \tag{13}$$

where HX and HY are the entropies of p_x and p_y, and

$$HXY = -\sum_i \sum_j p(i,j) \log(p(i,j)), \tag{14}$$

$$HXY1 = -\sum_i \sum_j p(i,j) \log(p_x(i)p_y(j)), \text{ and} \tag{15}$$

$$HXY2 = -\sum_i \sum_j p_x(i)p_y(j) \log(p_x(i)p_y(j)). \tag{16}$$

In Equations (1)–(13), to calculate the features, the following notation was used:

$$p(i,j), \tag{17}$$

which is the (i,j)th pixel in a gray sub-image matrix.

$$p_x(i), \tag{18}$$

which is the ith entry in the sub-image matrix, obtained by summing the rows of p.

$$p_y(j), \tag{19}$$

which is the jth entry in the sub-image matrix, obtained by summing the columns of p.

$$P_{x+y}(k) = \sum_{i=1}^{n} \sum_{j=1}^{n} p(i,j), \tag{20}$$

where $i + j = k$ and $k = 2, 3, \ldots, 2N$.

$$P_{x-y}(k) = \sum_{i=1}^{n} \sum_{j=1}^{n} p(i,j), \tag{21}$$

where $|i - j| = k$ and $k = 0, 1, \ldots, N - 1$.

2.4. Feature Extraction

The feature extraction is performed using local and global scales. The feature vector is created by extracting features from each different image partition (local and global). An example is shown in Figure 7. The original image (a) is divided into five partitions, (b) four local and one global (the same original image), as shown in Figure 4. The feature vector is obtained from each partition as shown in (c). Since BQMP is applied once, we obtain different values for two representative colors for each sub-image. As in the example shown in Figure 7c, the first representative color is brown with R1 = 78, G1 = 62, and B1 = 39. The other color is pink with R2 = 215, G2 = 80, B2 = 119. Through binarization, we know that brown represents 41% of the image and pink 59%; therefore, H1 = 0.41 and H2 = 0.59.

To achieve the binarization, the three-dimensional RGB information was transformed in a four-dimensional quaternion. Those quaternions were used to obtain the moments of order 1, 2, and 3, and the moments were used to obtain the equations of momentum conservation, to obtain the representative colors (R1,G1,B1 and R2,G2,B2) and the representative histograms (H1 and H2), as Reference [6] described. The moments were computed using quaternion multiplication that is a four-dimensional operation. For example, in the case of two quaternions $a = a_1 + a_2 i + a_3 j + a_4 k$ and $b = b_1 + b_2 i + b_3 j + b_4 k$, the product will be equal to $ab = (a_1 b_1 - a_2 b_2 - a_3 b_3 - a_4 b_4) + (a_1 b_2 + a_2 b_1 + a_3 b_4 - a_4 b_3)i + (a_1 b_3 - a_2 b_4 + a_3 b_1 + a_4 b_2)j + (a_1 b_4 + a_2 b_3 - a_3 b_2 + a_4 b_1)k$. Therefore, even if a_1 and b_1 are equal to zero, the real part of the multiplication will not necessarily be zero. In the case of the example, this extra information is Q1 = −0.61 in the first color and Q2 = 0.43 in the second color.

Then, we extracted the 13 Haralick features (explained in Section 2.3) in each sub-image and each color channel, obtaining $13 \times 5 \times 3 = 195$ more features to concatenate into the final vector.

In the texture of Figure 7, we performed only one BQMP iteration in an image from Brodatz database, obtaining only two color representatives. In general, the BQMP method generates 2^n representatives from each image, when n iterations are used. In Figures 5 and 6, an example for a simple color pattern shows the representatives for two iterations, n = 2. In our preliminary experiments, the results did not improve significantly for $n \geq 3$, and computational time increased significantly. The feature extraction was performed in local and global scales, so that the representative colors would capture the diversity of the whole image in a local and global manner.

In the case of more complex textures, it is possible to use more iterations of the BQMP method obtaining more representative colors, histograms, and quaternions. Figure 8 shows the feature extraction from one sample image from the Vistex database (Food0007) using one, two, or three iterations of the method.

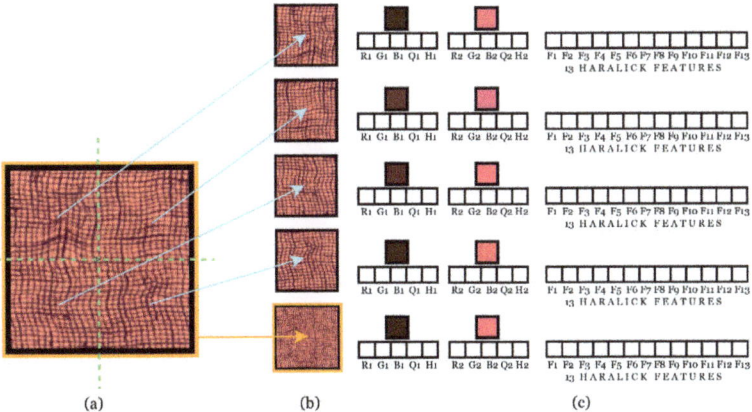

Figure 7. Feature vector extracted from one image from the Brodatz Database. (**a**) The original image is divided into five partitions (**b**): four local partitions and one global (the same image, bottom). (**c**) The feature vectors obtained from each partition. The color feature vector and texture feature vector are concatenated.

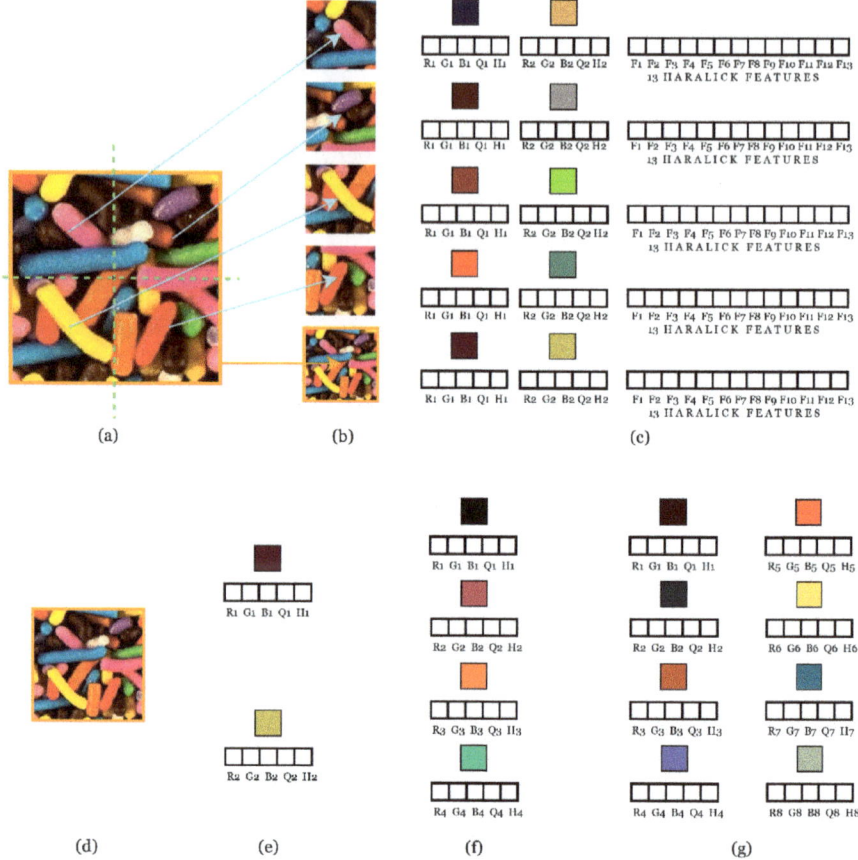

Figure 8. Feature vector extracted from one image of the Vistex Database (Food0007). (**a**) The original image is divided into five partitions (**b**): four local partitions and one global (the same image, bottom). (**c**) The feature vectors obtained from each partition. The color feature vectors and texture feature vectors are concatenated. (**d**–**g**) The global sub-image as a real example using one iteration (**e**), two iterations (**f**), or three iterations (**g**) of the BQMP method.

As in the example shown in Figure 8c, the first representative color is dark blue with R1 = 30, G1 = 32, and B1 = 69. The other color is cream with R2 = 217, G2 = 172, and B2 = 106. Through binarization, we know that dark blue represents 74% of the image and cream 26%; therefore, H1 = 0.74 and H2 = 0.26. Q1 and Q2 are 1.49 and −2.43, respectively. The Haralick features computed from Equations (1)–(13) for the first sub-image are the following: F1 = 1.09×10^{-4}, F2 = 2.66×10^{3}, F3 = 9.90×10^{8}, F4 = 4.75×10^{3}, F5 = 2.60×10^{-2}, F6 = 1.24×10^{2}, F7 = 1.65×10^{4}, F8 = 5.28, F9 = 9.32, F10 = 2.43×10^{-5}, F11 = 4.63, F12 = -6.37×10^{-2}, and F13 = 6.78×10^{-1}.

2.5. SVM Classifier and Post-Processing

After features were extracted from each image, an SVM classifier was used to determine each texture class. The SVM became very popular within the machine learning community due to its great classification potential [41,42]. The SVM maps input vectors in a non-linear transformation to a high-dimensional space where a linear decision hyperplane is constructed for class separation.

A Gaussian SVM kernel was used, and a coarse exhaustive search over the remaining SVM parameters was performed to find the optimal configuration on the training set.

A grid search with cross-validation was used to find the best parameters for the multiclass SVM cascade. We used half of the training set to determine the SVM parameters, and the other half in validation. For testing, we used a different set as explained in Section 3.1. In the case of bagging, we took repeated samples from the original training set for balancing the class distributions to generate new balanced datasets. Two parameters were tuned: the number of decision trees voting in the ensemble, and the complexity parameter related to the size of the decision tree. The method was trained for texture classification using the training sets as they are specified for each database.

In order to have a fair comparison between our obtained classification rates and those previously published, we employed the same partitions used for training and testing as in Diaz-Pernas et al., 2011 [1], Khan et al., 2015 [7], Arvis et al., 2004 [9], Mäenpää et al., 2004 [27], Qazi et al., 2011 [28], Losson et al., 2013 [4], and Couto et al., 2017 [50]. The training and test sets came from separate sub-images, and the methods never used the same sub-image for both training and testing.

In general, combining multiple classification models increases predictive performance [51]. In the post-processing stage, a bagging predictive model composed of a weighted combination of weak classifiers was performed with the results of the SVM model [52]. Bagging is a technique which uses bootstrap sampling to reduce the variance and improve the accuracy of a predictor [51]. It may be used in classification and regression. We created a bagging ensemble for classification using deep trees as weak learners. The bagging predictor was trained with new images taken from the training set of each database. This result was assigned as the final classification for each image.

We compared our results with those published previously on the same databases.

2.6. Databases

It is important to validate the method on standard colored texture databases with previously published results [53]. Therefore, we chose four colored texture databases that were used recently for this purpose: the Colored Brodatz Texture (CBT) [20], Vistex [4], Outex [47], and KTH-TIPS2b [48] databases.

The Brodatz Colored Texture (CBT) database has 111 images of 640 × 640 pixels. Each image in the database has a different texture. The Vistex Database was developed at Massachusetts Institute of Technology (MIT). It has 54 images of 512 × 512 pixels. Each image in the database has a natural color texture. The Outex Database was developed at the University of Oulu, Finland. We used 68 color texture images of 746 × 538, to obtain 1360 images of 128 × 128 with 68 different textures. Each image in the database has a natural color texture. Finally, the KTH-TIPS2b database contains images of 200 × 200 pixels. It has four samples of 108 images of 11 materials at different scales. Each image in the database has a natural color texture.

3. Results

3.1. Experiments

In order to have a fair comparison between our obtained classification rates and those previously published, we used the same databases and partitions used for training and testing as in Diaz-Pernas et al., 2011 [1], Khan et al., 2015 [7], Arvis et al., 2004 [9], Mäenpää et al., 2004 [27], Qazi et al., 2011 [28], Losson et al., 2013 [4], and Couto et al., 2017 [50]. The training and test sets came from separate sub-images, and the methods never used the same sub-image for both training and testing.

3.1.1. Brodatz Database

The methodology, as in Reference [1], used four different sets of images from the same database: the first one consisted of 10 images, the second of 30 images, the third of 40 images, and the fourth of all 111 images. The classification results in previously published articles reached 91.03% in Reference [1], 89.71% in Reference [43], and 88.15% in Reference [44] for the fourth set of the Brodatz database. We used the same partition used for training and testing as in Diaz-Pernas et al., 2011 [1]. The Brodatz image database consists of 111 images of size 640 × 640 pixels. Partitioning each image into nine non overlapping sub-images of 213 × 213 pixels, we obtained 999 sub-images from 111 texture classes. Diaz-Pernas et al., 2011 [1], using the (2 × 2) center sub-image as training and the other to test, reached the best classification results (see Figure 2).

In training, each training sub-image with a size of 213 × 213 pixels was subdivided into a number n of images. Features were extracted from each subdivided image. The parameter n changed from 2 to 7 in the first three experiments, and from 2 to 9 in the last one. Once the feature vector was computed, each vector was assigned to a texture class using an SVM as a classifier.

3.1.2. Vistex Database

The methodology used 54 images that were subdivided into 864 sub-images; 432 were used in training and the other 432 as testing images, as in Arvis et al., 2004 [9], Mäenpää et al., 2004 [27], Qazi et al., 2011 [28], Losson et al., 2013 [4], and Couto et al., 2017 [50]. Previously published results reached 98.61% and 99.07% [4] on the same database using color filter array (CFA) chromatic co-occurrence matrices (CCM). We chose the same 54 texture images to be able to compare our results with those previously published.

3.1.3. Outex Database

The methodology used 68 images that were subdivided into 1360 sub-images; 680 were used in training and the other 680 as testing images, as in Arvis et al., 2004 [9]. The same partition was performed by Mäenpää et al., 2004 [27], Qazi et al., 2011 [28], Losson et al., 2013 [4], and Couto et al., 2017 [50]. Previously published results reached 94.85% and 94.41% [4] on the same database using CFA chromatic co-occurrence matrices. We chose the same 68 texture images and partition to be able to compare our results with those previously published.

3.1.4. KTH-TIPS2b Database

The KTH-Tips2b database consists of four sets of 1,188 images. Each set has 11 different classes. The methodology used four sets of 108 images, each one with 11 images, making a total of 1188 images. We followed the same protocol described in Reference [7], where the average classification performance was reported over four test runs. In each run, all images from one sample were used for training, while all the images from the remaining three samples were used for testing as in Khan et al., 2015 [7]. Previously published results reached 70.6% [7] and 91.3% [8] on the same database using Divisive Information Theoretic Clustering (DITC) and three-dimensional adaptive sum and difference (3D-ASDH) methods, respectively.

3.2. Results

3.2.1. Brodatz Database

Table 1 shows the classification results for the Brodatz database. In this experiment, we used the features extracted in the first set (10 images), varying the size of the images in training $r \times n^2$. The last column shows the results of using the post-processing stage applied on the column with the best performance. The best result for the first experiment on the Brodatz database using 10 images was 100%. In this case, 27 (3×3^2) images with size 71 × 71 (213/3) pixels were used for training. It can also be observed that the use of the post-processing step improved the results up to 100% in all cases. These results were higher than 98.23%, the best result previously published for this experiment on the Brodatz database [1]. Also, Table 1 shows the classification results for the second, third, and fourth experiments on the Brodatz database with 30, 40, and 111 images, respectively. The best result reached using 30 images was 99.84%. In this case, 64 (4×4^2) random images with size 53 × 53 (213/4) pixels were used for training. It can also be observed that the use of the post-processing step improved the results up to 100% in all cases. These results are higher than the best result, 97.54%, previously published for this experiment on the Brodatz database [1].

Table 1. Classification results of the experiments on the Brodatz database for sets of 10, 30, 40, and 111 images. The best results reached with and without post-processing are highlighted by bold text.

	$r \times n^2$ Random Images [1] in Training with							
	No Post-Processing				Post-Processing			
	S = 10, r = 3	S = 30, r = 4	S = 40, r = 3	S = 111, r = 4	S = 10, r = 3	S = 30, r = 4	S = 40, r = 4	S = 111, r = 4
n = 2	99.68%	97.71%	97.03%	87.56%	100%	100%	100%	97.07%
n = 3	100%	99.21%	99.38%	94.81%	100%	100%	100%	99.32%
n = 4	99.92%	99.84%	99.67%	96.24%	100%	100%	100%	99.66%
n = 5	99.8%	99.73%	99.71%	96.71%	100%	100%	100%	99.88%
n = 6	99.83%	99.56%	99.70%	97.30%	100%	100%	100%	99.77%
n = 7	99.51%	99.60%	99.61%	97.63%	100%	100%	100%	99.77%
n = 8				90.05%				99.77%
n = 9				80.31%				96.85%

[1] n is the number of images per side in the training stage; r is the times we iterated the method in each image.

The best result reached using 40 images was 99.71%. In this case, 100 (4×5^2) random images of size 42×42 (213/5) pixels were used for training. It can also be observed that the use of the post-processing step improved the results up to 100% in all cases. These results are higher than 95.5%, the best previously published result for this experiment on the Brodatz database [1]. The best result reached using all the 111 images was 97.63%. In this case, 196 (4×7^2) random images of size 30×30 (213/7) pixels were used for training. It can also be observed that the use of the post-processing step improved the results up to 99.88%. These results are higher than the bests result of 98.25% and 99.5% previously published for this experiment on the Brodatz database [30,50].

The size of the smaller images reached an optimum for n = 7 with an image size of 30×30. We performed an exhaustive search varying from n = 2 to n = 9, reaching an optimum at n = 7. An explanation is that n = 7 is optimal for the complete method using texture and color features. Table 2 compares the results previously published in the literature and our results on the Brodatz database for the four experiments which included 10, 30, 40, and 111 images, respectively.

Table 2. Best results of global–local Haralick– binary quaternion-moment-preserving (BQMP) classification for the Brodatz database for the four sets of 10, 30, 40, and 111 images compared to previously published studies. SVM—support vector machine.

Best Results vs. Number of Images	10	30	40	111
Lazebnick et al., 2005 [43]	-	-	-	88.15%
Mellor et al., 2008 [44]	-	-	-	89.71%
Diaz-Pernas et al., 2011 [1]	98.23%	97.54%	95.5%	91.03%
Couto el al, 2017 [50]	-	-	-	98.25%
Kim et al., 2017 [31]	-	-	-	97.84%
SVM global–local Haralick BQMP method [1]	100%	99.84%	99.71%	97.63%
KNN [2] global–local Haralick BQMP method + post-processing [1]	100%	99.94%	99.98%	94.46%
SVM global–local Haralick BQMP method + post-processing [1]	100%	100%	100%	99.88%

[1] The last three methods are our results; [2] k-nearest neighbors.

It can be seen in Table 2 that our method, with post-processing, reached the highest results. The most significant improvement was reached on the complete Brodatz database that includes 111 images.

3.2.2. Vistex Database

Table 3 shows the classification results of our method applied to the Vistex database (54 images). Each image in the training set was partitioned randomly, and the number of windows per side was varied from two to four in each image chosen for training, with the number of random images from $4 \times n^2$ to $10 \times n^2$. The first column shows the best results reached by our method, and the second column shows the results after the post-processing stage. Table 4 compares the results published previously in the literature and our results for the Vistex database with 54 images. It can be observed in Table 4 that our post-processed method reached the highest results with 100%.

Table 3. Classification results of the experiment on the Vistex database for the set of 54 images. The best results reached with and without post-processing are highlighted by bold text.

	$r \times n^2$ Random Images [1] in Training					
	Without Post-Processing			With Post-Processing		
	r = 2	r = 3	r = 4	r = 2	r = 3	r = 4
n = 4	95.52%	94.83%	94.13%	99.54%	100%	100%
n = 5	95.68%	95.29%	94.64%	99.54%	100%	100%
n = 6	96.13%	95.88%	95.22%	100%	100%	100%
n = 7	96.45%	96.26%	95.36%	100%	100%	100%
n = 8	96.72%	96.42%	96.42%	100%	100%	100%
n = 9	95.67%	97.13%	95.88%	100%	100%	100%
n = 10	96.96%	96.32%	95.62%	100%	100%	100%

[1] n is the number of images per side in the training stage; r is the times we iterated the method in each image.

Table 4. Best results of global–local Haralick–BQMP classification for the Vistex database with 54 images and best results published previously on the same database.

Paper	Method	Result
Arvis et al., 2004 [9]	Multispectral	97.9%
Mäenpää et al., 2004 [27]	Color histogram $I_1 I_2 I_3 32^3$	100%
Qazi et al., 2011 [28]	Improved hue, luminance, and saturation color space (IHLS) B = 16	100%
Losson et al., 2013 [4]	Color filter array - chromatic co-occurrence matrices	98.61%
Losson et al., 2013 [4]	Chromatic co-occurrence matrices	99.07%
Couto et al., 2017 [50]	Deterministic walks' direction histogram	99.65%
Cernadas et al., 2017 [30]	Parallel vectors	99.5%
Neiva et al., 2018 [23]	Smoothed morphological operators (SMO)	99.54%
Kalakech et al., 2018 [38]	Adapted Laplace score	94.9%
Global–local Haralick BQMP method [1]		97.13%
KNN global–local Haralick BQMP method + post-processing [1]		97.25%
Global–local Haralick BQMP method + post-processing [1]		100%

[1] The last three methods are our results.

3.2.3. Outex Database

Table 5 shows the classification results of our method applied to the Outex database (68 images). Each image in the training set was partitioned randomly, varying the number of windows per side from two to four in each training image, and the number of random images from $5 \times n^2$ to $18 \times n^2$. The first column shows the best results reached by our method, and the second column shows the results after the post-processing stage. The best results are highlighted by bold text.

Table 5. Classification results of the experiment on the Outex database for the set of 68 images. The best results reached with and without post-processing are highlighted by bold text.

	r × n² Random Images [1] in Training					
	Without Post-Processing			With Post-Processing		
	r = 2	r = 3	r = 4	r = 2	r = 3	r = 4
n = 5	87.43%	86.46%	85.40%	97.50%	97.79%	97.79%
n = 6	88.49%	86.72%	85.72%	97.50%	98.08%	98.08%
n = 7	88.84%	86.37%	86.15%	97.94%	97.50%	98.08%
n = 8	89.19%	87.44%	86.27%	98.23%	97.79%	98.38%
n = 9	89.38%	87.73%	86.51%	98.67%	97.79%	98.23%
n = 10	89.61%	87.96%	86.62%	97.94%	98.23%	97.94%
n = 11	90.12%	88.20%	86.85%	97.94%	98.23%	97.94%
n = 12	90.35%	88.37%	87.04%	**98.97%**	97.94%	98.52%
n = 13	90.52%	88.50%	87.22%	98.82%	98.38%	97.05%
n = 14	90.58%	88.21%	87.38%	98.67%	97.20%	97.20%
n = 15	90.78%	88.06%	86.53%	98.67%	97.79%	96.32%
n = 16	90.19%	86.92%	86.24%	97.94%	96.62%	96.03%
n = 17	89.98%	86.02%	85.10%	97.50%	96.32%	95.44%
n = 18	90.18%	85.54%	84.16%	97.64%	95.59%	94.85%

[1] n is the number of images per side in the training stage; r is the times we iterate the method in each image.

Table 6 compares the results published previously in the literature and our results for the Outex database with 68 images. It can be observed in Table 6 that our post-processed method reached the highest results with 98.97%.

Table 6. Best results of global–local Haralick–BQMP classification for the Outex database with 68 images. The best previously published results are compared to our results.

Paper	Method	Result
Arvis et al., 2004 [9]	Multispectral	94.9%
Mäenpää et al., 2004 [27]	Color histogram HSV16³	95.4%
Bianconi et al., 2011 [11]	Gabor and chromatic features	90.0%
Qazi et al., 2011 [28]	IHLS color space B = 16	94.5%
Losson et al., 2013 [4]	Color filter array - chromatic co-occurrence matrices	94.41%
Losson et al., 2013 [4]	Chromatic co-occurrence matrices	94.85%
Sandid et al., 2016 [8]	three-dimensional adaptive sum and difference histograms (3D-ASDH)	95.8%
Couto el al, 2017 [50]	Deterministic walks direction histogram	97.28%
Neiva et al., 2018 [23]	Smoothed morphological operators (SMO)	86.47%
Cernadas et al., 2017 [30]	Parallel vectors	90.6%
Global–local Haralick BQMP method [1]		90.78%
KNN global–local Haralick BQMP method + post-processing [1]		96.72%
Global–local Haralick BQMP method + post-processing [1]		98.97%

[1]; [1] the last three methods are our results.

3.2.4. KTH-TIPS2b Database

Table 7 shows the classification results of our method applied to the KTH-TIPS2b database (1188 × 4 images). In each test, all the images from one sample were used for training, while the images from the remaining three samples were used as a test set. The first column shows the best results reached by our method, and the second column shows the results after the post-processing stage.

Table 7. Classification results of the experiment on the KTH-TIPS2b database for the four sets of 1188 images.

	n ^{2} Random Images [1] in Training			
	Without Post-Processing		With Post-Processing	
	n = 1	n = 2	n = 1	n = 2
S = 1	98.73%	99.16%	99.73%	99.65%
S = 2	92.72%	91.26%	95.79%	95.59%
S = 3	90.74%	88.99%	94.53%	93.71%
S = 4	89.42%	87.88%	92.93%	92.89%
Mean	92.90%	91.82%	95.75%	95.46%

[1] n is the number of images per side in the training stage; S is the set used for training, using the other three sets as test.

Table 8 compares the results published previously in the literature and our results for the KTH-TIPS2b database with 1188 × 4 images. It can be observed in Table 8 that our post-processing method reached the highest results with 95.75%.

Table 8. Best results of global–local Haralick–BQMP classification for the KTH-TIPS2b database with 1188 images per set and the best previously published results on the same database.

Paper	Method	Result
Khan et al., 2015 [7]	Divisive information theoretic clustering (DITC)	70.6%
Sandid et al., 2016 [8]	3D-ASDH	91.3%
El Merabet et al., 2018 [34]	Local concave/convex micro-structure pattern	84.44%
El Merabet et al., 2019 [35]	Attractive-and-Repulsive Center-Symmetric -LBP	93.61%
Global–local Haralick BQMP method [1]		92.90%
KNN global–local Haralick BQMP method + post-processing [1]		92.72%
Global–local Haralick BQMP method + post-processing [1]		95.75%

[1] The last three methods are in this paper.

3.2.5. Color or Texture vs. Color and Texture

Table 9 compares the results of color features, texture features, and of the combination of both, for texture classification measured on the Brodatz, Vistex, Outex, and KTH-TIPS2b databases. It can be observed that both types of features, color and texture, contribute to the overall results, with maximum performance when both types of features are combined. Comparing these results to those previously published on the same databases, it can be observed that, although the method reached 100% on the Vistex database in Reference [28], the results yielded on Outex were only 94.5%.

Table 9. Best results of global–local Haralick–BQMP classification for the all databases with the contribution of each part of the model (color and spatial structure).

Database	Global–Local Haralick BQMP Method with All Features	Global–Local Haralick BQMP Method with Only Color Features	Global–Local Haralick BQMP Method with Only Texture Features
Brodatz	99.88%	96.28%	87.84%
Outex	98.97%	91.86%	88.57%
Vistex	100%	95.67%	90.84%
KTH-TIPS2b	95.75%	94.95%	92.59%

3.2.6. Computation Time

Table 10 displays the computational time required for feature extraction (FE), classification time with the SVM, and post-processing (PP) time performed on the Vistex database. All implementations were carried out using Python 3 on an Intel (R) Core (TM) i7-7700HM 3.6 GHz, with 64 GB of random-access memory (RAM).

Table 10. Computational time of the proposed method on feature extraction (FE), SVM classification time, postprocessing (PP) time and total time. Experiments were conducted on the Vistex database (54 images).

Vistex	FE Time (s)	SVM Time (s)	PP Time (s)	Total Time (s)
Training	57.05 s	305.19 s	44.96 s	102.86 s
Test	55.88 s	41.22 s	5.76 s	407.2 s
Total	112.93 s	346.41 s	50.72 s	510.06 s

4. Discussion

The idea of combining color and texture was proposed previously, but the proposed feature extraction process allows the method to preserve the information available in the original image, yielding significantly better results than those previously published. A possible drawback of previous texture classification methods is that important information is lost from the original image with the feature extraction method, hampering its ability to improve texture classification results. The feature extraction process that includes global and local features is something new from the point of view of combining color with texture. Sub-dividing the training partition of the database into sub-images, and trying to obtain all the information present in the image using various image sizes or a different number of images is something that was not reported in previous publications.

Color and texture features are extracted in order to classify complex colored textures. However, the feature extraction process loses part of the information present in the image because the two-dimensional (2D) information is transformed into a reduced space. By using global and local features extracted from many different partitions of the image, the information needed for colored texture classification is preserved better. Sub-dividing the training data into sub-images (local–global) and trying to obtain all the information present using different image sizes is a new approach.

Although the BQMP method was proposed several years ago [6], it was used in color data compression, color edge detection, and multiclass clustering of color data. The reduction of an image into representative colors and a histogram that indicates which part of the image is represented by these colors achieves excellent results. In addition, local and global features are extracted from each image. The results of our method were compared with those of several other feature extraction implementations on the Brodatz database with those published in References [1,27,28,31,43,44,50], on the Vistex database with those published in References [4,9,13,23,27,28,30,38,45,46,50], on the Outex database with those published in References [4,8,9,11,27,28,30,38,45,46,50], and on the KTH-TIPS2b with those published in References [7,8,31,34,35] (please see Tables 2, 4, 6 and 8). The proposed method generated better results than those that were published previously.

The databases Brodatz, Vistex, Outex, and KTH2b-Tips are available for comparing the results of different texture classification methods. Tests should be performed under the same conditions. We compared our results with those of References [1,7] under the same conditions using the same training/test distribution, and an SVM as a classifier. We also compared our results with those of Reference [4] in which they used a nearest-neighbor classifier (KNN) and, therefore, we tested our method with KNN instead of SVM. The results with KNN are shown in Tables 2, 4, 6 and 8, corroborating that SVM achieves better results. The proposed method achieved better results than those previously published.

5. Conclusions

In this paper, we presented a new method for classifying complex patterns of colored textures. Our proposed method includes four main steps. Firstly, the image is divided into local and global images. This image sub-division allows feature extraction in different spatial scales, as well as adding spatial information to the extracted features. Therefore, we capture global and local features from the texture. Secondly, texture and color features are extracted from each divided image using the BQMP and Haralick algorithms. Thirdly, a support vector machine is used to classify each image with the extracted features as inputs. Fourthly, a post-processing stage using bagging is employed.

The method was tested on four databases, the Brodatz, VisTex, Outex, and KTH-TIPS2B databases, yielding correct classification rates of 97.63%, 97.13%, 90.78%, and 92.90% respectively. The post-processing stage improved the results to 99.88%, 100%, 98.97%, and 95.75%, respectively, for the same databases. We compared our results on the same databases to the best previously published results finding significant improvements of 8.85%, 0.93% (to 100%), 4.12%, and 4.45%.

The partition of the image into local and global images provides information about features at different scales and spatial locations within each image, which is useful in color/texture classification. The above, combined with the use of a post-processing stage using a bagging predictive model, allows achieving such results.

Author Contributions: Conceptualization, C.F.N. and C.A.P.; methodology, C.F.N. and C.A.P.; software, C.F.N.; validation, C.F.N. and C.A.P.; formal analysis, C.F.N. and C.A.P.; investigation, C.F.N. and C.A.P.; resources, C.A.P.; data curation, C.F.N.; Writing—Original Draft preparation, C.F.N. and C.A.P.; Writing—Review and Editing, C.F.N. and C.A.P.; visualization, C.F.N.; supervision, C.A.P.; project administration, C.A.P.; funding acquisition, C.A.P.

Funding: This research was funded by FONDECYT, grant number 1191610 from CONICYT, the Department of Electrical Engineering, and the Advanced Mining Technology Center, Universidad de Chile.

Conflicts of Interest: The authors declare no conflict of interest.

References

1. Díaz-Pernas, F.J.; Antón-Rodríguez, M.; Perozo-Rondón, F.J.; González-Ortega, D. A multi-scale supervised orientational invariant neural architecture for natural texture classification. *Neurocomputing* **2011**, *74*, 3729–3740. [CrossRef]
2. Drimbarean, A.; Whelan, P.F. Experiments in colour texture analysis. *Pattern Recognit. Lett.* **2001**, *22*, 1161–1167. [CrossRef]
3. Haralick, R.M.; Shanmugam, K.S.; Dinstein, I. Textural Features for Image Classification. *IEEE Trans. Syst. Man Cybern.* **1973**, *3*, 610–621. [CrossRef]
4. Losson, O.; Porebski, A.; Vandenbroucke, N.; Macaire, L. Color texture analysis using CFA chromatic co-occurrence matrices. *Comput. Vis. Image Underst.* **2013**, *117*, 747–763. [CrossRef]
5. Ojala, T.; Pietikainen, M.; Harwood, D. Performance evaluation of texture measures with classification based on Kullback discrimination of distributions. In Proceedings of the 12th International Conference on Pattern Recognition, Jerusalem, Israe, 9–13 October 1994; Volume 1, pp. 582–585.
6. Pei, S.C.; Cheng, C.M. Color image processing by using binary quaternion-moment-preserving thresholding technique. *IEEE Trans. Image Process.* **1999**, *8*, 614–628.
7. Khan, F.S.; Anwer, R.M.; van de Weijer, J.; Felsberg, M.; Laaksonen, J. Compact color-texture description for texture classification. *Pattern Recognit. Lett.* **2015**, *51*, 16–22. [CrossRef]
8. Sandid, F.; Douik, A. Robust color texture descriptor for material recognition. *Pattern Recognit. Lett.* **2016**, *80*, 15–23. [CrossRef]
9. Arvis, V.; Debain, C.; Berducat, M.; Benassi, A. Generalization of the cooccurrence matrix for colour images: Application to colour texture classification. *Image Anal. Stereol.* **2011**, *23*, 63–72. [CrossRef]
10. Bianconi, F.; González, E.; Fernández, A.; Saetta, S.A. Automatic classification of granite tiles through colour and texture features. *Expert Syst. Appl.* **2012**, *39*, 11212–11218. [CrossRef]
11. Bianconi, F.; Harvey, R.W.; Southam, P.; Fernández, A. Theoretical and experimental comparison of different approaches for color texture classification. *J. Electron. Imaging* **2011**, *20*, 043006. [CrossRef]

12. Cament, L.A.; Galdames, F.J.; Bowyer, K.W.; Perez, C.A. Face recognition under pose variation with local Gabor features enhanced by Active Shape and Statistical Models. *Pattern Recognit.* **2015**, *48*, 3371–3384. [CrossRef]
13. Hiremath, P.S.; Shivashankar, S.; Pujari, J. Wavelet based features for color texture classification with application to cbir. Int. J. Comput. Sci. Netw. Secur. 2006; 6, pp. 124–133.
14. Perez, C.A.; Aravena, C.M.; Vallejos, J.I.; Estevez, P.A.; Held, C.M. Face and iris localization using templates designed by particle swarm optimization. *Pattern Recognit. Lett.* **2010**, *31*, 857–868. [CrossRef]
15. Perez, C.A.; Cament, L.A.; Castillo, L.E. Methodological improvement on local Gabor face recognition based on feature selection and enhanced Borda count. *Pattern Recognit.* **2011**, *44*, 951–963. [CrossRef]
16. Perez, C.A.; Saravia, J.A.; Navarro, C.F.; Schulz, D.A.; Aravena, C.M.; Galdames, F.J. Rock lithological classification using multi-scale Gabor features from sub-images, and voting with rock contour information. *Int. J. Miner. Process.* **2015**, *144*, 56–64. [CrossRef]
17. Perez, C.A.; Estévez, P.A.; Vera, P.A.; Castillo, L.E.; Aravena, C.M.; Schulz, D.A.; Medina, L.E. Ore grade estimation by feature selection and voting using boundary detection in digital image analysis. *Int. J. Miner. Process.* **2011**, *101*, 28–36. [CrossRef]
18. Nan, B.; Mu, Z.; Chen, L.; Cheng, J. A Local Texture-Based Superpixel Feature Coding for Saliency Detection Combined with Global Saliency. *Appl. Sci.* **2015**, *5*, 1528–1546. [CrossRef]
19. Zeng, D.; Zhu, M.; Zhou, T.; Xu, F.; Yang, H. Robust Background Subtraction via the Local Similarity Statistical Descriptor. *Appl. Sci.* **2017**, *7*, 989. [CrossRef]
20. Abdelmounaime, S.; Dong-Chen, H. New Brodatz-Based Image Databases for Grayscale Color and Multiband Texture Analysis. *Isrn Mach. Vis.* **2013**, *2013*, 1–14. [CrossRef]
21. Ilea, D.E.; Whelan, P.F. Image segmentation based on the integration of colour–texture descriptors—A review. *Pattern Recognit.* **2011**, *44*, 2479–2501. [CrossRef]
22. Wang, Y.; Yang, J.; Peng, N. Unsupervised color–texture segmentation based on soft criterion with adaptive mean-shift clustering. *Pattern Recognit. Lett.* **2006**, *27*, 386–392. [CrossRef]
23. Barros Neiva, M.; Vacavant, A.; Bruno, O.M. Improving texture extraction and classification using smoothed morphological operators. *Digit. Signal Process.* **2018**, *83*, 24–34. [CrossRef]
24. Andrearczyk, V.; Whelan, P.F. Using filter banks in convolutional neural networks for texture classification. *Pattern Recognit. Lett.* **2016**, *84*, 63–69. [CrossRef]
25. Zhu, Q.; Zhong, Y.; Liu, Y.; Zhang, L.; Li, D. A Deep-Local-Global Feature Fusion Framework for High Spatial Resolution Imagery Scene Classification. *Remote Sens.* **2018**, *10*, 568.
26. Basu, S.; Mukhopadhyay, S.; Karki, M.; DiBiano, R.; Ganguly, S.; Nemani, R.; Gayaka, S. Deep neural networks for texture classification—A theoretical analysis. *Neural Netw.* **2018**, *97*, 173–182. [CrossRef]
27. Mäenpää, T.; Pietikäinen, M. Classification with color and texture: Jointly or separately? *Pattern Recognit.* **2004**, *37*, 1629–1640. [CrossRef]
28. Qazi, I.-U.-H.; Alata, O.; Burie, J.-C.; Moussa, A.; Fernandez-Maloigne, C. Choice of a pertinent color space for color texture characterization using parametric spectral analysis. *Pattern Recognit.* **2011**, *44*, 16–31. [CrossRef]
29. Shi, L.; Funt, B. Quaternion color texture segmentation. *Comput. Vis. Image Underst.* **2007**, *107*, 88–96. [CrossRef]
30. Cernadas, E.; Fernández-Delgado, M.; González-Rufino, E.; Carrión, P. Influence of normalization and color space to color texture classification. *Pattern Recognit.* **2017**, *61*, 120–138. [CrossRef]
31. Kim, N.C.; So, H.J. Directional statistical Gabor features for texture classification. *Pattern Recognit. Lett.* **2018**, *112*, 18–26. [CrossRef]
32. Wang, M.; Gao, L.; Huang, X.; Jiang, Y.; Gao, X. A Texture Classification Approach Based on the Integrated Optimization for Parameters and Features of Gabor Filter via Hybrid Ant Lion Optimizer. *Appl. Sci.* **2019**, *9*, 2173. [CrossRef]
33. Backes, A.R.; de Mesquita Sá Junior, J.J. LBP maps for improving fractal based texture classification. *Neurocomputing* **2017**, *266*, 1–7. [CrossRef]
34. El merabet, Y.; Ruichek, Y. Local Concave-and-Convex Micro-Structure Patterns for texture classification. *Pattern Recognit.* **2018**, *76*, 303–322. [CrossRef]
35. El merabet, Y.; Ruichek, Y.; El idrissi, A. Attractive-and-repulsive center-symmetric local binary patterns for texture classification. *Eng. Appl. Artif. Intell.* **2019**, *78*, 158–172. [CrossRef]

36. Hiremath, P.S.; Bhusnurmath, R.A. Multiresolution LDBP descriptors for texture classification using anisotropic diffusion with an application to wood texture analysis. *Pattern Recognit. Lett.* **2017**, *89*, 8–17. [CrossRef]
37. Jeena Jacob, I.; Srinivasagan, K.G.; Jayapriya, K. Local Oppugnant Color Texture Pattern for image retrieval system. *Pattern Recognit. Lett.* **2014**, *42*, 72–78. [CrossRef]
38. Kalakech, M.; Porebski, A.; Vandenbroucke, N.; Hamad, D. Unsupervised Local Binary Pattern Histogram Selection Scores for Color Texture Classification. *J. Imaging* **2018**, *4*, 112. [CrossRef]
39. Shivashankar, S.; Kudari, M.; Hiremath, P.S. Galois Field-based Approach for Rotation and Scale Invariant Texture Classification. *Int. J. Image Graph. Signal Process. (Ijigsp)* **2018**, *10*, 56–64.
40. Cortes, C.; Vapnik, V. Support-Vector Networks. *Mach. Learn.* **1995**, *20*, 273–297. [CrossRef]
41. Meyer, D.; Leisch, F.; Hornik, K. The support vector machine under test. *Neurocomputing* **2003**, *55*, 169–186. [CrossRef]
42. Shang, C.; Barnes, D. Support vector machine-based classification of rock texture images aided by efficient feature selection. In Proceedings of the 2012 International Joint Conference on Neural Networks (IJCNN), Brisbane, QLD, Australia, 10–15 June 2012; pp. 1–8.
43. Lazebnik, S.; Schmid, C.; Ponce, J. A sparse texture representation using local affine regions. *IEEE Trans. Pattern Anal. Mach. Intell.* **2005**, *27*, 1265–1278. [CrossRef]
44. Mellor, M.; Hong, B.; Brady, M. Locally Rotation, Contrast, and Scale Invariant Descriptors for Texture Analysis. *IEEE Trans. Pattern Anal. Mach. Intell.* **2008**, *30*, 52–61. [CrossRef]
45. Hayati, S.; Ahmadzadeh, M.R. WIRIF: Wave interference-based rotation invariant feature for texture description. *Signal Process.* **2018**, *151*, 160–171. [CrossRef]
46. Pham, M.-T.; Mercier, G.; Bombrun, L. Color Texture Image Retrieval Based on Local Extrema Features and Riemannian Distance. *J. Imaging* **2017**, *3*, 43. [CrossRef]
47. Ojala, T.; Maenpaa, T.; Pietikainen, M.; Viertola, J.; Kyllonen, J.; Huovinen, S. Outex-new framework for empirical evaluation of texture analysis algorithms. In Proceedings of the Object Recognition Supported by User Interaction for Service Robots, Quebec City, QC, Canada, 11–15 August 2002; Volume 1, pp. 701–706.
48. Caputo, B.; Hayman, E.; Mallikarjuna, P. Class-specific material categorisation. In Proceedings of the Tenth IEEE International Conference on Computer Vision (ICCV'05) Volume 1, Beijing, China, 17–21 October 2005; Volume 2, pp. 1597–1604.
49. Kashyap, R.L.; Khotanzad, A. A Model-based Method for Rotation Invariant Texture Classification. *IEEE Trans. Pattern Anal. Mach. Intell.* **1986**, *8*, 472–481. [CrossRef]
50. Couto, L.N.; Backes, A.R.; Barcelos, C.A. Texture characterization via deterministic walks' direction histogram applied to a complex network-based image transformation. *Pattern Recognit. Lett.* **2017**, *97*, 77–83. [CrossRef]
51. Breiman, L. Bagging predictors. *Mach. Learn.* **1996**, *24*, 123–140. [CrossRef]
52. Zhang, C.; Ma, Y. *Ensemble Machine Learning: Methods and Applications*; Springer: Cham, Switzerland, 2012; ISBN 1-4419-9325-8.
53. Hossain, S.; Serikawa, S. Texture Databases—A Comprehensive Survey. *Pattern Recogn. Lett.* **2013**, *34*, 2007–2022. [CrossRef]

© 2019 by the authors. Licensee MDPI, Basel, Switzerland. This article is an open access article distributed under the terms and conditions of the Creative Commons Attribution (CC BY) license (http://creativecommons.org/licenses/by/4.0/).

Article

Partial Order Rank Features in Colour Space

Fabrizio Smeraldi [1,2,*], Francesco Bianconi [2], Antonio Fernández [3] and Elena González [3]

1. School of Electronic Engineering and Computer Science, Queen Mary University of London, Mile End Road, London E1 4NS, UK
2. Department of Engineering, Università degli Studi di Perugia, Via G. Duranti 93, 06125 Perugia, Italy; bianco@ieee.org
3. Department of Engineering Design, Universidade de Vigo, Rúa Maxwell s/n, 36310 Vigo, Spain; antfdez@uvigo.es (A.F.); elena@uvigo.es (E.G.)
* Correspondence: f.smeraldi@qmul.ac.uk

Received: 26 November 2019; Accepted: 31 December 2019; Published: 10 January 2020

Abstract: Partial orders are the natural mathematical structure for comparing multivariate data that, like colours, lack a natural order. We introduce a novel, general approach to defining rank features in colour spaces based on partial orders, and show that it is possible to generalise existing rank based descriptors by replacing the order relation over intensity values by suitable partial orders in colour space. In particular, we extend a classical descriptor (the Texture Spectrum) to work with partial orders. The effectiveness of the generalised descriptor is demonstrated through a set of image classification experiments on 10 datasets of colour texture images. The results show that the partial-order version in colour space outperforms the grey-scale classic descriptor while maintaining the same number of features.

Keywords: mathematics of colour and texture; hand-designed image descriptors; rank features; partial orders

1. Introduction

It is, at first sight, peculiar that one of the most robust tools for image description, namely rank features, have only seen limited application to colour images. The problem is, of course, that while they are very effective at dealing with noise, rank features run afoul of the main theoretical difficulty associated with colour spaces—that is the absence of a natural order.

In recent years there has been a revival of interest in ranking of colour pixels. Notably, Ledoux et al. [1] published an extensive comparative study in the use of total orders as rank features for texture recognition. However, interest has been keenest in the field of colour morphology, where several solutions have been proposed—from adaptive orders that work around the 'false colour problem' to the natural mathematical structure for ordering higher-dimensional sets—that is partial orders [2–6].

In partially-ordered sets we simply admit that there will be couples of elements incomparable to each other. Partial orders are therefore particularly suitable for dealing with colour spaces, where statements like "yellow is greater than green" make little or no sense at all.

The objective of this work is to introduce a novel category of rank features based on partial orders. In the remainder, after providing some background on partial orders (Section 2), we detail the ways in which rank features can be defined (Section 2.5) and extend a classical descriptor (the Texture Spectrum) to work with partial orders (Section 3.1). We demonstrate the feasibility of the method through a set of experiment on 10 datasets of colour texture images (Section 3.2) and show that partial orders in colour space can outperform grey-scale total ordering (Section 4).

2. Background

2.1. Rank Features

Rank features are a well established technique for dealing with noise in images, enforcing invariance to all sorts of contrast or illumination variations and sensor nonlinearities [7]. Because of their robustness, they were first developed in the context of wide-baseline stereo matching—see for instance the census and rank transforms [8]. More recently, descriptors in the popular Local Binary Pattern (LBP) family, including Texture Spectrum, Binary Gradient Contours, etc. [9,10] have turned rank features into a general purpose tool, with applications—among others—in texture classification, face recognition, surface inspection and content-based image retrieval [11]. The descriptive power of rank features has been expanded to explicitly capture orientation (Ranklets [12]) and second-order stimuli (Variance Ranklets [13]), all types of information that were seen as the preserve of linear filters or ad-hoc algorithms.

Common to all rank features is the fact that they are defined in terms of ordinal information between pixels only, with the actual pixel values being discarded. This can be done in terms of pairwise pixel comparisons (rank and census transform, LBP), pixel ranks (Ranklets) or a permutation of ranks (Variance Ranklets), but it is easy to see that the two approaches are equivalent [12] and that all descriptors rely on the natural order relation ($\leq$) between pixel values.

Before proceeding to definitions it is worth noting that, notwithstanding the trend towards the use of convolutional neural networks as feature extractors [14], rank features are still competitive in texture applications [15]. In the following section we recall the axioms for an order relation.

2.2. Order Relations

An order relation is an abstraction of the common notion of "greater than" used to compare numerical values, in our case pixel values in P (typically the set of 8-bit intensity values). In order to be called a (total) order, a binary relation $\leq$ needs to satisfy the following four conditions:

Definition 1 (Order axioms). *For all* $(x, y, z) \in P^3$,

1. $x \leq x$ *(reflexivity)*.
2. *if* $x \leq y$ *and* $y \leq x$ *then* $y = x$ *(antisymmetry)*.
3. *if* $x \leq y$ *and* $y \leq z$ *then* $x \leq z$ *(transitivity)*.
4. *either* $x \leq y$ *or* $y \leq x$ *(totality)*.

The last condition guarantees that we know how to compare any pair of pixel values.

2.3. Ordering High-Dimensional Data

The application of rank features to multi-channel images or higher dimensional data is hindered by the fact that there is no natural way of ordering multivariate data. It is certainly possible to provide a total order for a colour space; for instance, one could order RGB data lexicographically using the R channel as the primary sorting key, followed by G and finally by B. However, like other similar options, this has a disadvantage, namely, there are colours that are very close to each other in colour space, but very far in the order—and is therefore of limited practical interest (a sub-relation of the lexicographical order, the *product order*, is indeed of practical interest and will be discussed in detail in this paper, see Section 2.5.1). In general, it is best to resort to some sort of sub-ordering principle. These can be broadly divided in four categories [16]:

- Marginal ordering (*M-ordering*).
- Reduced (aggregate) ordering (*R-ordering*).
- Conditional (sequential) ordering (*C-ordering*).
- Partial ordering (*R-ordering*).

In marginal ordering, ranking is carried out on one or more components (*marginals*) of the multivariate data. Ranking colour data in the RGB space by the value of red is an example of M-ordering; lexicographical ordering is another one. Reduced (aggregate) ordering relies on converting multivariate data to univariate through suitable transformations. A common way to do this consists of establishing a reference point in the data space and using the distance from that point to rank the data. Conditional ordering occurs when we sort a random multivariate sample based on the corresponding (usually marginally-sorted) values of another sample. C-ordering is closely related to the concept of *concomitants* in Statistics [17]. Partial ordering will be discussed in detail in Section 2.5.

Interestingly, many of the common ways of dealing with order in colour space fall under the first two categories, i.e., marginal and reduced (aggregate) ordering. For instance, ranking based on intensity can be seen as a marginal ordering of the HSV space along the V axis; or as a reduced or aggregate ordering over the RGB space, where the aggregating function is the grey-level intensity. Other examples of aggregating orders will be given in Section 3.1.

In this paper, we will focus on rank features based on *partial orders*. Before introducing these, we review recent approaches to using multivariate orders on colour images.

2.4. Rank-Based Approaches to Colour Processing

Previous approaches to rank-based colour features typically extend grey-scale rank-based methods to the colour domain by considering either the colour channels separately (*intra*-channel features) and/or in pairwise combination (*inter*-channel features). Mäenpää and Pietikäinen [18] for instance extended classic LBP by applying it both to each R, G and B colour channel separately and pairwise between each of the R–G, R–B and G–B pairs. Bianconi et al. [19] adopted the same approach for extending grey-scale ranklets [12] to the colour domain. Lee et al. [20] defined Local Colour Vector Binary Patterns (LCVBP) by decomposing the colour triplets into a norm and angular component and by computing LBP on each of them. More recently, Cusano et al. [21] introduced Local Angular Patterns (LAP) which consider the angular component only and discard the norm part altogether.

Another possible strategy consists of establishing some sort of *a priori* total ordering on the colour data. This approach is not uncommon in colour morphology—see for instance Angulo [4], van De Gronde and Roerdink [6]—and has been advocated for extending LBP to colour images by Barra [22]. Of late, this family of methods has been extensively investigated by Ledoux et al. [1] and Bello-Cerezo et al. [23]. The problem is that imposing a total ordering on the colour data inevitably entails a certain degree of arbitrariness, with the consequence that the results tend to be dataset-dependent. On the other hand, morphology for tensor-valued images (that arise from certain magnetic resonance techniques) has relied on the Loewner order, that is in fact a partial order (see for instance Burgeth et al. [24]; more on this in Section 2.5.2). More recently, this approach has been extended to morphology for colour images Burgeth and Kleefeld [5]. Partial ordering circumvents the problem of ordering multivariate data totally, at the expense of not allowing comparisons between some colour values.

2.5. Partial Orders

A partial order differs from a total order in that the fourth axiom in Definition 1 is waived, i.e., there are pairs of elements in the set that are incomparable. In order to distinguish this from a total order we use the notation $x \preceq y$. If the elements x and y are incomparable, we shall write $x \nsim y$.

In the following, we describe two types of partial orders that are applicable to colour spaces with Cartesian and polar coordinates respectively.

2.5.1. Product Order

By product order we mean the relation obtained from the component-wise comparison of colour values. Given $\mathbf{u} = (c_{1u}, c_{2u}, c_{3u})$ and $\mathbf{v} = (c_{1v}, c_{2v}, c_{3v})$ two triplets representing colours in a generic space we write:

$$\begin{cases} \mathbf{u} \preceq_\times \mathbf{v} & \text{if } c_{1u} \leq c_{1v}, c_{2u} \leq c_{2v}, c_{3u} \leq c_{3v} \\ \mathbf{u} \nsim_\times \mathbf{v} & \text{if neither } \mathbf{u} \preceq_\times \mathbf{v} \text{ nor } \mathbf{v} \preceq_\times \mathbf{u}. \end{cases} \quad (1)$$

Note that this is a subset of the lexicographical order introduced in Section 2.3; it is however of higher practical interest as it treats all three channels symmetrically. In the RGB space, for instance, a given colour $\mathbf{u}$ weakly dominates the rectangular parallelepiped $C(\mathbf{u})$ with three edges along the axes and a vertex in the colour itself (see Figure 1). For any colour $\mathbf{v}$ that does not dominate all of $C(\mathbf{u})$, $\mathbf{v} \nsim \mathbf{u}$.

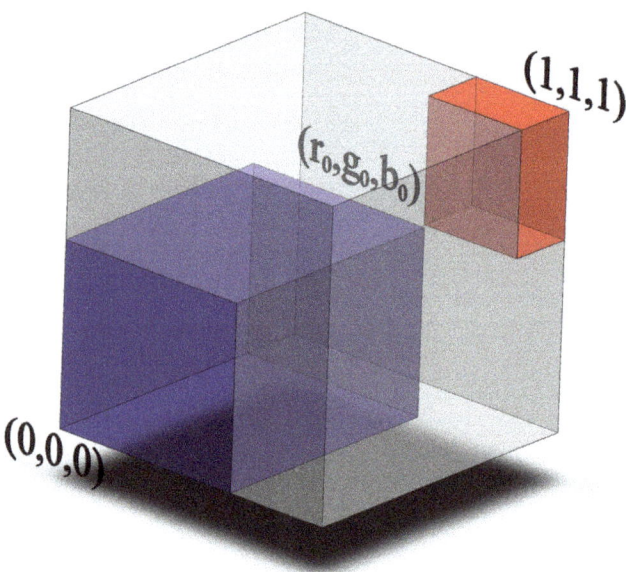

Figure 1. Product order in the RGB space: A generic colour (r_0, g_0, b_0) dominates all the colours in the blue volume and is dominated by all the colours in the red volume.

The product order can of course be applied to any colour space, giving relations of various degree of interpretability and effectiveness for pattern recognition (see Sections 3.1 and 4).

2.5.2. Loewner Order

The Loewner (partial) order is defined on symmetric matrices. Given two symmetric matrices $\mathbf{A}$, $\mathbf{B}$ we write:

$$\begin{cases} \mathbf{A} \preceq_\mathcal{L} \mathbf{B} & \text{if } (\mathbf{B} - \mathbf{A}) \in \mathcal{S}_+ \\ \mathbf{A} \nsim_\mathcal{L} \mathbf{B} & \text{if neither } \mathbf{A} \preceq_\mathcal{L} \mathbf{B} \text{ nor } \mathbf{B} \preceq_\mathcal{L} \mathbf{A} \end{cases} \quad (2)$$

where $\mathcal{S}_+$ indicates the set of positive semi-definite matrices. Applying this to a colour space requires mapping colour values to symmetric matrices. Following [5], we start from a modified colour space HCL̃ obtained from HSL ([25], Section 4.6) by setting $\tilde{L} = 2L - 1$ for the (modified) luminance and replacing saturation with chroma $C = \max\{R, G, B\} - \min\{R, G, B\}$. The resulting colour gamut fills a bicone with axis $\tilde{L}$ and opening angle 90°. We isometrically map colours to the space $Sym(2)$ of symmetric 2×2 matrices by setting [5]:

$$\mathbf{M}(h, c, \tilde{l}) = \frac{1}{\sqrt{2}} \begin{pmatrix} \tilde{l} - c & h \\ h & \tilde{l} + c \end{pmatrix}. \tag{3}$$

For two colours $\mathbf{u} = (h_u, c_u, \tilde{l}_u)$ and $\mathbf{v} = (h_v, c_v, \tilde{l}_v)$ in the HC$\tilde{L}$ space we therefore write:

$$\begin{cases} \mathbf{u} \preceq_L \mathbf{v} & \text{if } \mathbf{M}(h_v, c_v, \tilde{l}_v) \preceq_{\mathcal{L}} \mathbf{M}(h_u, c_u, \tilde{l}_u) \\ \mathbf{u} \nsim_L \mathbf{v} & \text{if neither } \mathbf{u} \preceq_L \mathbf{v} \text{ nor } \mathbf{v} \preceq_L \mathbf{u} \end{cases} \tag{4}$$

where $\preceq_{\mathcal{L}}$ is defined in Equation (2). Geometrically (Figure 2), a given colour $\mathbf{v}$ weakly dominates all colours of lower luminance that fall in a cone with its vertex in $\mathbf{v}$ and its axis parallel to the $\tilde{L}$ axis.

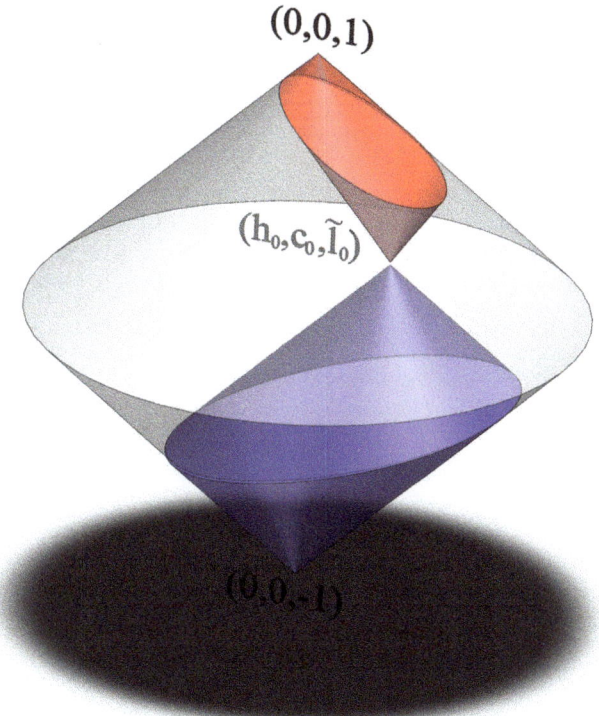

Figure 2. Loewner order in the HC$\tilde{L}$ space: A generic colour $(h_0, c_0, \tilde{l}_0)$ dominates all the colours in the blue volume and is dominated by all the colours in the red volume.

3. Materials and Methods

3.1. Rank Features on Partial Orders

In this section we show how to generalise existing rank-based descriptors by replacing total order in grey-scale with suitable partial orders in colour space. In the remainder we shall use the Texture Spectrum [26] as our reference model—though other descriptors such as Local Binary Patterns and Local Ternary Patterns are amenable to the same procedure with virtually no effort.

In Texture Spectrum, a local image pattern $\mathcal{P} = \{\mathbf{p}_0, \mathbf{p}_1, \ldots, \mathbf{p}_n\}$ is assigned a unique decimal code as follows:

$$f_{\text{TS}}(\mathcal{P}) = \sum_{i=1}^{n} 3^i \tau \left[g(\mathbf{p}_0), g(\mathbf{p}_i) \right] \tag{5}$$

where $\mathbf{p}_0$ represents the central pixel and $\mathbf{p}_i$, $i \in \{1, \ldots, n\}$ the peripheral pixels, which we assume to be arranged on a circle around the central pixel. We also assume that $\mathbf{p}$ represents a point in a 3D colour space, though again extension to multi-spectral data is straightforward. In Equation (5) the function $g(\mathbf{p}_0)$ stands for a generic conversion from colour into grey-scale, whereas $\tau(u, w)$ indicates the ternary thresholding function:

$$\tau(w, z) = \begin{cases} 0 & \text{if } w < z \\ 1 & \text{if } w = z \\ 2 & \text{if } w > z. \end{cases} \tag{6}$$

An image is represented by the dense, orderless statistical distribution over the set of possible codes. For Texture Spectrum, the number of (directional) features generated by the method is clearly 3^n. Invariance under rotations and/or reflections can be obtained by grouping together all those codes that represent patterns which can be transformed into one another by such transforms. The corresponding mathematical structures are *necklaces* and *bracelets*, respectively for invariance under rotations (i.e., cyclic group of order n; C_n) and under rotations + reflections (i.e., dihedral group of order n; D_n). For general formulas about the number of resulting C_n- and D_n-invariant features and for other mathematical details please refer to González et al. [27], Zelenyuk and Zelenyuk [28]. Specifically, for $n = 8$ (which is the case considered herein—see below) the number of features is respectively 834 and 498.

A ternary rank feature for partially ordered data analogous to Texture Spectrum—the Partial Order Texture Spectrum (POTS)—can easily be defined in the following way:

$$f_{\text{POTS}}(\mathcal{P}) = \sum_{i=1}^{n} 3^i \varphi(\mathbf{p}_0, \mathbf{p}_i) \tag{7}$$

$$\varphi(\mathbf{u}, \mathbf{v}) = \begin{cases} 0 & \text{if } \mathbf{u} \prec \mathbf{v} \\ 1 & \text{if } \mathbf{u} \succeq \mathbf{v} \\ 2 & \text{if } \mathbf{u} \nsim \mathbf{v} \end{cases} \tag{8}$$

where $\preceq$ indicates a generic partial order relation in the colour space (see Section 2.5). Notably, the number of features generated by this formulation is the same as generated by the Texture Spectrum.

In the experiments we considered the following partial order/colour space combinations: product order (Section 2.5.1) in the RGB, Ohta's and opponent spaces [25]; Loewner order (Section 2.5.2) in the HCL̃ space. When reporting experimental results we use subscripts 'RGB', 'ohta' and 'opp' to indicate the colour spaces, and superscripts $\preceq_\times$ and $\preceq_L$ respectively for the product and Loewner orders (see Equations (1) and (4)). No superscript was used to indicate the natural total order on greyscale values.

Conversion from RGB to grey-scale was also performed in three different ways: (1) through the standard PAL/NTSC formula ([25], Section 4.3.1); (2) by computing the average of the three channels; and (3) by determining, for each image, the principal axes of the colour distribution in the RGB space and projecting each (r, g, b) triplet onto the first axis. In the remainder we denote the corresponding variations of Texture Spectrum respectively as TS_{grey}, TS_μ and TS_{p1}.

Finally, we computed C_n- and D_n-invariant features over 3×3, non-interpolated, square neighbourhoods of radius 1px and 2px. The overall feature vector was obtained by concatenating the feature vectors obtained at each resolution—see also González et al. [27] for details. These settings respectively generates $834 \times 2 = 1668$ and $498 \times 2 = 996$ features.

3.2. Experiments

To test the effectiveness of the partial-order rank features described in Section 3.1 we ran a set of supervised image classification experiments. Datasets, classification strategy and accuracy estimation are described in the following subsections.

3.3. Datasets

We used ten datasets of colour texture images from different sources as described below. The main properties of each dataset are summarised in Table 1.

Table 1. Datasets used in the experiments: round-up table.

ID	Name	No. of Classes	No. of Samples per Class	Variations in Imaging Conditions	Sample Images
1	Epistroma	2	825/551	Unspecified	
2	KTH-TIPS	10	81	illumination, rotation, scale	
3	KTH-TIPS2b	11	432	illumination, rotation, scale	
4	Kylberg-Sintorn	25	6	None	
5	MondialMarmi	25	16	None	
6	Outex-13	68	20	None	
7	Outex-14	68	60	illumination	
8	Pap smear	2	204	Unspecified	
9	PlantLeaves	20	60	None	
10	RawFooT	68	184	illumination	

KEY TO SYMBOLS: 💡 = illumination, ↻ = rotation, 🔍 = scale.

3.3.1. Epistroma

Contains 1376 histopathological images from colorectal cancer representing either *epithelium* (825 images) or *stroma* (551 images). The image size ranges from 93 px to 2372 px in width and from 94 px to 2373 px in height. Further details about tissue preparation and digitisation procedure are available in Linder et al. [29].

3.3.2. KTH-TIPS

Includes 10 classes of common materials (e.g., *aluminum foil, bread, corduroy,* etc.) with 81 image samples for each class [30,31]. Each material was acquired under nine scales, three rotation angles and three lighting directions.

3.3.3. KTH-TIPS2b

Features 11 classes of materials (432 sample images per class) and is actually an extension of KTH-TIPS. The image acquisition settings were the same as in KTH-TIPS, but four rather than three illumination conditions were used in this case [32].

3.3.4. Kylberg–Sintorn

Is composed of 25 classes of heterogeneous materials, such as food (e.g., *lentils, oatmeal* and *sugar*), fabric (e.g., *knitwear* and *towels*) and tiles [33,34]. For each class one sample image was acquired using invariable illumination conditions and under nine different rotation angles—of which only the images at 0° were included in our experiments. Each image was further subdivided into six non-overlapping sub-images of dimension 1728 × 1728 px.

3.3.5. MondialMarmi

Comprises 25 classes of marble and granite products identified by their commercial denominations, e.g., *Azul Platino, Bianco Sardo, Rosa Porriño* and *Verde Bahía* [35]. Each class is represented by four tiles; ten images for each tile were acquired under steady illumination conditions and at rotation angles from 0° deg to 90° in steps of 10°. In the experiments we only used the images at 0°; moreover, we subdivided each image into four non-overlapping sub-images therefore obtaining 16 image samples for each class.

3.3.6. OUTEX-13 and OUTEX-14

Are based on the same sets of images that respectively make up the OUTEX_ TC_00013 and OUTEX_TC_00014 test suites—see Ojala et al. [36] for details. Specifically, OUTEX-13 features 68 classes of materials with 20 images per class acquired under invariable illumination conditions; OUTEX-14 contains the same classes—but in this case the image samples were acquired under three different illumination conditions—therefore there are 60 samples per class. Please notice, however, that in order to maintain the same evaluation protocol for all the datasets considered here (see Section 3.4), the subdivisions into train and test sets used in our experiments were not the same as in the OUTEX_TC_00013 and OUTEX_TC_00014 test suites.

3.3.7. Pap Smear

Consists of 917 PAP-stained images of variable dimension representing cells from the cervix [37]. The images represent either *abnormal* cases—675 samples or *normal* cases—242 samples. The dataset also comes with a further subdivision into seven sub-classes which was not considered in our experiments. The image size ranges from 84 × 88 px to 392 × 262 px. In our experiments we considered a balanced sub-set containing 204 samples for each of the two classes.

3.3.8. Plant Leaves

Includes a total of 1200 samples of plant leaves from 20 different classes with 60 samples per class [38]. The images were acquired using a planar scanner and have a dimension of 128 × 128 px.

3.3.9. RawFooT

Comprehends 68 classes of raw food and grains such as *corn, chicken breast, pomegranate, salmon* and *tuna* [39,40]. The materials were acquired under 46 different illumination conditions resulting in as many image samples for each class. We further subdivided the images into four non-overlapping sub-images, thus obtaining 184 samples for each class. The dimension of the resulting image tiles was 400 × 400 px.

3.4. Classification and Accuracy Estimation

For each dataset described in Section 3.3 we performed supervised classification using a nearest neighbour classifier (1-NN) with the L_1 ('Manhattan') distance. In detail, after extracting a feature vector from all images according to one of the descriptors tested, we computed the distance between such vectors as the sum of the absolute differences between components. We then assigned each test vector to the class of the closest training vector. The absence of tuning parameters, the ease of implementation and other desirable asymptotic properties make the 1-NN particularly appealing for comparison purposes. Its use in related works is indeed customary: see for instance Cusano et al. [39], Kandaswamy et al. [41], Liu et al. [42].

Accuracy estimation was based on split-half validation with stratified sampling—for each dataset we used half of the samples of each class to train the classifier (*train set*) and the other half (*test set*) to compute the accuracy. This was defined as the ratio between of number of samples of the test set correctly classified (N_c) and the total number of samples of the test set (N):

$$a = \frac{N_c}{N}. \qquad (9)$$

For a stable estimation we averaged the above value over a hundred different subdivisions into train and test set:

$$\hat{a} = \frac{\sum_{i=1}^{100} a_i}{100}, \qquad (10)$$

where a_i indicates the accuracy achieved in the i-th subdivision into train and test set. In Table 2 we report the 95% confidence intervals for $\hat{a}$ (computed under the simplifying assumption of normal distribution).

Table 2. Overall classification accuracy: confidence intervals for the cross-validated accuracy $\hat{a}$. Best results highlighted for grey-level (orange) and colour space features (blue). Boldface figures indicate statistically significant differences.

Descriptor	Inv.	Dataset									
		1	2	3	4	5	6	7	8	9	10
LBP_{grey}	C_n	91.3–91.7	93.2–93.8	92.6–92.8	93.4–94.6	79.5–79.9	82.5–82.7	97.4–97.7	81.5–82.2	72.9–73.5	94.3–94.5
LBP_{grey}	D_n	91.5–91.9	93.3–93.9	92.5–92.7	93.4–94.6	79.7–80.1	82.4–82.7	97.3–97.6	81.1–81.9	73.0–73.6	94.3–94.5
TS_{grey}	C_n	91.3–91.8	92.4–93.0	94.3–94.5	95.2–96.2	79.5–79.9	82.5–82.7	97.7–98.1	83.5–84.3	74.5–75.1	96.1–96.2
TS_{grey}	D_n	91.8–92.2	92.3–92.9	94.2–94.3	95.3–96.3	79.3–79.8	82.4–82.6	97.8–98.2	84.0–84.9	74.8–75.4	96.1–96.2
TS_μ	C_n	91.6–92.0	92.7–93.3	93.7–93.9	92.8–94.2	79.2–79.6	79.5–79.7	97.1–97.5	83.1–83.8	73.9–74.4	95.8–96.0
TS_μ	D_n	91.9–92.2	92.8–93.4	93.5–93.7	92.8–94.2	78.8–79.3	79.5–79.7	97.0–97.4	83.7–84.5	74.3–74.8	95.9–96.0
TS_{p1}	C_n	91.3–91.6	92.9–93.4	93.0–93.2	94.4–95.6	78.6–79.0	80.0–80.2	96.8–97.3	80.6–81.4	73.2–73.8	95.1–95.2
TS_{p1}	D_n	91.5–91.8	92.9–93.5	92.8–93.0	94.6–95.8	78.4–78.9	80.0–80.3	96.8–97.3	80.9–81.7	73.0–73.6	95.1–95.3
$POTS_{ohta}^{\leq \times}$	C_n	86.9–87.3	89.3–89.8	94.7–94.9	93.3–94.4	82.2–82.6	82.0–82.2	97.6–98.0	77.6–78.6	69.3–69.9	94.5–94.6
$POTS_{ohta}^{\leq \times}$	D_n	87.0–87.3	89.4–89.9	94.9–95.1	93.3–94.4	82.4–82.8	82.3–82.5	97.7–98.1	78.3–79.3	69.3–69.9	94.5–94.6
$POTS_{opp}^{\leq \times}$	C_n	90.3–90.6	89.9–90.5	95.1–95.3	95.3–96.5	80.1–80.5	78.9–79.2	97.5–97.9	73.7–74.7	64.1–64.7	93.1–93.3
$POTS_{opp}^{\leq \times}$	D_n	90.2–90.6	89.7–90.3	95.2–95.3	94.9–96.0	80.4–80.9	79.0–79.2	97.5–97.9	74.6–75.6	64.4–65.0	93.2–93.4
$POTS_{HCL}^{\leq L}$	C_n	91.4–91.8	93.6–94.1	95.3–95.5	90.7–92.5	73.8–74.3	73.4–73.7	96.1–96.7	82.9–83.7	67.7–68.3	95.3–95.5
$POTS_{HCL}^{\leq L}$	D_n	92.0–92.4	93.5–94.0	95.2–95.4	90.7–92.5	74.3–74.7	74.0–74.3	96.2–96.7	83.9–84.7	68.1–68.6	95.4–95.5
$POTS_{RGB}^{\leq \times}$	C_n	90.4–90.8	94.1–94.6	94.9–95.1	87.9–90.0	72.6–73.0	74.5–74.7	96.0–96.6	82.6–83.4	75.2–75.8	96.4–96.5
$POTS_{RGB}^{\leq \times}$	D_n	91.3–91.7	94.0–94.5	94.8–95.0	87.9–90.0	73.0–73.5	74.8–75.0	96.2–96.8	83.6–84.5	75.9–76.4	96.3–96.5

4. Results and Discussion

Table 2 reports the confidence intervals for the means of the overall classification accuracy (see Section 3.4). For each dataset we highlighted in orange the best result obtained by total-order grey-scale rank features; in blue the best result obtained by partial order rank features (POTS) in colour space. When there was a statistically significant difference between the two, the best figure was indicated in boldface. As can be seen, partial order rank features in colour space performed significantly better in five datasets out 10, whereas the reverse occurred in one dataset only (dataset six). In the remaining four datasets there was no significant difference between the two methods.

As for grey-scale rank features, the results show that in most cases (i.e., 8 datasets out of 10) the best performance was obtained using standard PAL/NTSC grey-scale conversion. By contrast, partial-order rank features denoted a higher dependence on the colour space used.

The computational cost of all the descriptors considered is roughly equivalent, as the number of features is the same and the complexity of computing a partial or total order comparison in colour space is comparable to the cost of a colour space transformation. Indeed, as we have just described, even the traditional TS requires a grey-scale conversion, the choice of which can be seen as an integral part of the descriptor.

In Table 3 we compare our results to published results obtained using rank-based descriptors in conjunction with other ordering methods in colour spaces. As can be seen, in most cases our partial-order based approach improves significantly over previous results. We should here emphasise that the computational requirements of our partial-order descriptors are not higher than those of the other ordering methods cited.

Table 3. Comparison with the results obtained by other ordering methods as reported in the references indicated. Key to symbols: 'cvn' = colour vector norm, 'lex' = lexicographic ordering, 'rcl' = preorder based on white as reference colour. Please refer to the cited works for further details.

Dataset	Best Result (Literature)			Best Result (This Paper)
	LBP_{cvn}	LBP_{lex}	LBP_{rcl}	
KTH-TIPS	94.3 [23]	94.3 [23]	94.0 [23]	94.1–94.6 ($POTS_{RGB}^{\preceq \times}/C_n$)
KTH-TIPS2b	92.3 [23]	92.3 [23]	92.1 [23]	95.3–95.5 ($POTS_{HCL}^{\preceq \times}/C_n$)
Kylberg-Sintorn	N/A	99.1 [43]	N/A	95.3–96.5 ($POTS_{RGB}^{\preceq \times}/C_n$)
Outex-13	85.3 [1]	86.3 [1]	85.9 [1]	82.3–82.5 ($POTS_{opp}^{\preceq \times}/D_n$)
Outex-14	74.3 [1]	73.4 [1]	72.3 [1]	97.7–98.1 ($POTS_{ohta}^{\preceq \times}/D_n$)
PapSmear	N/A	N/A	66.2 [43]	83.6–84.5 ($POTS_{RGB}^{\preceq \times}/D_n$)
PlantLeaves	69.9 [23]	65.2 [23]	71.9 [23]	75.9–76.4 ($POTS_{RGB}^{\preceq \times}/D_n$)
RawFoot	N/A	N/A	80.5 [43]	96.4–96.5 ($POTS_{RGB}^{\preceq \times}/C_n$)

5. Conclusions and Future Work

The lack of a natural order among colours represents an intrinsic impediment to the definition of rank features in colour space. In this paper we have introduced a novel and general approach based on partial orders. Partial orders overcome the problems inherent to ordering multivariate data at the expense of admitting that not all pairs of colours can be compared to each other. We showed that this scheme fits in well with existing grey-scale local image descriptors, that are amenable to extension to the colour domain with little effort. Taking the Texture Spectrum as a model, we showed that its partial-order version in colour space (POTS) can outperform the grey-scale classic descriptor while maintaining the same number of features and with comparable computational complexity. Previous studies have also demonstrated that the use of colour can improve texture discrimination, but at the expense of employing a higher number of features [44–46]. Notably, our approach improves on published results that use descriptors based specifically on (total) colour space ordering (see Table 3).

To the best of our knowledge this is the first time that partial orders have been used to define rank features for pattern recognition. The method is conceptually simple, fairly general and shows potential for application in a wide number of computer vision tasks. Future studies will be focussed on extending the approach to the broader class of descriptors known as Histograms of Equivalent Patterns [9]. The effect of the colour space on the performance of rank features based on partial orders is also an important topic for further investigation. Finally, the insertion of partial order based algorithms in more involved image processing pipelines (e.g., convolutional neural networks) also represents an interesting opportunity for future research; integration at the level of matching [47] has so far been successful.

Author Contributions: Conceptualization, F.S. and F.B.; Formal analysis, F.S., F.B. and A.F.; Methodology, F.S., F.B., A.F. and E.G.; Software, F.S. and F.B.; Validation, F.S., F.B., A.F. and E.G.; Visualization, F.B. and A.F.; Writing—original draft, F.S. and F.B.; Writing—review & editing, A.F. and E.G. All authors have read and agreed to the published version of the manuscript.

Funding: This work was partially supported by the Spanish Government under projects AGL2014-56017-R and TIN2014-56919-C3-2-R, and by the Department of Engineering at the Università degli Studi di Perugia (UniPG Eng), Italy, under project *Machine learning algorithms for the control of autonomous mobile systems and the automatic classification of industrial products and biomedical images* (Fundamental resarch grants 2017). F.S. performed part of this work as a

Visiting Researcher at UniPG Eng. He gratefully acknowledges the support of UniPG under international mobility grant 'D.R. n.2270/2015'.

Conflicts of Interest: The authors declare no conflict of interest.

References

1. Ledoux, A.; Losson, O.; Macaire, L. Color local binary patterns: Compact descriptors for texture classification. *J. Electron. Imaging* **2016**, *25*, 061404. [CrossRef]
2. Hanbury, A.; Serra, J. Mathematical morphology in the CIELAB space. *Image Anal. Stereol.* **2002**, *21*, 201–206. [CrossRef]
3. Aptoula, E.; Lefèvre, S. A comparative study on multivariate mathematical morphology. *Pattern Recognit.* **2007**, *40*, 2914–2929. [CrossRef]
4. Angulo, J. Morphological colour operators in totally ordered lattices based on distances: Application to image filtering, enhancement and analysis. *Comput. Vis. Image Underst.* **2007**, *107*, 56–73. [CrossRef]
5. Burgeth, B.; Kleefeld, A. Morphology for color images via Loewner order for matrix fields. In *Lecture Notes in Computer Science (Including Subseries Lecture Notes in Artificial Intelligence and Lecture Notes in Bioinformatics)*; Springer: Berlin/Heidelberg, Germany, 2013; Volume 7883 LNCS, pp. 243–254.
6. van De Gronde, J.; Roerdink, J. Group-invariant colour morphology based on frames. *IEEE Trans. Image Process.* **2014**, *23*, 1276–1288. [CrossRef] [PubMed]
7. Hodgson, R.; Bailey, D.; Naylor, M.; Ng, A.; McNeill, S. Properties, implementations and applications of rank filters. *Image Vis. Comput.* **1985**, *3*, 3–14. [CrossRef]
8. Zabih, R.; Woodfill, J. Non-parametric Local Transforms for Computing Visual Correspondence. In *European Conference on Computer Vision*; Springer: Stockholm, Sweden, 1994; pp. 151–158.
9. Fernández, A.; Álvarez, M.X.; Bianconi, F. Texture description through histograms of equivalent patterns. *J. Math. Imaging Vis.* **2013**, *45*, 76–102. [CrossRef]
10. Liu, L.; Fieguth, P.; Wang, X.; Pietikäinen, M.; Hu, D. Evaluation of LBP and deep texture descriptors with a new robustness benchmark. In Proceedings of the 14th European Conference on Computer Vision (ECCV 2016), Amsterdam, The Netherlands, 11–14 October 2016; Springer: Amsterdam, The Netherlands, 2016; Volume 9907, pp. 69–86.
11. Brahnam, S.; Jain, L.; Nanni, L.; Lumini, A. (Eds.) *Local Binary Patterns: New Variants and Applications*; Studies in Computational Intelligence; Springer: Berlin/Heidelberg, Germany, 2014; Volume 506.
12. Smeraldi, F. Ranklets: Orientation selective non-parametric features applied to face detection. In Proceedings of the 16th International Conference on Pattern Recognition (ICPR'02), Quebec City, QC, Canada, 11–15 August 2002; Volume 3, pp. 379–382.
13. Azzopardi, G.; Smeraldi, F. Variance Ranklets: Orientation-selective rank features for contrast modulations. In Proceedings of the British Machine Vision Conference, BMVC 2009, London, UK, 7–10 September 2009.
14. Liu, L.; Chen, J.; Fieguth, P.; Zhao, G.; Chellappa, R.; Pietikäinen, M. From BoW to CNN: Two decades of texture representation for texture classification. *Int. J. Comput. Vis.* **2019**, *127*, 74–109. [CrossRef]
15. Bello-Cerezo, R.; Bianconi, F.; Di Maria, F.; Napoletano, P.; Smeraldi, F. Comparative Evaluation of Hand-Crafted Image Descriptors vs. Off-the-Shelf CNN-Based Features for Colour Texture Classification under Ideal and Realistic Conditions. *Appl. Sci.* **2019**, *9*, 738. [CrossRef]
16. Barnett, V. The ordering of multivariate data. *J. R. Stat. Soc. Ser. A (Gen.)* **1976**, *139*, 318–355. [CrossRef]
17. Yang, S. Distribution Theory of the Concomitants of Order Statistics. *Ann. Stat.* **1977**, *5*, 996–1002. [CrossRef]
18. Mäenpää, T.; Pietikäinen, M. Texture analysis with local binary patterns. In *Handbook of Pattern Recognition and Computer Vision*, 3rd ed.; Chen, C., Wang, P., Eds.; World Scientific: Singapore, 2005; pp. 197–216.
19. Bianconi, F.; Fernández, A.; González, E.; Armesto, J. Robust color texture features based on ranklets and discrete Fourier transform. *J. Electron. Imaging* **2009**, *18*, 043012.
20. Lee, S.; Choi, J.; Ro, Y.; Plataniotis, K. Local color vector binary patterns from multichannel face images for face recognition. *IEEE Trans. Image Process.* **2012**, *21*, 2347–2353. [CrossRef] [PubMed]
21. Cusano, C.; Napoletano, P.; Schettini, R. Local angular patterns for color texture classification. In *Lecture Notes in Computer Science (Including Subseries Lecture Notes in Artificial Intelligence and Lecture Notes in Bioinformatics)*; Springer: Berlin/Heidelberg, Germany, 2015; Volume 9281, pp. 111–118.

22. Barra, V. Expanding the local binary pattern to multispectral images using total orderings. *Commun. Comput. Inf. Sci.* **2011**, *229 CCIS*, 67–80.
23. Bello-Cerezo, R.; Fieguth, P.; Bianconi, F. LBP-Motivated Colour Texture Classification. In Proceedings of the 2nd International Workshop on Compact and Efficient Feature Representation and Learning in Computer Vision (in Conjunction with ECCV 2018), Munich, Germany, 9 September 2018; Volume 11132, pp. 517–533.
24. Burgeth, B.; Welk, M.; Feddern, C.; Weickert, J. Mathematical morphology on tensor data using the Loewner ordering. In *Visualization and Processing of Tensor Fields*; Mathematics and Visualization; Springer: Berlin/Heidelberg, Germany, 2006; pp. 357–368.
25. Palus, H. Representations of colour images in different colour spaces. In *The Colour Image Processing Handbook*; Sangwine, S.J., Horne, R.E.N., Eds.; Springer: Berlin/Heidelberg, Germany, 1998; pp. 67–90.
26. He, D.C.; Wang, L. Texture Unit, Texture Spectrum, And Texture Analysis. *IEEE Trans. Geosci. Remote. Sens.* **1990**, *28*, 509–512.
27. González, E.; Bianconi, F.; Fernández, A. An investigation on the use of local multi-resolution patterns for image classification. *Inf. Sci.* **2016**, *361–362*, 1–13. [CrossRef]
28. Zelenyuk, Y.; Zelenyuk, Y. Counting symmetric bracelets. *Bull. Aust. Math. Soc.* **2014**, *89*, 431–436. [CrossRef]
29. Linder, N.; Konsti, J.; Turkki, R.; Rahtu, E.; Lundin, M.; Nordling, S.; Haglund, C.; Ahonen, T.; Pietikäinen, M.; Lundin, J. Identification of tumor epithelium and stroma in tissue microarrays using texture analysis. *Diagn. Pathol.* **2012**, *7*, 22. [CrossRef]
30. Hayman, E.; Caputo, B.; Fritz, M.; Eklundh, J.O. On the Significance of Real-World Conditions for Material Classification. In Proceedings of the 8th European Conference on Computer Vision (ECCV 2004), Prague, Czech Republic, 11–14 May 2004; Springer: Prague, Czech Republic, 2004; Volume 3024, pp. 253–266.
31. The KTH-TIPS and KTH-TIPS2 Image Databases. 2004. Available online: http://www.nada.kth.se/cvap/databases/kth-tips/ (accessed on 21 September 2016).
32. Caputo, B.; Hayman, E.; Mallikarjuna, P. Class-specific material categorisation. In Proceedings of the Tenth IEEE International Conference on Computer Vision (ICCV'05), Beijing, China, 17–20 October 2005; Volume II, pp. 1597–1604.
33. Kylberg, G. Automatic Virus Identification Using TEM. Image Segmentation and Texture Analysis. Ph.D. Thesis, Faculty of Science and Technology, University of Uppsala, Uppsala, Sweden, 2014.
34. Kylberg Sintorn Rotation Dataset. 2013. Available online: http://www.cb.uu.se/gustaf/KylbergSintornRotation/ (accessed on 6 January 2016).
35. Bello-Cerezo, R.; Bianconi, F.; Fernández, A.; González, E.; Di Maria, F. Experimental comparison of color spaces for material classification. *J. Electron. Imaging* **2016**, *25*, 061406. [CrossRef]
36. Ojala, T.; Pietikäinen, M.; Mäenpää, T.; Viertola, J.; Kyllönen, J.; Huovinen, S. Outex—New Framework for Empirical Evaluation of Texture Analysis Algorithms. In Proceedings of the 16th International Conference on Pattern Recognition (ICPR'02), Quebec, QC, Canada, 11–15 August 2002; Volume 1, pp. 701–706.
37. Jantzen, J.; Noras, J.; Dounias, G.; Bjerregaard, B. Pap-smear Benchmark Data For Pattern Classification. In *Nature Inspired Smart Information Systems (NiSIS 2005)*; NiSIS: Albufeira, Portugal, 2005.
38. Casanova, D.; de Mesquita Sá Junior, J.J.; Bruno, O.M. Plant leaf identification using Gabor wavelets. *Int. J. Imaging Syst. Technol.* **2009**, *19*, 236–243. [CrossRef]
39. Cusano, C.; Napoletano, P.; Schettini, R. Evaluating color texture descriptors under large variations of controlled lighting conditions. *J. Opt. Soc. Am. A* **2016**, *33*, 17–30. [CrossRef] [PubMed]
40. RawFooT DB: Raw Food Texture Database. 2015. Available online: http:projects.ivl.disco.unimib.it/rawfoot/ (accessed on 22 September 2016).
41. Kandaswamy, U.; Schuckers, S.A.; Adjeroh, D. Comparison of Texture Analysis Schemes Under Nonideal Conditions. *IEEE Trans. Image Process.* **2011**, *20*, 2260–2275. [CrossRef] [PubMed]
42. Liu, L.; Zhao, L.; Long, Y.; Kuang, G.; Fieguth, P.W. Extended local binary patterns for texture classification. *Image Vis. Comput.* **2012**, *30*, 86–99. [CrossRef]
43. Fernández, A.; Lima, D.; Bianconi, F.; Smeraldi, F. Compact Color Texture Descriptor Based on Rank Transform and Product Ordering in the RGB Color Space. In Proceedings of the IEEE International Conference on Computer Vision Workshops, ICCVW 2017, Venice, Italy, 22–29 October 2017; Institute of Electrical and Electronics Engineers Inc.: Venice, Italy, 2017; pp. 1032–1040.
44. Drimbarean, A.; Whelan, P.F. Experiments in colour texture analysis. *Pattern Recognit. Lett.* **2001**, *22*, 1161–1167. [CrossRef]

45. Mäenpää, T.; Pietikäinen, M. Classification with color and texture: Jointly or separately? *Pattern Recognit.* **2004**, *37*, 1629–1640. [CrossRef]
46. Bianconi, F.; Harvey, R.; Southam, P.; Fernández, A. Theoretical and experimental comparison of different approaches for color texture classification. *J. Electron. Imaging* **2011**, *20*, 043006. [CrossRef]
47. Abdollahyan, M.; Cascianelli, S.; Bellocchio, E.; Costante, G.; Ciarfuglia, T.A.; Bianconi, F.; Smeraldi, F.; Fravolini, M.L. Visual Localization in the Presence of Appearance Changes Using the Partial Order Kernel. In Proceedings of the 2018 26th European Signal Processing Conference (EUSIPCO), Roma, Italy, 3–7 September 2018; pp. 697–701.

© 2019 by the authors. Licensee MDPI, Basel, Switzerland. This article is an open access article distributed under the terms and conditions of the Creative Commons Attribution (CC BY) license (http://creativecommons.org/licenses/by/4.0/).

Article

Detection of Tampering by Image Resizing Using Local Tchebichef Moments

Dengyong Zhang [1,†] , Shanshan Wang [1,†], Jin Wang [1,*], Arun Kumar Sangaiah [2], Feng Li [1] and Victor S. Sheng [3]

1. Hunan Provincial Key Laboratory of Intelligent Processing of Big Data on Transportation, School of Computer and Communication Engineering, Changsha University of Science and Technology, Changsha 410114, China
2. School of Computing Science and Engineering, Vellore Institute of Technology (VIT), Vellore 632014, India
3. Department of Computer Science, Texas Tech University, Lubbock, TX 79409, USA
* Correspondence: jinwang@csust.edu.cn; Tel./Fax: +86-(0731)85603438
† These authors contributed equally to this work.

Received: 30 June 2019; Accepted: 23 July 2019; Published: 26 July 2019

Abstract: There are many image resizing techniques, which include scaling, scale-and-stretch, seam carving, and so on. They have their own advantages and are suitable for different application scenarios. Therefore, a universal detection of tampering by image resizing is more practical. By preliminary experiments, we found that no matter which image resizing technique is adopted, it will destroy local texture and spatial correlations among adjacent pixels to some extent. Due to the excellent performance of local Tchebichef moments (LTM) in texture classification, we are motivated to present a detection method of tampering by image resizing using LTM in this paper. The tampered images are obtained by removing the pixels from original images using image resizing (scaling, scale-and-stretch and seam carving). Firstly, the residual is obtained by image pre-processing. Then, the histogram features of LTM are extracted from the residual. Finally, an error-correcting output code strategy is adopted by ensemble learning, which turns a multi-class classification problem into binary classification sub-problems. Experimental results show that the proposed approach can obtain an acceptable detection accuracies for the three content-aware image re-targeting techniques.

Keywords: image resizing; local Tchebichef moments (LTM); scaling; scale-and-stretch; seam carving

1. Introduction

As image editing tools and various mobile devices are easily acquired and conveniently used, maximizing the viewing experience of end users on small devices becomes very important. Compared to traditional image re-targeting methods, such as linear scaling and cropping, many content-aware image resizing methods can preserve salient areas, avoiding serious distortions or loss of significant information[1–3]. Meanwhile, many content-aware resizing algorithms have been adopted using image editing tools, such as photoshop and GIMP. An ordinary user can very easily create tampered images for malicious purposes using image editing tools. Moreover, it is impossible to distinguish those tampered images from authentic images with the naked eye. Therefore, how to detect tampered images is a hot topic in the field of image content security.

In recent years, a few approaches have existed about the detection of content-aware image re-targeting. Moreover, most of the detection methods are for the seam carved images. Lu et al. adopted a forensic hash to distinguish whether an image is subjected to a seam carving operation [4]. However, it is an active forensics approach; moreover, a falsifier might remove the forensic hash. For passive forensic detection, Sarkar et al. used 324D Markov features to detect image seam carving [5]. Later,

Fillion et al. exploited a series of intrinsic features to expose the trace of seam carved images [6]. In Wei et al. [7], an approach based patch analysis was adopted to distinguish whether an image is original or not. According to noise and energy distribution in seam carved images, Ryu et al. [8] exploited the features based on noise and energy bias to detect seam carved images. Local binary pattern (LBP) was adopted to detect seam carved images [9] in our recent work. Inspired by image entropy with the ability of capturing the intrinsic information of an image, we exploited multi-scale spectral and spatial entropies to detect seam carved images with low resizing ratios [10]. Web Local Descriptor (WLD) and LBP were adopted to distinguish whether an image is original or not [11]. In [12], a large feature mining approach was proposed to detect image seam carving under recompression in joint photographic experts group (JPEG) images.

However, most existing detections of image resizing are designed for a specific content-aware resizing. Much less has been done to distinguish different content-aware resizing approaches in the process of image re-targeting. In practice, the best re-targeting method relies on an image itself. For example, scaling images in horizontal or vertical direction can be performed in real time using interpolation and will preserve the global visual effects and retarget images with medium perceptual quality. However, scaling will introduce some shape deformation into the retarget image. Seam carving [1] supports various visual saliency measures for defining the energy of an image. Nevertheless, seam carving can excessively carve less important parts of an image and result in unwanted visual distortions. scale-and-stretch [3]can preserve the aspect ratios of local objects. However, if there are many quads in the image, the approach will fail to preserve the aspect ratio of the whole image [2]. Therefore, there are different image resizing methods depending on the image content to achieve image change in size while preserving the saliency region. it is necessary to propose a universal detection of image resizing.

The rest paper is organized as follows: Section 2 summarizes several common methods of image re-targeting and analyzes their artifacts. Section 3 briefly introduces the proposed detection approach. Our experimental results are described and analyzed in detail in Section 4, and conclusions are made in Section 5.

2. Image Resizing and Their Possible Artifacts

2.1. Several Common Methods of Image Resizing

Among the methods of content-aware image resizing, scaling, seam carving [1], and scale-and-stretch [3] are three common approaches to re-target an image. Seam carving is defined by forward energy. The intensity gradient magnitude in L_1 metric is defined as an importance map. The contiguous chains of pixels that pass through the regions of the least importance in the image are deleted or duplicated by seam carving to obtain image resizing. Dynamic programming is adopted to compute seams. Scale-and-stretch is defined warping. Both image dimensions are processed by warping at once. Moreover, an objective function is optimized to allow important regions uniformly scaling in order to preserve their shapes. A saliency and combination of L_2 gradient magnitude (defined by [13]) is defined as the importance map. Scaling implemented image resizing by simply bi-cubic interpolation and non-uniform scaling.

According to the above description of the three resizing methods, it is clearly found that scale-and-stretch can keep significant regions in an image, which is consistent with its original image after the image is re-targeted. However, seam carving implements image re-targeting by deleting or inserting pixels within a minimal energy; therefore, it may cause the salient object distortion.

2.2. Analysis of Image Resizing Artifacts

There exist three kinds of artifacts for the image processed by the content-aware resizing method [13], such as geometric deformation, information loss and local texture distortion. Figure 1 shows these artifacts caused by content-aware re-targeting. Figure 1b shows line or edge distortion after re-targeting. However, salient areas of an image, such as people and building, do not significantly change. The shape distortion of an image is shown in Figure 1c. This further explains that removed pixels might exist in salient areas of the image when a pixel with a minimum energy is deleted in the process of the image seam carving. Therefore, it is easy to deform important objects of the image in the process of image resizing. Figure 1d shows the artifact of information loss. A scaling method is bi-cubic interpolation and re-targets the entire image in the process of image re-targeting. To better show the image distortion in the process of image resizing, we adopt an LTM histogram to identify the distortion caused by different resizing methods in this paper. Figure 2 shows the residual LTM diagram for three different resizing methods. It can be found from Figure 2b that the image distortions of a non-subject area are not easily perceived in Figure 1b. However, these distortions can be clearly found in the residual LTM diagram. It can be clearly showed that the distortions are found in the process of image re-targeting in Figure 2d.

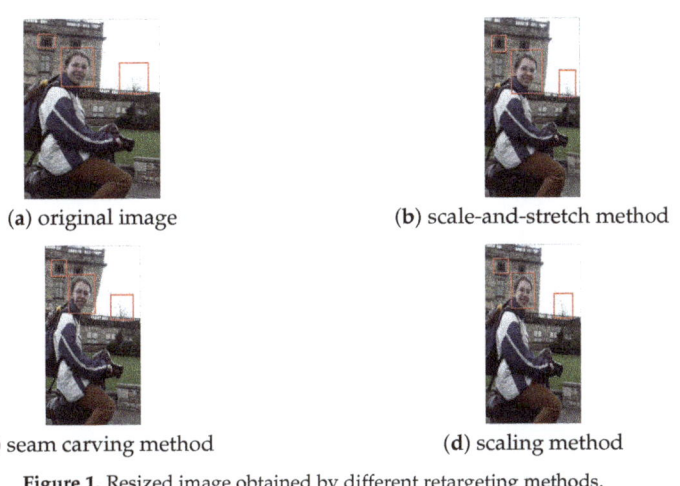

Figure 1. Resized image obtained by different retargeting methods.

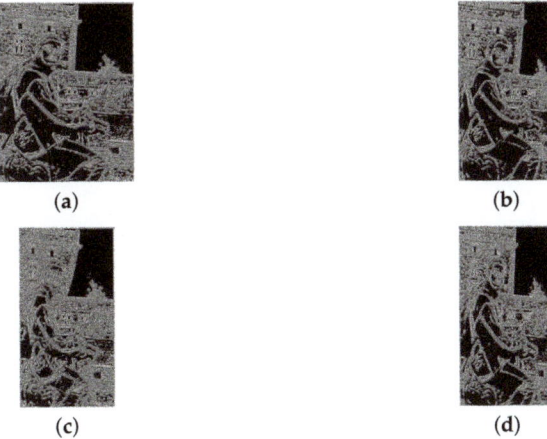

Figure 2. Residual LTM (local Tchebichef moments) diagram obtained by different re-targeting methods: (a–d) correspond to the residual LTM diagram of Figure 1a–d, respectively.

3. Proposed Method

A passive detection method is presented for image resizing forgery detection in this paper. Figure 3 shows the implementation process diagram of our presented algorithm. Similar to the process of most existing forensics methods, our proposed method consists of two parts, i.e., a training part and a testing part. In the training process, tampered images and their corresponding original images are adopted as data sets. First, all the training images are preprocessed. Second, the LTM histogram features are extracted from preprocessed images. Finally, a training model is obtained by using ensemble learning based on extracted features. In the testing process, the LTM histogram features are extracted according to the same steps in the training part. Finally, the extracted features are used by the trained ensemble classifier to distinguish which resizing method a tested image is re-targeted.

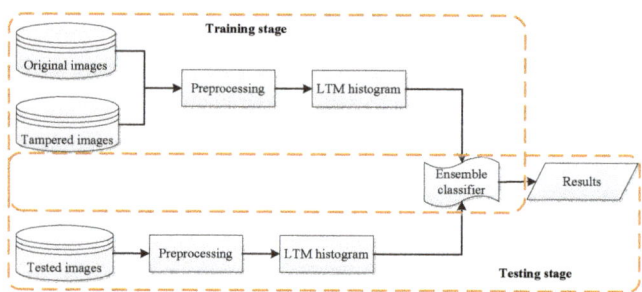

Figure 3. A block diagram of our proposed approach.

3.1. Preprocessing

An image obtained by content-aware resizing methods usually has a good visual effect. Furthermore, it is impossible for users to distinguish from authentic photographs using the naked eye. However, the correlation between adjacent pixels will inevitably change after an image is resized. Therefore, it is necessary to preprocess re-targeted images. Image residuals can efficiently capture the change of adjacent pixels in the process of image re-targeting. In this paper, a one-dimensional low-pass filter is adopted to calculate residuals along the horizontal and the vertical directions. The formula is shown as Equation (1):

$$R(x,y) = I(x,y) - I(x,y) * L(u), \qquad (1)$$

where $I(x,y)$ is an image and $L(u)$ is low-pass filter. Figure 4 shows the residuals of the preprocessed content-aware image resizing. Through the residuals, it can be clearly found that tampered traces are caused by different content-aware image resizing methods.

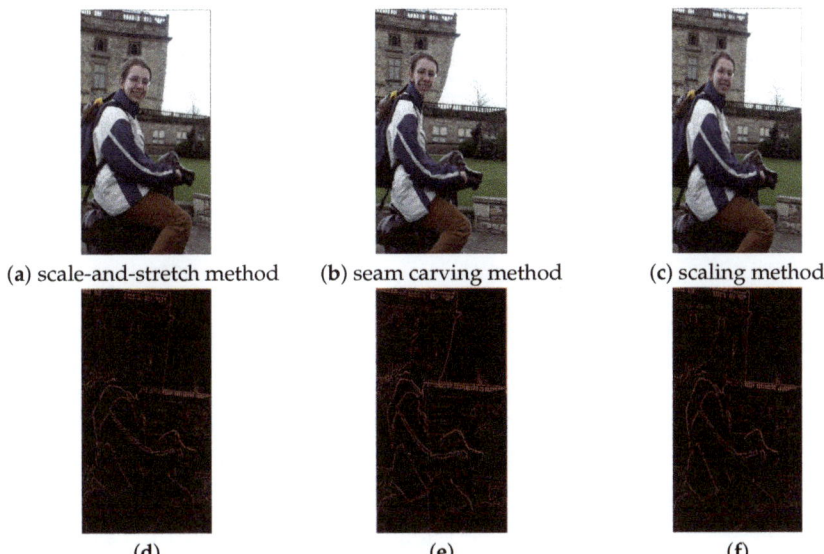

Figure 4. Tampered images and correspond residual images obtained by different re-targeting methods: (**d–f**) corresponds to the residual LTM diagram of (**a–c**), respectively.

3.2. Features of LTM

After images preprocessing, orthogonal Tchebichef moments are adopted to construct feature vectors on 5 × 5 neighbor pixels. In addition, the texture information is encoded with Lehmer to represent the relative strength of moments. The extracted feature vectors are called LTM. A byte value for each pixel is provided, and an LTM diagram is generated by the encoding scheme. Therefore, the histogram features of LTM are adopted to identify whether an image is subjected to image resizing. Figure 5 shows the histogram features of LTM after preprocessing.

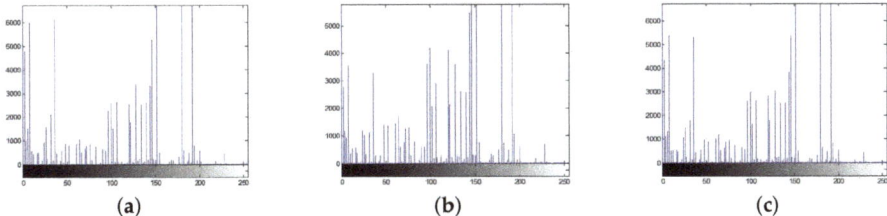

Figure 5. The histogram features of LTM: (**a–c**) corresponds to the LTM histogram of Figure 4d–f, respectively.

3.3. Ensemble Learning for Blind Forensics

In this paper, an error-correcting output codes (ECOC) strategy [14] based on ensemble learning is adopted, which transforms multi-class classification problems into binary classification sub-problems. This is because ECOC is an excellent multi-class classification tool, and the ensemble learning performs well in terms of computational complexity and detection accuracy. The tamper of three different resizing methods, such as scale-and-stretch, seam carving and scaling, is identified. Therefore, for this three-class classification problem, a pair coupling strategy [15] is adopted. Specifically, a discrete matrix (coding matrix) is defined first. In addition, then, the problem is decomposed into $n = 3$ binary classification sub-problems according to the sequence of 0 and 1 in the coding matrix, namely dichotomies. After that, ensemble learning is adopted to train these dichotomies and test the extracted histogram of LTM to obtain binary vectors. Finally, the class is identified by the minimum hamming distance between the encoded word and the vector.

4. Results

4.1. Experimental Environment

To verify the performance of our proposed algorithm, we conduct a number of experiments in our personal computer. The passive forensics approach is implemented in Matlab2012b. The ensemble learning can be downloaded from [16]. In this paper, the Uncompressed Colour Image Database (UCID) [17] is adopted as the original images. The image database contains 1338 images, which are composed of people, buildings, animals and landscapes. Since there is no publicly available image database of image resizing available, we construct an image library from UCID for resizing carving detection. According to different resizing ratios, three resizing methods are used to produce tampered images. The resizing ratios vary from 10% to 50% with a step size of 10%. That is, for every resizing method, the resizing ratios of tampered images are 10%, 20%, 30%, 40%, and 50%. Therefore, we have 1338 original images and $3 \times 5 \times 1338$ tampered images. To verify our proposed approach, we perform the performance evaluation for the following cases: (1) tamper detection for a single resizing method; (2) tamper detection for multiple resizing methods; and (3) tamper detection without preprocessing. In all experiments, the ECOC based on an ensemble learning strategy is adopted to test the effect of our proposed method. The image dataset is divided into two groups, 50% for training and 50% for testing. The training and testing are repeated ten times, and the average detection accuracy is reported in this paper.

4.2. Experimental Discussions

4.2.1. Tamper Forensics on a Single Resizing Method

In this experiment, we test the detection performance of our proposed method for a single resizing method. The tampered images with scaling ratios from 10% to 50% are adopted to test the performance of our proposed approach. Table 1 shows the detection results. From Table 1, we can see that the detection accuracy is improved with the increment of the scaling ratio. When the scaling ratio is less than 20%, our proposed approach has a higher detection accuracy for the scale-and-stretch resizing method. Since the optimal local scaling ratio of each local block is calculated iteratively and the warped image is updated simultaneously to match the scaling ratios as much as possible, the entire image is resized in the process of scale-and-stretch re-targeting. However, the seam carving method resizes an image by deleting the seams with the lowest energy once. Therefore, the tampered images obtained by the seam carving method are difficult to be distinguished from the authentic images when the scaling ratio is low. With the increment of the scaling ratio, the algorithm will cause a global structure distortion. It can also be reflected in Table 1. Our proposed approach can get a higher accuracy rate for images obtained using the seam carving method than that using other resizing methods with the increment of the resizing ratio.

Table 1. Comparisons in terms of accuracy for tampered images with single re-targeting methods.

Scaling Ratio (%)	Accuracy (%)		
	Scale-and-Stretch	Seam Carving	Scaling
10	82.51	75.26	74.33
20	88.45	91.67	87.37
30	95.51	97.87	95.55
40	98.95	99.77	98.87
50	99.85	100	99.81

4.2.2. Identifying Images Obtained by Different Content-Aware Resizing Methods

In this experiment, the tampered images with scaling ratios from 10% to 50% are obtained by different content-aware resizing methods. They are adopted to test the performance of our proposed algorithm. The average detection accuracies of different content-aware re-targeting methods, where the average detection accuracy is the average value of diagonal elements in the confusion matrix, are summarized in Table 2. Note that "mixed" represents the mixed test set of tampered images with the scaling ratios from 10% to 50%. There are three content-aware resizing methods in this paper. Therefore, this is a four-class classification problem (the original images as a special class), according to the ECOC strategy.

Table 2. The average detection accuracies of our proposed approach with preprocessing.

Uncompressed (%)						Compressed (%)					
10%	20%	30%	40%	50%	mixed	10%	20%	30%	40%	50%	mixed
82.04	90.00	93.30	94.61	95.94	92.04	76	81.13	84.08	87.98	88.34	78.33

Table 2 shows that the average accuracy is improved with the increment of the scaling ratio. However, the detection accuracy is apparently decreased for highly compressed images with Quality factor (QF) being equal to 75. Through careful analysis of our experimental results, it is found that the main reason for the decrement of the detection accuracy in the compressed condition is that the traces of the tampered images are weakened when images are compressed. Therefore, the detection accuracy is decreased in this case. We have also completed the experiment of the "mixed" tampered uncompressed and compressed images and get the confusion matrix. Table 3 shows our experimental results, where CMOMTUI represents the confusion matrix of "mixed" tampered uncompressed images, CMOMTCI represents the confusion matrix of "mixed" tampered compressed images, "*" represents the classified accuracy less than 1%, OR represents original images, SNS represents a scale-and-stretch method, SC represents a seam carving method, and SL represents a scaling method. From Table 3, we can see that our proposed method can get a high accuracy for the three content-aware resizing methods mentioned in this paper. However, it can't obtain a good detection accuracy on the tampered images with JPEG compression. In addition, its false positive rate is relatively high for the seam carving method (SC) and the scaling method (SL).

Table 3. CMOMTUI and CMOMTCI with preprocessing.

Uncompressed	Detection Accuracy (%)				Compressed	Detection Accuracy (%)			
	OR	SNS	SC	SL		OR	SNS	SC	SL
OR	95.93	*	*	*	OR	93.45	*	*	*
SNS	*	90.14	11.12	3.61	SNS	*	80.69	11.75	10.49
SC	*	*	83.71	6.01	SC	*	1.49	68.43	18.77
SL	3.26	8.90	5.17	90.38	SL	5.56	17.82	19.82	70.73

4.2.3. The Detection Accuracy without Preprocessing

Tables 4 and 5 report the identified results for the uncompressed and compressed tampered images without preprocessing, respectively, where CMOMTUI represents the confusion matrix of "mixed" tampered uncompressed images, CMOMTCI represents the confusion matrix of "mixed" tampered compressed images, "*" represents the classified accuracy less than 1%, OR represents original images, SNS represents the scale-and-stretch method, SC represents the seam carving method, and SL represents the scaling method. It can be found from Tables 4 and 5 that our proposed approach has a sightly higher detection accuracy on the uncompressed images when the features of LTM are extracted from the images without preprocessing. However, when the images are compressed by QF = 75, the accuracy of our proposed approach is significantly lower than that of the images with preprocessing. The main reason is that the residual may weaken the tampered trace of the images without compression when it is used in the process of preprocessing. However, the images with compression are preprocessed by residuals, and the changes of these images will be highlighted.

Table 4. The average detection accuracies of our proposed approach without preprocessing.

Uncompressed (%)						Compressed (%)					
10%	20%	30%	40%	50%	mixed	10%	20%	30%	40%	50%	mixed
84.04	91.02	94.32	95.41	96.04	93.04	70.93	74.59	79.63	83.18	85.34	76.09

Table 5. CMOMTUI and CMOMTCI without preprocessing.

Uncompressed	Detection Accuracy (%)				Compressed	Detection Accuracy (%)			
	OR	SNS	SC	SL		OR	SNS	SC	SL
OR	96.39	*	*	*	OR	87.98	*	*	*
SNS	*	91.54	9.21	3.21	SNS	*	77.25	16.29	12.49
SC	*	*	85.71	5.41	SC	1.35	1.73	65.14	17.07
SL	3.60	8.40	5.08	91.38	SL	10.67	21.02	18.57	70.73

5. Conclusions

Content-aware image re-targeting methods are widely adopted to resize images to display on all kinds of terminals. However, they can be also used to make fake images, which don't have any perceptual annoying distortions. By the principle analysis of the three image resizing methods, we found that the correlation between adjacent pixels can be destroyed in the process of image resizing. Tchebichef moments have been extensively applied in field of image/vedio such as information security [18], pattern recognition [19] and image quality assessment [20], and so on. Inspired by this, it can be found from experiments that LTM can effectively reflect the correlation changes between adjacent pixels. We proposed a passive forensics algorithm based on LTM to identify tampered images obtained by content-aware image resizing methods in this paper. Our experimental results showed that our proposed method can obtain a better accuracy. It is verified that it has a good performance for the image resizing with high scaling ratios. In the future, we will try to evaluate the detection accuracy on the tampered images obtained by image resizing with low scaling ratios. In addition, our proposed method could not obtain a satisfied detection accuracy on the re-targeted images with JPEG compression. We will further analyze the tampered trace of the resized images with JPEG compression. In addition, since there are still a few image resizing methods [21,22] besides the three methods proposed in this paper, we will attempt to distinguish these image resizing techniques for re-targeted images by applying other multi-class classifiers [23–28] and designing more general features from image/video processing methods [29–38]. In view of the importance of social media digital images in practical applications, research on their authenticity, integrity and traceability has been one of the hot and challenging research topics in the field of information security. We will adopt network optimization methods [39–48] to improve the real-time and high efficiency performance of the feature extraction phases.

Author Contributions: Conceptualization: D.Z. and S.W.; investigation: J.W.; methodology: D.Z. and S.W.; software: D.Z. and S.W.; supervision: F.L.; validation: A.K.S. and V.S.S.; writing—original draft: D.Z. and S.W.; writing—review and editing: J.W. and A.K.S.

Funding: This research was funded by the National Natural Science Foundation of China (61772454, 61811530332, 61811540410, 61772087, 61232016), the Scientific Research Fund of Hunan Provincial Education Department of China (14C0029) and the "Double First-class" International Cooperation and Development Scientific Research Project of Changsha University of Science and Technology (No. 2018IC25).

Conflicts of Interest: The authors declare no conflict of interest.

References

1. Avidan, S.; Shamir, A. Seam carving for content-aware resizing. *ACM Trans. Graphics* **2007**, *26*, 10. [CrossRef]
2. Vaquero, D.; Turk, M.; Pulli, K. A survey of image retargeting techniques. *Int. Soc. Opt. Photonics* **2010**, *7898*, 789814.
3. Wang, Y.S.; Tai, C.L.; Sorkine, O. Optimized scale-and-stretch for image resizing. *ACM Trans. Graphics (TOG)* **2008**, *27*, 118. [CrossRef]
4. Lu, W.; Wu, M. Seam carving estimation using forensic hash. In Proceedings of the Thirteenth ACM Multimedia Workshop on Multimedia and Security, Buffalo, NY, USA, 29–30 September 2011; pp. 9–14.
5. Sakar, A.; Nataraj, L.; Manjunath, B.S. Detection of seam carving and localization of seam insertions in digital images. In Proceedings of the 11th ACM Workshop on Multimedia and Security, Princeton, NJ, USA, 7–8 September 2009; pp. 107–116.
6. Fillion, C.; Sharma, G. Detecting content adaptive scaling of images for forensic applications. *SPIE Electron. Imaging Int. Soc. Opt. Photonics* **2010**, *7541*, 75410Z.
7. Wei, J.D.; Lin, Y.J.; Wu, Y.J. A patch analysis method to detect seam carved images. *Pattern Recognit. Lett.* **2014**, *36*, 100–106. [CrossRef]
8. Seung-Jin, R.Y.U.; Hae-Yeoun, L.E.E.; Heung-Kyu, L.E.E. Detecting trace of seam carving for forensic analysis. *IEICE Trans. Inf. Syst.* **2014**, *97*, 1304–1311.
9. Yin, T.; Yang, G.; Li, L. Detecting seam carving based image resizing using local binary patterns. *Comput. Secur.* **2015**, *55*, 130–141. [CrossRef]
10. Zhang, D.Y.; Yin, T.; Yang, G. Detecting image seam carving with low scaling ratio using multiscale spatial and spectral entropies. *J. Vis. Commun. Image Represent.* **2017**, *48*, 281–291. [CrossRef]
11. Zhang, D.Y.; Li, Q.; Yang, G. Detection of image seam carving by using weber local descriptor and local binary patterns. *J. Inf. Secur. Appl.* **2017**, *36*, 135–144. [CrossRef]
12. Liu, Q. An approach to detecting JPEG down-recompression and seam carving forgery under recompression anti-forensics. *Pattern Recognit.* **2017**, *65*, 35–46. [CrossRef]
13. Itti, L.; Koch, C.; Niebur, E. A model of saliency-based visual attention for rapid scene analysis. *IEEE Trans. Pattern Anal. Mach. Intell.* **1998**, *20*, 1254–1259. [CrossRef]
14. Dietterich, T.G.; Bakiri, G. Solving multiclass learning problems via error-correcting output codes. *J. Artif. Intell. Res.* **1995**, *2*, 263–286. [CrossRef]
15. Hastie, T.; Tibshirani, R. Classification by pairwise coupling. In *Advances in Neural Information Processing Systems*; MIT Press: Cambridge, UK, 1998; pp. 507–513.
16. The Ensemble Learning. Available online: http://dde.binghamton.edu/download/ensemble (accessed on 25 May 2019).
17. Schaefer, G.; Stich, M. UCID: An uncompressed color image database. *Int. Soc. Opt. Photonics* **2003**, *5307*, 472–480.
18. Chen, B.; Coatrieux, G.; Wu, J. Fast computation of sliding discrete Tchebichef moments and its application in duplicated regions detection. *IEEE Trans. Signal Process.* **2015**, *63*, 5424–5436. [CrossRef]
19. Zhang, H.; Dai, X.; Sun, P. Symmetric image recognition by Tchebichef moment invariants. In Proceedings of the 2010 IEEE International Conference on Image Processing, Hong Kong, China, 26–29 September 2010; pp. 2273–2276.
20. Li, L.; Zhu, H.; Yang, G. Referenceless measure of blocking artifacts by Tchebichef kernel analysis. *IEEE Signal Process. Lett.* **2014**, *21*, 122–125. [CrossRef]
21. Niu, Y.; Liu, F.; Li, X. Image resizing via non-homogeneous warping. *Multimed. Tools Appl.* **2012**, *56*, 485–508. [CrossRef]
22. Lin, S.S.; Yeh, I.C.; Lin, C.H. Patch-based image warping for content-aware retargeting. *IEEE Trans. Multimed.* **2013**, *15*, 359–368. [CrossRef]
23. Zhang, J.; Lu, C.; Li, X.; Kim, H.J.; Wang, J. A full convolutional network based on DenseNet for remote sensing scene classification. *Math. Biosci. Eng.* **2019**, *16*, 3345–3367. [CrossRef]
24. Yu, J.; Zhang, B.; Kuang, Z.; Lin, D.; Fan, J. iPrivacy: Image Privacy Protection by Identifying Sensitive Objects via Deep Multi-Task Learning. *IEEE Trans. Inf. Forensics Secur.* **2017**, *12*, 1005–1016. [CrossRef]
25. Tirkolaee, E.B.; Hosseinabadi, A.A.R.; Soltani, M. A Hybrid Genetic Algorithm for Multi-trip Green Capacitated Arc Routing Problem in the Scope of Urban Services. *Sustainability* **2018**, *10*, 1366. [CrossRef]

26. Chen, Y.T.; Wang, J.; Chen, X.; Zhu, M.; Yang, K.; Wang, Z.; Xia, R. Single-Image Super-Resolution Algorithm Based on Structural Self-Similarity and Deformation Block Features. *IEEE Access* **2019**, *7*, 58791–58801. [CrossRef]
27. Zhang, J.; Jin, X.; Sun, J.; Wang, J.; Sangaiah, A.K. Spatial and semantic convolutional features for robust visual object tracking. *Multimed. Tools Appl.* **2018**. [CrossRef]
28. Zhang, J.; Jin, X.; Sun, J.; Wang, J.; Li, K. Dual model learning combined with multiple feature selection for accurate visual tracking. *IEEE Access* **2019**, *7*, 43956–43969. [CrossRef]
29. Yun, S.; Gaobo, Y.; Hongtao, X. Residual domain dictionary learning for compressed sensing video recovery. *Multimed. Tools Appl.* **2017**, *76*, 10083C10096.
30. Xiang, L.; Shen, X.; Qin, J.; Hao, W. Discrete Multi-Graph Hashing for Large-scale Visual Search. *Neural Process. Lett.* **2019**, *49*, 1055–1069. [CrossRef]
31. Yu, J.; Rui, Y.; Tang, Y.Y.; Tao, D. High-Order Distance-Based Multiview Stochastic Learning in Image Classification. *IEEE Trans. Cybern.* **2014**, *44*, 2431–2442. [CrossRef]
32. Li, Y.; Yang, G.; Zhu, Y.; Ding, X.; Gong, R. Probability model-based early Merge mode decision for dependent views in 3D-HEVC. *ACM Trans. Multimed. Comput. Commun. Appl. (TOMM)* **2018**, *14*, 8501–8515. [CrossRef]
33. Ding, X.; Yang, G.; Li, R.; Zhang, L.; Li, Y.; Sun, X. Identification of MC-FRUC based on spatial-temporal Markov features of residue signal. *IEEE Trans. Circuits Syst. Video Technol.* **2018**, *28*, 1497–1512. [CrossRef]
34. Xia, M.; Yang, G.; Li, L.; Li, R.; Sun, X. Detecting video frame rate up-conversion based on frame-level analysis of average texture variation. *Multimed. Tools Appl.* **2017**, *76*, 8399–8421. [CrossRef]
35. He, J.; Yang, G.; Song, J.; Ding, X.; Li, R. Hierarchical prediction-based motion vector refinement for video frame-rate up-conversion. *J. Real-Time Image Process.* **2018**, 1–15. [CrossRef]
36. Tang, D.; Zhou, S.; Yang, W. Random-filtering based sparse representation parallel face recognition. *Multimed. Tools Appl.* **2019**, *78*, 1419–1439. [CrossRef]
37. Pan, J.S.; Kong, L.; Sung, T.W.; Tsai, P.W.; Snášel, V. A Clustering Scheme for Wireless Sensor Networks Based on Genetic Algorithm and Dominating Set. *J. Internet Technol.* **2018**, *19*, 1111–1118.
38. Meng, Z.; Pan, J.-S.; Tseng, K.-K. PaDE: An enhanced Differential Evolution algorithm with novel control parameter adaptstion schemes for numerical optimization. *Knowl.-Based Syst.* **2019**, *168*, 80–99. [CrossRef]
39. Nguyen, T.-T.; Pan, J.-S.; Dao, T.-K. An Improved Flower Pollination Algorithm for Optimizing Layouts of Nodes in Wireless Sensor Network. *IEEE Access* **2019**, *7*. [CrossRef]
40. Wang, J.; Gao, Y.; Liu, W.; Sangaiah, A.K.; Kim, H.J. An Intelligent Data Gathering Schema with Data Fusion Supported for Mobile Sink in WSNs. *Int. J. Distrib. Sens. Netw.* **2019**, *15*. [CrossRef]
41. Wang, J.; Cao, J.; Sherratt, R.S.; Park, J.H. An improved ant colony optimization-based approach with mobile sink for wireless sensor networks. *J. Supercomput.* **2018**, *74*, 6633–6645. [CrossRef]
42. Wang, J.; Cao, J.; Ji, S.; Park, J.H. Energy Efficient Cluster-based Dynamic Routes Adjustment Approach for Wireless Sensor Networks with Mobile Sinks. *J. Supercomput.* **2017**, *73*, 3277–3290. [CrossRef]
43. Wang, J.; Gao, Y.; Yin, X.; Li, F.; Kim, H.J. An Enhanced PEGASIS Algorithm with Mobile Sink Support for Wireless Sensor Networks. *Wirel. Commun. Mob. Comput.* **2018**, *2018*. [CrossRef]
44. Wang, J.; Gao, Y.; Liu, W.; Wu, W.; Lim, S.J. An Asynchronous Clustering and Mobile Data Gathering Schema based on Timer Mechanism in Wireless Sensor Networks. *Comput. Mater. Contin.* **2019**, *58*, 711–725. [CrossRef]
45. Pan, J.S.; Lee, C.Y.; Sghaier, A.; Zeghid, M.; Xie, J. Novel Systolization of Subquadratic Space Complexity Multipliers Based on Toeplitz Matrix-Vector Product Approach. *IEEE Trans. Very Large Scale Integr. Syst.* **2019**, *27*, 1614–1622. [CrossRef]
46. He, Y.; Xiang, S.; Li, K.; Liu, Y. Region-Based Compressive Networked Storage with Lazy Encoding. *IEEE Trans. Parallel Distrib. Syst.* **2019**, *30*, 1390–1402.
47. Pan, J.S.; Kong, L.P.; Sung, T.W.; Tsai, P.W.; Snasel, V. Alpha-Fraction First, Strategy for Hierarchical Wireless Sensor Networks. *J. Internet Technol.* **2018**, *19*, 1717–1726.
48. He, S.; Xie, K.; Chen, W.; Zhang, D.; Wen, J. Energy-aware Routing for SWIPT in Multi-hop Energy-constrained Wireless Network. *IEEE Access* **2018**, *6*, 17996–18008. [CrossRef]

© 2019 by the authors. Licensee MDPI, Basel, Switzerland. This article is an open access article distributed under the terms and conditions of the Creative Commons Attribution (CC BY) license (http://creativecommons.org/licenses/by/4.0/).

Article
A Bounded Scheduling Method for Adaptive Gradient Methods

Mingxing Tang [1], **Zhen Huang** [1,*], **Yuan Yuan** [2], **Changjian Wang** [2] **and Yuxing Peng** [1]

[1] Science and Technology on Parallel and Distributed Laboratory, National University of Defense Technology, Changsha 410073, China
[2] College of Computer, National University of Defense Technology, Changsha 410073, China
* Correspondence: huangzhen@nudt.edu.cn

Received: 22 July 2019; Accepted: 28 August 2019; Published: 1 September 2019

Abstract: Many adaptive gradient methods have been successfully applied to train deep neural networks, such as Adagrad, Adadelta, RMSprop and Adam. These methods perform local optimization with an element-wise scaling learning rate based on past gradients. Although these methods can achieve an advantageous training loss, some researchers have pointed out that their generalization capability tends to be poor as compared to stochastic gradient descent (SGD) in many applications. These methods obtain a rapid initial training process but fail to converge to an optimal solution due to the unstable and extreme learning rates. In this paper, we investigate the adaptive gradient methods and get the insights on various factors that may lead to poor performance of Adam. To overcome that, we propose a bounded scheduling algorithm for Adam, which can not only improve the generalization capability but also ensure the convergence. To validate our claims, we carry out a series of experiments on the image classification and the language modeling tasks on several standard benchmarks such as ResNet, DenseNet, SENet and LSTM on typical data sets such as CIFAR-10, CIFAR-100 and Penn Treebank. Experimental results show that our method can eliminate the generalization gap between Adam and SGD, meanwhile maintaining a relative high convergence rate during training.

Keywords: deep neural networks; adaptive gradient methods; stochastic gradient descent; bounded scheduling method; image classification; language modeling

1. Introduction

Deep neural networks (DNNs) [1] have achieved great successes in many applications, such as image recognition [2], object detection [3], speech recognition [4,5], face recognition [6] and machine translation [7]. How to train DNNs quickly and accurately has attracted the attention of many researchers. Training neural networks is equivalent to solving the following non-convex optimization problems:

$$\min_{w \in R^d} F(w) = \frac{1}{n} \sum_{i=1}^{n} f_i(w), \qquad (1)$$

where w is the parameter to train, n is the number of instances, $f_i(\cdot) : \mathbb{R}^d \to \mathbb{R}$ is a loss function defined on the instance with d dimensions and indexed i. Training algorithms need to search parameters to minimize the loss function.

Stochastic gradient descent (SGD) [8] has become the dominant training algorithm for DNNs. Simple as it is, SGD performs well in many applications. SGD obtains a smaller loss by moving the parameters of the model in the negative direction of gradient evaluated on a minibatch. The iteration of SGD can be described as follows:

$$w_k = w_{k-1} - \eta \nabla f_{i_k}(w_{k-1}), \qquad (2)$$

where η is learning rate, i_k is the instance index at the k-th iteration, $\nabla f_{i_k}(w_{k-1})$ denotes the stochastic gradient computed at w_{k-1}.

There are two main drawbacks of SGD. The first one is SGD needs to find an appropriate learning rate, which means that excessive learning rate will cause the loss function unable to converge to the optimal value and exceptionally small learning rate will slow down the convergence speed of loss function. The other one is SGD scales the gradient uniformly in all directions, which leads that the ill-scaled or sparse problems cannot be solved well [9].

To train DNNs, SGD uses a standard decreasing learning rate scheme, where the learning rate is initialized as a large value at the beginning and decreases gradually with iteration. However, a suitable initial learning rate is difficult to tune. Linear search [10] and grid search are often used to find the optimal learning rate, but the computational overhead is high. Cyclical learning rates method [11] changes the learning rate periodically within a fixed bound, which can practically eliminate the need to experimentally find the best values and schedule for the global learning rates. Then a super-convergence method [12] is proposed to train networks with one learning rate cycle and a large maximum learning rate, which can achieve an increase in performance compared with standard methods. However, the uniformly scaled gradient still makes these methods perform poorly when the data set is sparse or ill-scaled.

In recent years, a series of adaptive gradient methods have been proposed. These methods scale the gradient by some form of squared past gradients, which can achieve a rapid training speed with an element-wise scaling term on learning rates [13]. Adagrad [9] is the first popular algorithm to use an adaptive gradient, which has obviously better performance than SGD when the gradients are sparse. However, the learning rate of Adagrad will drop rapidly because of its accumulation of the squared gradients in the denominator, which may lead to deterioration in the case that the loss functions are non-convex or gradients are dense. Then Adadelta [14], RMSprop [15], Adam [16], Nadam [17] are proposed to fix this issue, which use the exponential moving averages of squared past gradients to avoid the rapid drop of learning rate. These algorithms have been successfully applied to a variety of practical problems, especially Adam has become the default algorithm for training neural networks.

When training DNNs with adaptive gradient methods, the loss function decreases rapidly in the early stage of training, but the final training loss and test loss are worse than that of SGD in many applications. Moreover, since the learning rate of Adam does not decrease monotonously, the training process will diverge in some applications [18]. Some work has proposed a hybrid scheme of Adam and SGD to solve these problems. SWATS [19] proposes a strategy that Adam can be switched to SGD when a triggering condition is satisfied, which can close the generalization gap between Adam and SGD. ADABOUND [13] can achieve a gradual and smooth transition from adaptive methods to SGD by employing dynamic bounds on learning rates. For these hybrid algorithms, the switching time of Adam and SGD and the learning rate of SGD after switching still have a great impact on the performance of the algorithm, which should be tuned elaborately.

In this paper, we study the adaptive gradient algorithms and propose a bounded scheduling method for Adam, called Bsadam, to improve the performance when training neural networks. The major contributions of this paper include:

1. We investigate the factors that lead to the poor performance of Adam while training complex neural networks.
2. We set effective bounds for the learning rate of Adam without manual tuning, which can improve the generation capability.
3. We schedule the bounds of learning rate to improve the performance of Adam. Firstly we fix the upper bound and increase the lower bound gradually to find wide, flat minima. Then we fix the lower bound and decrease the upper bound gradually to ensure the convergence of training. At last, a fixed learning rate is used to make the algorithm converge to the optimal solution.

4. We train multiple tasks on several models to evaluate the algorithm. MNIST [20] is trained on simple neural networks, CIFAR-10 [21] and CIFAR-100 [21] are trained on ResNet [22], DenseNet [23] and SENet [24], Penn Treebank [25] is trained on LSTM [26]. All these experiments show that our method is capable of eliminating the generalization gap between Adam and SGD and maintaining a higher convergence speed in training.

The rest of our paper is organized as follows. In Section 2, the background of this paper is reviewed, where the traditional learning rate methods and adaptive gradient methods are described. In Section 3, we introduce the bounded scheduling scheme for Adam. In Section 4, we present a series of experiments to verify the effectiveness of our method. In Section 5, we summarize the paper.

2. Background

2.1. Traditional Learning Rate Methods

Learning rate is one of the most important hyper-parameters of gradient-based optimization methods, there have been many related works on it. Line search [10] is often used to find the learning rate of the full gradient. The line search method will set a large initial learning rate and try a learning rate at each iteration, if the loss function does not fall a certain distance than the current value, the learning rate will decrease proportionally and iterate again, until the learning rate satisfying the fall condition is found. Line search needs a large amount of computation and is often used when the data set is small. A line search method for SGD is also proposed [27]. This method uses random samples to do basic line search and estimates the Lipschitz constant L, then deduces the theoretical optimal learning rate based on L. However, the optimal learning rate, in theory, is different from that in practice and this method can not guarantee convergence.

Barzilai-Borwen method [28,29] is also often used to estimate the learning rate. The Barzilai-Borwen method is based on the quasi-Newton method and uses second-order derivative information to evaluate the learning rate, which requires little extra computational overhead. However, the learning rate estimated by the Barzilai-Borwen method may lead to the divergence of the training process. Yann Ollivier et al. proposed a method to view the whole performance of the learning trajectory as a function of the learning rate, then adapt the learning rate by performing a gradient descent on the learning rate itself [30]. Although these methods do not need to search the learning rate, their performance is not good enough compared with the manually tuned optimal learning rate. Cyclical learning rate method [11] does not need to use a certain learning rate, but makes the learning rate vary periodically in a certain range. Then super-convergence [12] is proposed to train DNNs with one cycle and a large maximum learning rate, which provides a boost in performance. Traditional learning rate methods scale the gradient uniformly in all directions, the performance of which will decrease when data sets are sparse or ill-scaled.

2.2. Adaptive Gradient Methods

The recently proposed adaptive gradient methods can provide an element-wise scaling term on learning rates without the need to tune the learning rate manually. These methods use historical information to estimate the curvature of the loss function and adopt different learning rates for each parameter, so the learning rate is a vector and each element for a parameter, which is different from the traditional learning rate methods. The representative adaptive gradient methods are Adagrad [9], RMSprop [15], Adam [16], AMSgrad [18], etc.

Adagrad [9] is the first proposed adaptive gradient method. Its main idea is to adopt a smaller learning rate for the parameters corresponding to frequent features and a larger learning rate for the parameters corresponding to infrequent features. Therefore, Adagrad is very suitable for training sparse data, which can improve the robustness of SGD. The update of Adagrad can be shown as follows:

$$w_k = w_{k-1} - \eta \frac{\nabla f(w_{k-1})}{\sqrt{v_k} + \epsilon}, \qquad (3)$$

where

$$v_k = \Sigma_{j=0}^{k-1} \nabla f(w_j)^2, \qquad (4)$$

ϵ is a smoothing term that avoids division by zero, η is general learning rate.

Adagrad uses the accumulation of the squared gradients and the squared gradients are positive, which will lead to a rapid decline in learning rate to infinite small and the standstill of loss function. RMSprop [15] was proposed to solve the problem of the rapid disappearance of the gradient for Adagrad. The update rule of RMSprop is the same as (3), but the updating of v_k adopts exponential decaying average of square gradients, which can be shown as follows:

$$v_k = \beta v_{k-1} + (1-\beta)\nabla f(w_{k-1})^2, \qquad (5)$$

where $\beta \in [0,1)$ is the hyper-parameter that controls the exponential decay rate of average. The use of the exponential moving averages of squared past gradients can prevent the rapid rise of v_k and the learning rate will not decline rapidly.

Adam [16] can also calculate adaptive learning rate for each parameter. As a complement to RMSprop, Adam preserves the exponential moving averages of squared past gradients, as well as the exponential moving averages of past gradients, which gives the gradient momentum. The update formula of Adam is shown as follows:

$$w_k = w_{k-1} - \eta \cdot \frac{\sqrt{1-\beta_2^k}}{1-\beta_1^k} \cdot \frac{m_k}{\sqrt{v_k}+\epsilon}, \qquad (6)$$

where

$$\begin{aligned} m_k &= \beta_1 m_{k-1} + (1-\beta_1)\nabla f(w_{k-1}), \\ v_k &= \beta_2 v_{k-1} + (1-\beta_2)\nabla f(w_{k-1})^2, \end{aligned} \qquad (7)$$

where $\beta_1, \beta_2 \in [0,1)$ are hyper-parameters that control the exponential decay rate of moving average.

Reddi et al. pinpoint that the use of exponential moving averages of squared past gradients may make Adam fail to converge to the optimal solution. As a result, AMSGrad was proposed [18]. Unlike Adam, AMSGrad uses the maximum of exponential moving averages of squared past gradients, the update rule of v_k is show as follows:

$$\begin{aligned} \widehat{v}_k &= \beta_2 \widehat{v}_{k-1} + (1-\beta_2)\nabla f(w_{k-1})^2, \\ v_k &= \max(\widehat{v}_k, v_{k-1}). \end{aligned} \qquad (8)$$

The adaptive gradient methods has low generalization ability in training complex models and its performance is worse than the optimal learning rate tuned manually.

3. Bsadam: Bounded Scheduling Method for Adam

3.1. Preliminaries

Firstly, we use an empirical study to illustrate the existence of the generalization gap in Adam. We use SGD and Adam to do image classification for CIFAR-10 data set on ResNet-34 architecture and present training accuracy and test accuracy in Figure 1. As can be seen from Figure 1, the training and test accuracy of Adam both increased faster than that of SGD in the early stage. However, when the learning rate is reduced by 10 after 100 epochs, the training and test accuracy of Adam are lower than that of SGD. Although the final training accuracy of Adam reaches the level of SGD, its test accuracy is still 1% to 2% lower than that of SGD, which means that its generalization gap is larger than SGD.

There may be various factors that may lead to the weakly empirical generalization capability of Adam. Based on previous researches [13,19,31–33], we summarize these factors and work to eliminate them. The main factors can be listed as follows.

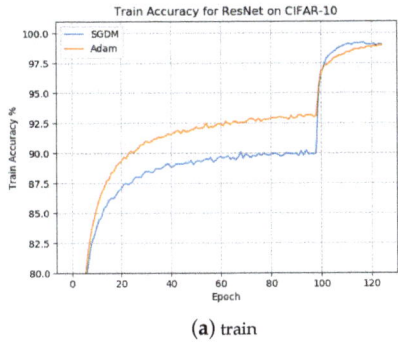
(a) train

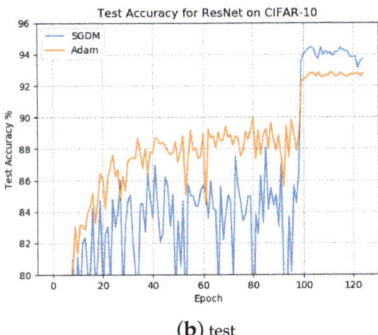
(b) test

Figure 1. Training the ResNet-34 architecture on the CIFAR-10 data set with stochastic gradient descent (SGD) and Adam. Adam has a faster initial convergence speed, but the final test accuracy is lower than that of SGD.

- The non-uniform scaling of the gradients will lead to the poor generalization performance of adaptive gradients methods. SGD is uniformly scaled and low training error will generalize well [19,31].
- The exponential moving average used in Adam can't make the learning rate monotonously decline, which will cause it to fail to converge to an optimal solution and arise the poor generalization performance [32,33].
- The learning rate learned by Adam may circumstantially be too small for effective convergence, which will make it fail to find a right path and converge to a suboptimal point [13].
- Adam may aggressively increase the learning rate, which is detrimental to the overall performance of the algorithm [13,18].

Taking all these factors into account, some improvements needs to be considered for Adam. Upper and lower bounds should be specified to avoid the side effect caused by extreme large and small learning rate. At the later stage of training, learning rate should be monotonous decreased to ensure the convergence and be uniformly scaled to improve generalization performance.

3.2. Specify Bounds for Adam

In this paper, we use the curve of loss function obtained by learning rate range test (LR range test) [11] to determine the upper and lower bounds of the learning rate for Adam. When training a new model or data set, the LR range test is a very effective way to find a reasonable learning rate range for SGD, although it can not find a specific learning rate. LR range test uses SGD to train the model for several epochs and makes the learning rate increase linearly from small to large, then the approximate range of reasonable learning rate can be estimated by the curve of the loss function. Specifically, when the loss decreases, it means that the current learning rate is reasonable when the loss rises, it means that the current learning rate is inappropriate.

However, as a result that Adam itself has the function of adjusting the learning rate, the standard of specifying the bounds for Adam is different from the classical LR range test, we need a wider range of bounds. Specifically, the lower bound can be set to the point where the loss function begins to decline and the upper bound can be set to the point where the loss function begins to rise. What is more, in order to get better generalization ability, the upper bound can be enlarged within five times.

For example, we use Resnet-34 architecture to perform the LR range test on CIFAR-10 and obtain the curve of loss function along with the learning rate. The result is shown in Figure 2. As can be seen from Figure 2, the loss begins to decline obviously when the learning rate is 0.001, so 0.001 can be used as the lower bound of the learning rate for Adam. When the learning rate is 0.1, the loss starts to rise and the training process starts to diverge, so 0.1 can be used as the upper bound of the learning rate for Adam. However, through experiments, we find that the algorithm can get better minima by increasing the upper bound appropriately, so the upper bound can be set to 0.5. The upper and lower bounds of learning rate are limited, the negative effects of too large or too small learning rate on Adam can be eliminated.

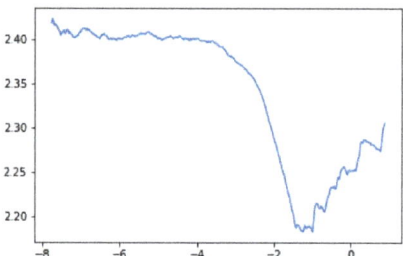

Figure 2. Learning rate range test. The x-axis is learning rate (log scale), the y-axis is training loss.

3.3. Schedule Bounds for Adam

We improve the empirical generalization capability of Adam by scheduling its lower and upper bounds, which can reduce the adverse effects of the non-uniform scaling of the gradients and the non-monotonically decreasing learning rate. According to the updated formula of Adam, we can regard $\frac{\sqrt{1-\beta_2^k}}{1-\beta_1^k} \cdot \frac{\eta}{\sqrt{v_k}+\epsilon}$ as the learning rate of Adam and m_k as gradients with momentum of Adam. Gradient clipping can constrain the learning rate to a certain range, which is an effective method to solve the problem of gradient disappearance or gradient explosion. We use gradient clipping to clip the learning rate of Adam which exceeds the threshold. Consider applying the following operations to Adam

$$Clip(\frac{\sqrt{1-\beta_2^k}}{1-\beta_1^k} \cdot \frac{\eta}{\sqrt{v_k}+\epsilon}, min\_lr, max\_lr), \tag{9}$$

which can clip the learning rate of Adam element-wisely such that each element of the learning rate is limited in the range of $[min\_lr, max\_lr]$, where min_lr and max_lr are lower bound and upper bound found in Section 3.2 respectively.

Then we will schedule the bounds of learning rate. The scheduling process is divided into three phases, which are finding minima, converging and uniform scaling. The scheduling details for each phase are described in detail below.

3.3.1. Finding Minima

In this phase, we use the concept of super-convergence, which implies that a large maximum learning rate can achieve better generalization capability. Using a relatively large learning rate in the early stage of training can make the loss function skip the suboptimal solution more easily and find wide, flat minima. Therefore, we fix the upper bound of learning rate and gradually increase the lower bound of learning rate, so that each element of learning rate can gradually rise to the upper bound. In this phase, gradient clipping can be expressed as follows:

$$Clip(\frac{\sqrt{1-\beta_2^k}}{1-\beta_1^k} \cdot \frac{\eta}{\sqrt{v_k}+\epsilon}, ascending(t), max\_lr), \tag{10}$$

where $ascending(t)$ is a function that lower bound increases gradually from min_lr to max_lr with iteration and t means the progress of epoch in this phase. $ascending(t)$ can be linear, exponential and trigonometric, which can be formulated as follows:

- linear rise:
$$ascending(t) = min\_lr + t \cdot \frac{max\_lr - min\_lr}{T}, \qquad (11)$$

- exponential rise:
$$ascending(t) = min\_lr \cdot \left(\frac{max\_lr}{min\_lr}\right)^{\frac{t}{T}}, \qquad (12)$$

- trigonometric rise:
$$ascending(t) = min\_lr + (max\_lr - min\_lr) \sin\left(\frac{t}{T} \cdot \frac{\pi}{2}\right), \qquad (13)$$

where T is the total epochs in this phase.

3.3.2. Converging

In this phase, to avoid the divergence or poor generalization performance caused by the non-monotonic decline of learning rate, we need to make sure that the learning rate of Adam is decreasing. Therefore, we fix the lower bound of learning rate and gradually decrease the upper bound of learning rate, so that each element of learning rate can gradually decrease to the lower bound. In this phase, gradient clipping can be expressed as follows:

$$Clip\left(\frac{\sqrt{1-\beta_2^k}}{1-\beta_1^k} \cdot \frac{\eta}{\sqrt{v_k}+\epsilon}, min\_lr, descending(t)\right), \qquad (14)$$

where $descending(t)$ is a function that upper bound decreases gradually from max_lr to min_lr with iteration and t means the progress of epoch in this phase. $descending(t)$ can be linear, exponential and trigonometric, which can be formulated as follows:

- linear decrease:
$$descending(t) = max\_lr - t \cdot \frac{max\_lr - min\_lr}{T}, \qquad (15)$$

- exponential decrease:
$$descending(t) = max\_lr \cdot \left(\frac{min\_lr}{max\_lr}\right)^{\frac{t}{T}}, \qquad (16)$$

- trigonometric decrease:
$$descending(t) = max\_lr - (max\_lr - min\_lr) \sin\left(\frac{t}{T} \cdot \frac{\pi}{2}\right), \qquad (17)$$

where T is the total epochs in this phase.

3.3.3. Uniform Scaling

There is a conventional phase for training neural networks, which is reducing the learning rate by 10 in the final stage of training, so that the algorithm will converge to the near minimum. In our algorithm, at the end of the converging phase, upper bound are reduced to min_lr, so the gradients are uniformly scaled. We use min_lr as a learning rate continuing training model. The training accuracy and test accuracy will be further improved and stabilized and the algorithm will eventually converge. In this phase, the gradients are uniformly scaled, which will help improve generalization performance.

3.4. Algorithm Overview

Based on the above analysis, in this subsection, we propose a new variant of the optimization methods, named Bsadam, which can maintain the fast convergence of the algorithm in the early stage and obtain a good finally generalization capacity.

Empirically, the number of epoch in the first phase is the same as that in the second phase and the number of epoch in the third phase should be less than that in the former two phases. Specifically, if the total number of training epochs is T, the allocation of the number of epochs for three phases are $\frac{2T}{5}$, $\frac{2T}{5}$ and $\frac{T}{5}$, respectively. The details of Bsadam are illustrated in Algorithm 1, where max_lr and min_lr can be found by the method mentioned in Section 3.2, β_1, β_2 and η is the hyper-parameters of Adam itself, $data\_loader()$ is a function that combines a data set and a sampler and provides an iterable process over the given data set.

Algorithm 1 Bsadam.

Parameters : total epochs T, max_lr, min_lr
Initialize : w_0, β_1, β_2, η

1: set $v_0 = 0$, $m_0 = 0$, $k = 0$
2: **for** $t = \{1, 2, ..., \frac{2T}{5}\}$ **do**
3:     $ascending\_lr = ascending(t)$
4:     **for** $data$ in $data\_loader()$ **do**
5:         $k = k + 1$
6:         compute gradient $g_k = \nabla f(w_{k-1})$ on $data$
7:         $m_k = \beta_1 m_{k-1} + (1 - \beta_1) g_k$
8:         $v_k = \beta_2 v_{k-1} + (1 - \beta_2) g_k^2$
9:         $lr = Clip(\frac{\sqrt{1-\beta_2^k}}{1-\beta_1^k} \cdot \frac{\eta}{\sqrt{v_k}+\epsilon}, ascending\_lr, max\_lr)$
10:        $w_k = w_{k-1} - lr \cdot m_k$
11:     **end for**
12: **end for**
13: **for** $t = \{1, 2, ..., \frac{2T}{5}\}$ **do**
14:     $ascending\_lr = descending(t)$
15:     **for** $data$ in $data\_loader()$ **do**
16:         $k = k + 1$
17:         compute gradient $g_k = \nabla f(w_{k-1})$ on $data$
18:         $m_k = \beta_1 m_{k-1} + (1 - \beta_1) g_k$
19:         $v_k = \beta_2 v_{k-1} + (1 - \beta_2) g_k^2$
20:         $lr = Clip(\frac{\sqrt{1-\beta_2^k}}{1-\beta_1^k} \cdot \frac{\eta}{\sqrt{v_k}+\epsilon}, min\_lr, descending\_lr)$
21:        $w_k = w_{k-1} - lr \cdot m_k$
22:     **end for**
23: **end for**
24: **for** $t = \{1, 2, ..., \frac{T}{5}\}$ **do**
25:     $lr = min\_lr$
26:     **for** $data$ in $data\_loader()$ **do**
27:         $k = k + 1$
28:         compute gradient $g_k = \nabla f(w_{k-1})$ on $data$
29:         $w_k = w_{k-1} - lr \cdot g_k$
30:     **end for**
31: **end for**

4. Experiments

To illustrate the effectiveness of our algorithm, we experimented with different models on different data sets to compare the new variant with other popular optimization methods, such as SGD with momentum (SGDM), Adagrad and Adam. We mainly consider two problems that are often solved by deep neural networks: image classification and language modeling. The models used in the experiment include feedforward neural network, convolutional neural network [34], deep convolutional neural network and recurrent neural network, The data sets used in the experiment are MNIST [20], CIFAR-10 [21], CIFAR-100 [21], Penn Treebank [25]. All these models or data sets are often encountered in deep learning.

4.1. Experimental Setup

We implemented these experiments on a server configured as 2 NVIDIA TITAN XP GPUs, 1 Intel I7-6800K CPU, 16G*8 DDR4, 240G SSD and 1T SATA. These experiments were coded in PyTorch, each experiment was run three times and we chose the best one.

The algorithms under consideration have many hyper-parameters and the setting of hyper-parameters has a great influence on the performance of the optimization algorithm. Here we will describe how we adjust hyper-parameters. We use a logarithmical grid search on a large space of learning rate and then fine-tune it, the results are shown in Table 1. Specifically, the learning rate of each algorithm is adjusted as follows:

- **SGD(M)** We used SGDM for image classification tasks and SGD for language modeling tasks. When using SGDM, we set the momentum parameter to the default value 0.9. we roughly tuned the learning rate for SGD(M) on a logarithmic scale from 10^{-3} to 10^2 first and then fine-tune the learning rate.
- **Adagrad** The general learning rates used for Adagrad are chosen from {0.0005, 0.001, 0.005, 0.01, 0.1}.
- **Adam** The general learning rates used for Adam were chosen from {0.0005, 0.001, 0.005, 0.01, 0.05}. We set $\{\beta_1, \beta_2\}$ as the recommended default value {0.9, 0.999}. The perturbation value ϵ is set to 10^{-8}.
- **Bsadam** We used the same hyper-parameter settings for Adam. The upper and lower bounds were determined by learning rate range test and the rise and decrease of bounds are linear.

Other hyper-parameters such as batch size and weight decay use the default values recommended by the model.

Table 1. Summarizing the models and the data sets utilized for our experiments. The optimal hyperparameters for stochastic gradient descent (SGD) with momentum (M), Adagrad, Adam and Bsadam for all experiments are also listed.

Data Set	Model	Network Type	SGD(M)	Adagrad	Adam	Bsadam
MNIST	1-Layer Perceptron	Feedforward	0.1	0.001	0.001	(0.01,0.5)
MNIST	1-Layer Convolutional	Convolutional	0.1	0.001	0.001	(0.01,0.5)
CIFAR-10	ResNet	Deep Convolutional	0.1	0.001	0.001	(0.01,0.5)
CIFAR-10	DenseNet	Deep Convolutional	0.1	0.001	0.001	(0.01,0.5)
CIFAR-10	SENet	Deep Convolutional	0.1	0.001	0.001	(0.01,0.5)
CIFAR-100	ResNet	Deep Convolutional	0.3	0.001	0.001	(0.05,1)
CIFAR-100	DenseNet	Deep Convolutional	0.1	0.001	0.001	(0.05,1)
CIFAR-100	SENet	Deep Convolutional	0.1	0.001	0.001	(0.01,0.5)
Penn Treebank	1-Layer LSTM	Recurrent	50	0.001	0.001	(5,100)
Penn Treebank	2-Layer LSTM	Recurrent	50	0.001	0.001	(5,100)

4.2. Image Classification

4.2.1. Simple Neural Network

The MNIST database of handwritten digits has a training set of 60,000 images, and a test set of 10,000 images, which can be divided into 10 classes. We train a simple fully connected neural network with one hidden layer and a one-layer convolutional network with one convolutional layer and one fully connected layer for the image classification problem on the MNIST dataset. We run 100 epochs and decay the learning rate by 10 at 80th epoch for fully connected neural network, then we run 75 epochs and decay the learning rate by 10 at 60th epoch for convolutional network.

Figure 3 shows the learning curve of each optimization method, which includes training accuracy and test accuracy. We find that all the optimization algorithms can achieve nearly 100% accuracy on the training set. However, the accuracy of each algorithm will be different on the test set. Among these algorithms, Adagrad converges fastest on the training set, but achieves lower accuracy on the test set and SGDM has a slightly better accuracy on the test set than Adam and Adagrad. Our proposed Bsadam has better convergence speed than SGDM in the early stage. Especially in the converging phase, the convergence speed of Bsadam is faster than all the compared algorithms on both training and test set. Moreover, the final test accuracy of Bsadam is as good as fine-tuned SGDM, which means that our algorithm can get faster training speed without sacrificing accuracy when training simple neural networks. We also run RMSProp and Nesterov with default setting on MNIST. We find that RMSProp has worst convergence speed and test accuracy, Nesterov has similar performance with SGD with momentum. So our method still has advantages over these methods.

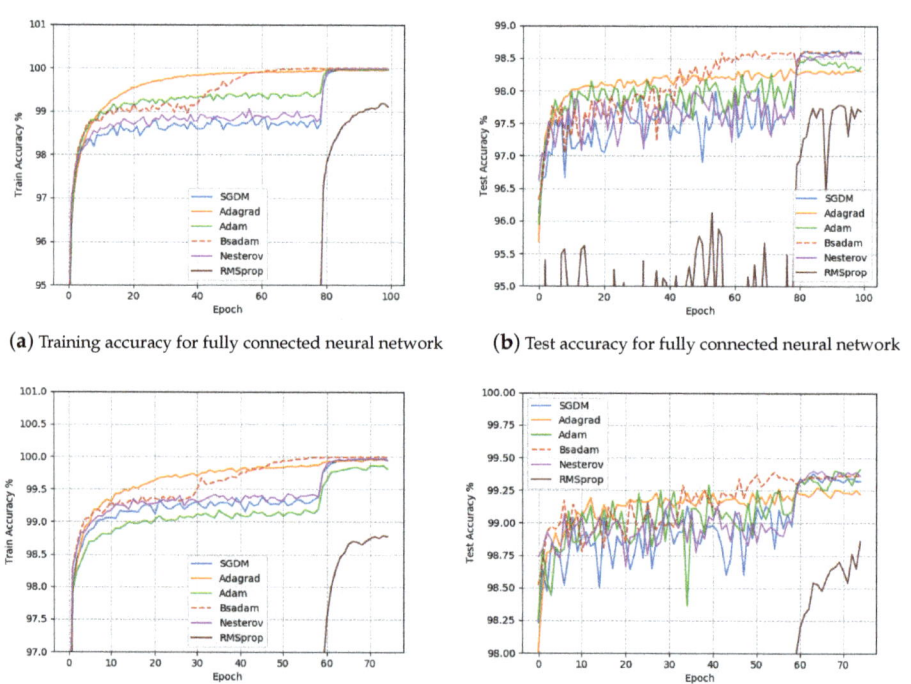

(**a**) Training accuracy for fully connected neural network (**b**) Test accuracy for fully connected neural network

(**c**) Training accuracy for one-layer convolutional neural Network (**d**) Test accuracy for one-Layer convolutional neural network

Figure 3. Training and test accuracy for fully connected neural network and one-layer convolutional neural network on MNIST.

4.2.2. Deep Convolutional Network

We evaluate our algorithm on a more complex deep convolutional network. Specifically, we perform experiments with three architectures: ResNet-34 [22], DenseNet-121 [23] and SENet-34 [24] on CIFAR-10 and CIFAR-100 data sets for image classification tasks. These data sets have a training set of 50,000 32 × 32 RGB images, and a test set of 10,000 images, which can be divided into 10 classes for CIFAR-10 and 100 classes for CIFAR-100. We ran 125 epochs for all the compared algorithms and decay the learning rate by 10 at 100th epoch.

Figure 4 shows the learning curve of each optimization method running on CIFAR-10, which includes training accuracy and test accuracy. As we can see, Adagrad had faster convergence speed and higher accuracy on training set, its accuracy is the lowest on test set, which indicates that its generalization gap is relatively large. Adam converges faster than SGDM in the early training, but the final test accuracy is lower than SGDM. SGDM has the slowest convergence speed on training set and test set, but its final test accuracy is higher than Adam and Adgrad, which means that its generalization capability is better than adaptive gradient methods. Our proposed Bsadam converges faster than fine-tuned SGDM in the early training. Especially in the converging phase, the convergence speed of Bsadam is faster than all the compared algorithms on both training and test set. The final training and test accuracy of Bsadam are the highest among all the compared algorithms, which indicates that our algorithm can accelerate the training process and improve the accuracy for complex deep neural networks.

Figure 5 shows the learning curve of each optimization method running on CIFAR-100, which includes training accuracy and test accuracy. The experimental results shown in Figure 5 are similar to Figure 4. The adaptive gradient methods often exhibit a relatively large generalization gap. Bsadam can achieve faster convergence speed and higher convergence accuracy on both training and test set.

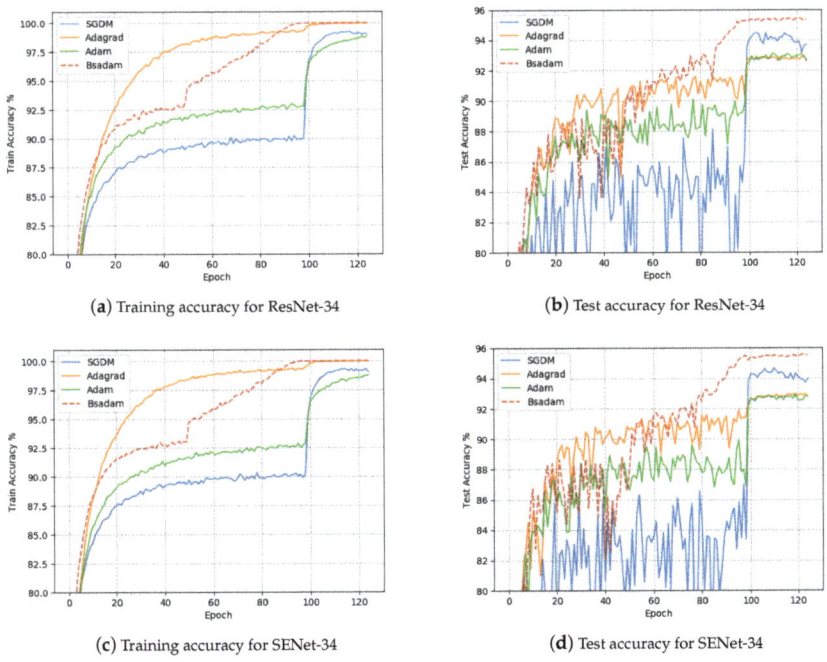

(a) Training accuracy for ResNet-34

(b) Test accuracy for ResNet-34

(c) Training accuracy for SENet-34

(d) Test accuracy for SENet-34

Figure 4. Cont.

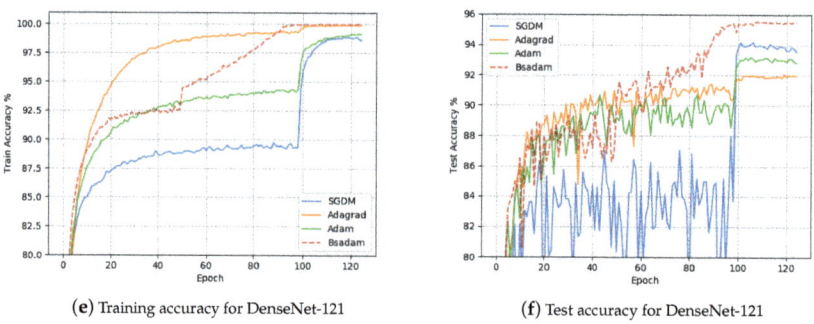

(e) Training accuracy for DenseNet-121

(f) Test accuracy for DenseNet-121

Figure 4. Training and test accuracy for ResNet-34, SENet-34 and DenseNet-121 on CIFAR-10.

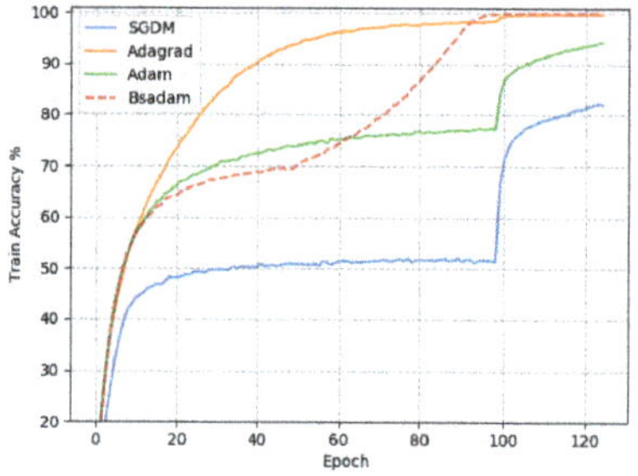

(g) Training accuracy for ResNet-34

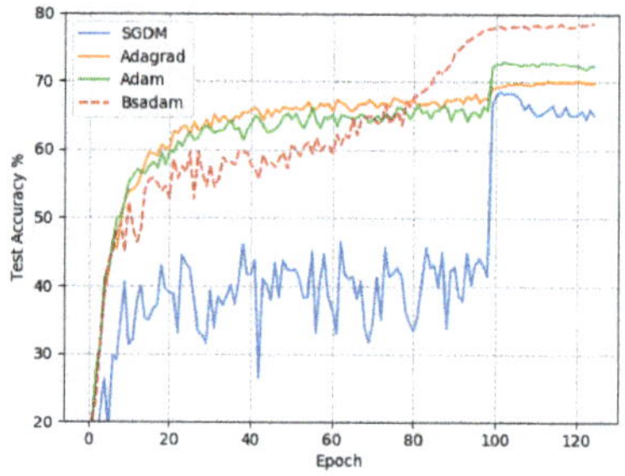

(h) Test accuracy for ResNet-34

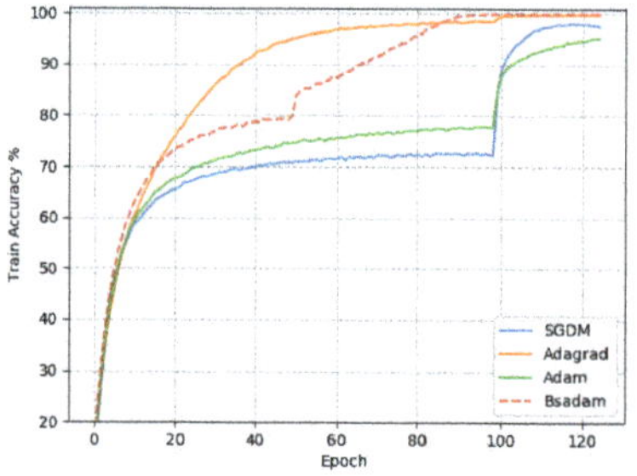

(i) Training accuracy for SENet-34

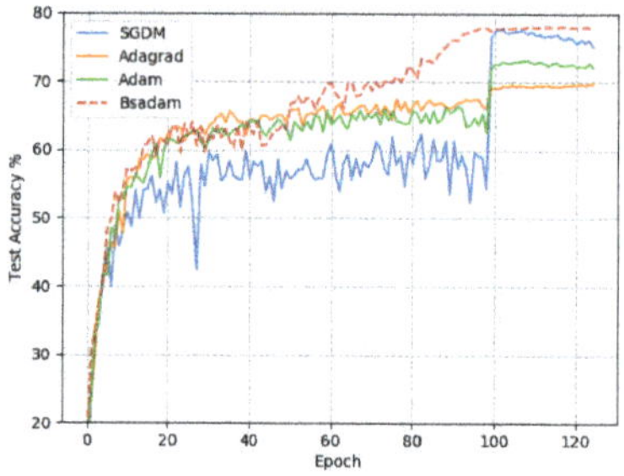

(j) Test accuracy for SENet-34

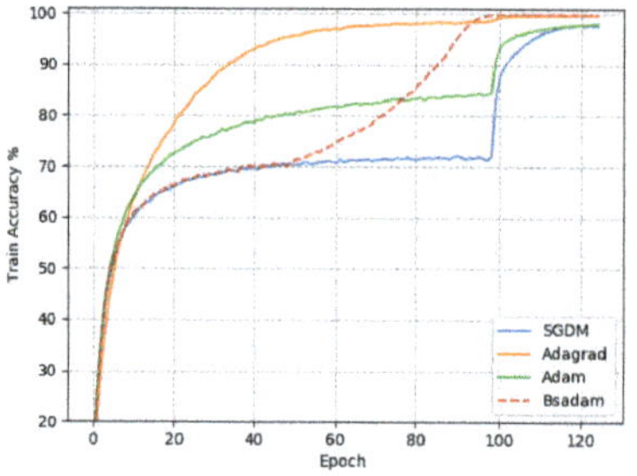

(k) Training accuracy for DenseNet-121

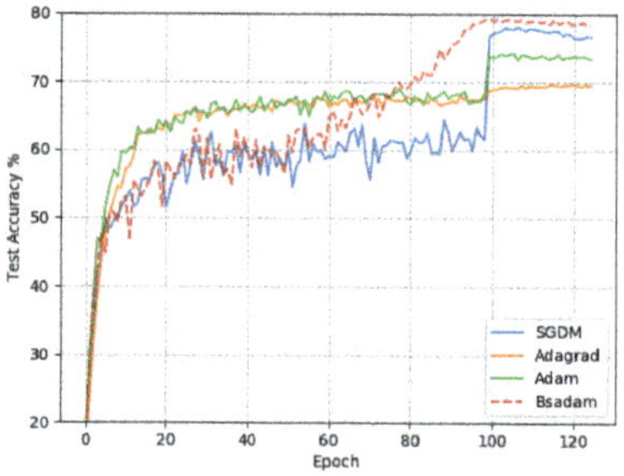

(l) Test accuracy for DenseNet-121

Figure 5. Training and test accuracy for ResNet-34, SENet-34 and DenseNet-121 on CIFAR-100.

4.3. Language Modeling

To illustrate the wide applicability of our algorithm, we also conduct experiments with the recurrent network. Specifically, we perform experiments with long short-term memory (LSTM) network [26] on Penn Treebank data set for word-level language modeling tasks. We compare our algorithm with Adam and SGD without the moment. We ran 125 epochs for all the compared algorithms and decay the learning rate by 10 at 100th epoch.

Figure 6 shows the learning curve of each optimization method running on Penn Treebank, which includes training accuracy and test accuracy. We find that the training perplexity of a two-layer LSTM is lower than a one-layer LSTM, but the valid perplexity is almost the same, which indicates that the complexity of the network may weaken the generalization capability of the algorithm. Although Adam achieves a lower perplexity on the training set, the final perplexity on a valid set is relatively high. SGD converges slowly in the early stage on a valid set, but the final perplexity is lower than Adam. Bsadam converges slowly in finding minima phase, but in converging phase, training and valid perplexity both decrease rapidly and the overall convergence speed is faster than SGD. What is more, Bsadam can get a similar or better final perplexity compared to fine-tuned SGD.

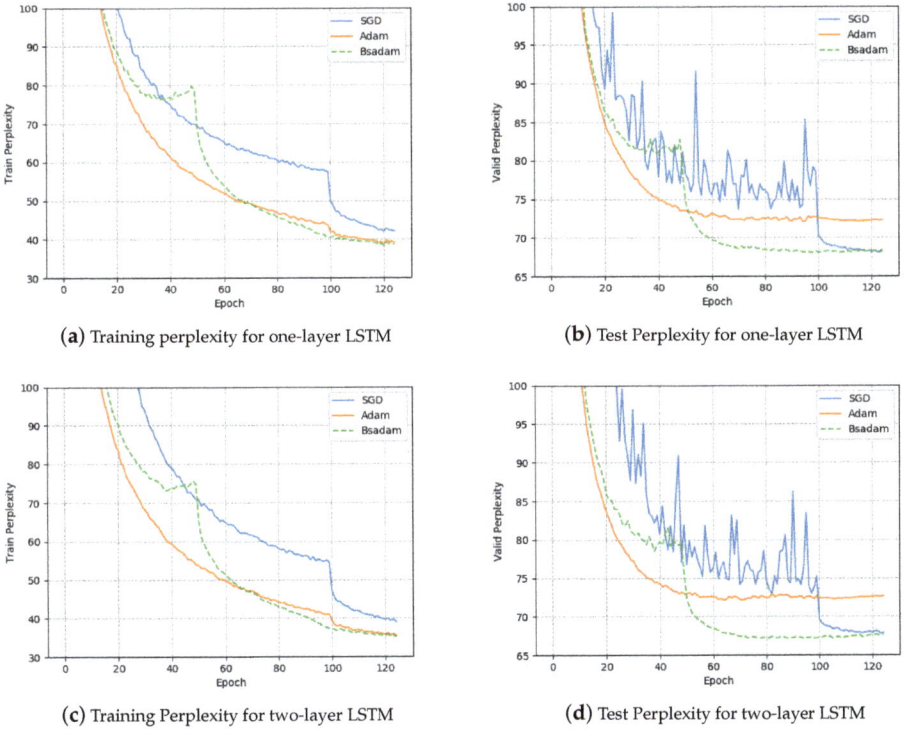

(a) Training perplexity for one-layer LSTM
(b) Test Perplexity for one-layer LSTM
(c) Training Perplexity for two-layer LSTM
(d) Test Perplexity for two-layer LSTM

Figure 6. Training and valid perplexity for long short-term memory (LSTM) with different layers on Penn Treebank.

4.4. Comparison of Different Scheduling Methods

In this paper, we propose three bounded scheduling methods: linear scheduling, exponential scheduling and trigonometric scheduling. We use these three bounded scheduling methods to train SENet-34 on CIFAR-10 and the learning curve is shown in Figure 7. As we can see, these scheduling methods have similar performance, but the details of the learning curve are slightly different. Exponential scheduling method has the fastest convergence speed among three scheduling methods, but the final test accuracy is lowest. Linear scheduling method has the highest final test accuracy, but the convergence speed is slowest among three scheduling methods.

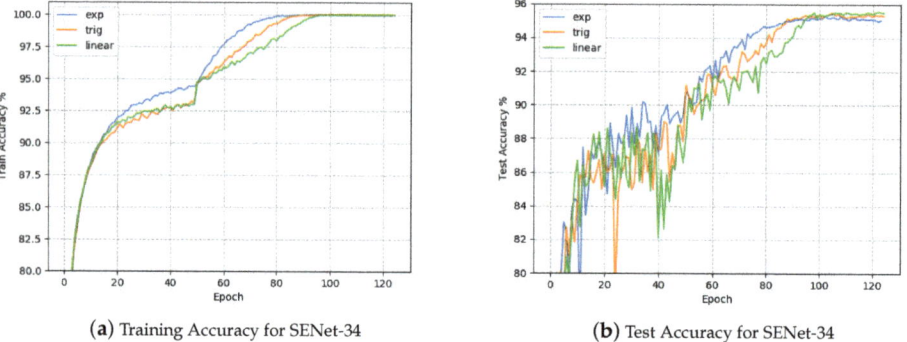

(a) Training Accuracy for SENet-34 (b) Test Accuracy for SENet-34

Figure 7. Training and test accuracy of different scheduling methods for SENet-34 on CIFAR-10.

5. Conclusions

Towards the poor generalization capability of adaptive gradient methods in training deep neural networks, a bounded scheduling method, called Bsadam, is proposed in this paper. We first find the upper and lower bound for Adam, then divide the training process into three phases: finding minima phase allows the algorithm to overcome the suboptimal solutions by raising the lower bound of Adam, converging phase ensures the convergence of the algorithm by decreasing the upper bound of Adam and uniform scaling phase allows the algorithm converge to the minimum. We evaluate our algorithm by using simple neural networks, deep convolution networks and recurrent network to perform image classification and language modeling tasks. The experimental results show that our algorithm outperforms SGD(M) and the adaptive gradient methods in convergence speed and accuracy.

Author Contributions: Conceptualization, M.T.; methodology, M.T.; software, M.T.; validation, Z.H., Y.Y. and C.W.; formal analysis, M.T. and Z.H.; investigation, M.T.; resources, Y.P.; data curation, M.T.

Funding: This research was funded by The National Key Research and Development Program of China grant number 2016YFB1000100.

Acknowledgments: All authors thank the referees for their valuable suggestions and help.

Conflicts of Interest: The authors declare no conflict of interest. The funders had no role in the design of the study; in the collection, analyses, or interpretation of data; in the writing of the manuscript, or in the decision to publish the results.

References

1. Schmidhuber, J. Deep learning in neural networks: An overview. *Neural Netw.* **2015**, *61*, 85–117. [CrossRef] [PubMed]
2. Seo, S.; Kim, J. Efficient Weights Quantization of Convolutional Neural Networks Using Kernel Density Estimation based Non-uniform Quantizer. *Appl. Sci.* **2019**, *9*, 2559. [CrossRef]
3. Song, K.; Yang, H.; Yin, Z. Multi-Scale Attention Deep Neural Network for Fast Accurate Object Detection. *IEEE Trans. Circuits Syst. Video Technol.* **2018**, 1. [CrossRef]
4. Maas, A.L.; Qi, P.; Xie, Z.; Hannun, A.Y.; Lengerich, C.T.; Jurafsky, D.; Ng, A.Y. Building DNN acoustic models for large vocabulary speech recognition. *Comput. Speech Lang.* **2017**, *41*, 195–213. [CrossRef]
5. Hinton, G.; Deng, L.; Yu, D.; Dahl, G.; Mohamed, A.R.; Jaitly, N.; Senior, A.; Vanhoucke, V.; Nguyen, P.; Kingsbury, B.; et al. Deep Neural Networks for Acoustic Modeling in Speech Recognition: The Shared Views of Four Research Groups. *IEEE Signal Process. Mag.* **2012**, *29*, 82–97. [CrossRef]

6. Violante, M.G.; Marcolin, F.; Vezzetti, E.; Ulrich, L.; Billia, G.; Di Grazia, L. 3D Facial Expression Recognition for Defining Users' Inner Requirements—An Emotional Design Case Study. *Appl. Sci.* **2019**, *9*, 2218. [CrossRef]
7. Zhang, J.; Zong, C. Deep Neural Networks in Machine Translation: An Overview. *IEEE Intell. Syst.* **2015**, *30*, 16–25. [CrossRef]
8. Robbins, H.; Monro, S. A Stochastic Approximation Method. *Ann. Math. Stat.* **1951**, *22*, 400–407. [CrossRef]
9. Duchi, J.; Hazan, E.; Singer, Y. Adaptive subgradient methods for online learning and stochastic optimization. *J. Mach. Learn. Res.* **2001**, *12*, 2121–2159.
10. Nocedal, J.; Wright, S. *Numerical Optimization*; Springer Science & Business Media: Berlin, Germany, 2006.
11. Smith, L.N. Cyclical learning rates for training neural networks. In Proceedings of the 2017 IEEE Winter Conference on Applications of Computer Vision (WACV), Santa Rosa, CA, USA, 24–31 March 2017; pp. 464–472.
12. Smith, L.N.; Topin, N. Super-convergence: Very fast training of neural networks using large learning rates. *Artif. Intell. Mach. Learn. Multi-Domain Oper. Appl.* **2019**, *11006*, 1100612.
13. Luo, L.; Xiong, Y.; Liu, Y.; Sun, X. Adaptive gradient methods with dynamic bound of learning rate. *arXiv* **2019**, arXiv:1902.09843.
14. Zeiler, M.D. ADADELTA: An adaptive learning rate method. *arXiv* **2012**, arXiv:1212.5701.
15. Tieleman, T.; Geoffrey, H. Lecture 6.5-rmsprop: Divide the gradient by a running average of its recent magnitude. *COURSERA Neural Netw. Mach. Learn.* **2012**, *4*, 26–31.
16. Kingma, D.P.; Ba, J.L. Adam: A Method for Stochastic Optimization. In Proceedings of the 3rd International Conference on Learning Representations, ICLR 2015, San Diego, CA, USA, 7–9 May 2015.
17. Dozat, T. Incorporating Nesterov Momentum into Adam. *ICLR Workshop* **2016**, *1*, 2013–2016.
18. Reddi, S.J.; Kale, S.; Kumar, S. On the Convergence of Adam and Beyond. *arXiv* **2019**, arXiv:1904.09237.
19. Nitish, S.K.; Richard, S. Improving generalization performance by switching from Adam to SGD. *arXiv* **2017**, arXiv:1712.07628.
20. LeCun, Y.; Bottou, L.; Bengio, Y.; Haffner, P. Gradient-based learning applied to document recognition. *Proc. IEEE* **1998**, *86*, 2278–2324. [CrossRef]
21. Krizhevsky, A.; Hinton, G. *Learning Multiple Layers of Features from Tiny Images*; Technical Report; University of Toronto: Toronto, ON, Canada, 2009.
22. He, K.; Zhang, X.; Ren, S.; Sun, J. Deep residual learning for image recognition. In Proceedings of the IEEE Conference on Computer Vision and Pattern Recognition, Las Vegas, NV, USA, 26 June–1 July 2016; pp. 770–778.
23. Iandola, F.; Moskewicz, M.; Karayev, S.; Girshick, R.; Darrell, T.; Keutzer, K. Densenet: Implementing efficient convnet descriptor pyramids. *arXiv* **2014**, arXiv:1404.1869.
24. Hu, J.; Shen, L.; Sun, G. Squeeze-and-excitation networks. In Proceedings of the IEEE Conference on Computer Vision and Pattern Recognition, Salt Lake City, UT, USA, 18–22 June 2018; pp. 7132–7141.
25. Mitchell, P.M.; Mary, A.M.; Beatrice, S. Building a large annotated corpus of english: The penn treebank. *Comput. Linguist.* **1993**, *19*, 313–330.
26. Hochreiter, S.; Schmidhuber, J. Long short-term memory. *Neural Comput.* **1997**, *9*, 1735–1780. [CrossRef]
27. Roux, N.L.; Schmidt, M.; Bach, F. A Stochastic Gradient Method with an Exponential Convergence Rate for Finite Training Sets. *Adv. Neural Inf. Process. Syst.* **2012**, *4*, 2663–2671.
28. Fletcher, R. On the barzilai-borwein method. In *Optimization and Control with Applications*; Springer: Boston, MA, USA, 2005; pp. 235–256.
29. Raydan, M. On the barzilai and borwein choice of steplength for the gradient method. *IMA J. Numer. Anal.* **1993**, *13*, 321-326. [CrossRef]
30. Massé, P.-Y.; Ollivier, Y. Speed learning on the fly. *arXiv* **2015**, arXiv:1511.02540.
31. Xiangyi, C.; Sijia, L.; Ruoyu, S.; Mingyi, H. On the convergence of a class of Adam-type algorithms for non-convex optimization. *arXiv* **2018**, arXiv:1808.02941.
32. Ashia, C.W.; Rebecca, R.; Mitchell, S.; Nati, S.; Benjamin, R. The marginal value of adaptive gradient methods in machine learning. *Adv. Neural Inf. Process. Syst.* **2017**, *30*, 4148–4158.

33. Hardt, M.; Recht, B.; Singer, Y. Train faster, generalize better: Stability of stochastic gradient descent. *arXiv* **2015**, arXiv:1509.01240.
34. Zhang, R. Making convolutional networks shift-invariant again. *arXiv* **2019**, arXiv:1904.11486.

 © 2019 by the authors. Licensee MDPI, Basel, Switzerland. This article is an open access article distributed under the terms and conditions of the Creative Commons Attribution (CC BY) license (http://creativecommons.org/licenses/by/4.0/).

Article

A Novel Bio-Inspired Method for Early Diagnosis of Breast Cancer through Mammographic Image Analysis

David González-Patiño [1], Yenny Villuendas-Rey [2], Amadeo-José Argüelles-Cruz [1,*] and Fakhri Karray [3]

1. Centro de Investigación en Computación, Instituto Politécnico Nacional, Ciudad de México 07738, Mexico; davidglezp-92@hotmail.com
2. Centro de Innovación y Desarrollo Tecnológico en Cómputo, Instituto Politécnico Nacional, Ciudad de México 07700, Mexico; yenny.villuendas@gmail.com
3. Department of Electrical and Computer Engineering, University of Waterloo, Ontario, ON N2L 3G1, Canada; karray@pami.uwaterloo.ca
* Correspondence: amadeomx@gmail.com; Tel.: +52-155-2855-8602

Received: 28 September 2019; Accepted: 17 October 2019; Published: 23 October 2019

Abstract: Breast cancer is a current problem that causes the death of many women. In this work, we test meta-heuristics applied to the segmentation of mammographic images. Traditionally, the application of these algorithms has a direct relationship with optimization problems; however, in this study, its implementation is oriented to the segmentation of mammograms using the Dunn index as an optimization function, and the grey levels to represent each individual. The update of grey levels during the process results in the maximization of the Dunn's index function; the higher the index, the better the segmentation will be. The results showed a lower error rate using these meta-heuristics for segmentation compared to a well-adopted classical approach known as the Otsu method.

Keywords: mammogram; meta-heuristics; optimization; breast cancer; segmentation; detection

1. Introduction

Breast cancer is an acute health problem all over the world and one of the most common cancers that cause death among women. This cancer has caused over 1.7 million deaths in 2012 and nearly 5 million cases were diagnosed. The estimation for 2030 is that breast cancer cases will continue to grow in developing countries [1].

To reduce these cases and to provide treatments, it is necessary to have better ways of diagnosing breast cancer early on. Using algorithms to help in the early diagnosis would be desirable [2]. The algorithms can use the result images given by other tests like mammography studies or other screening techniques to obtain an easier image to analyze. These algorithms can obtain a simplified representation of the image only by extracting the region of interest from the image for later analysis.

Physicians use mammography studies to obtain an image of the breast, using X-rays to detect breast cancer followed by a detailed analysis of the generated image. These evaluations have resulted in a reduction of breast cancer mortality [3]. When cancer is in the early phases, the group of abnormal cells is found in the same region, and hence it can be easily detected using mammography [4]. The experts highly recommend a mammography study every one or two years for women aged 39 to 69 [5,6]. After the generation of the mammography image, it is necessary to process it to identify the Region of Interest (ROI) of the image for later analysis. This process receives the name of segmentation.

However, this is not the only way to carry out these studies; as an example, a different way is seen in the work carried out by Yu et al. in 2017 [7]. They present an improvement in the images taken by ultrasound tomography, which improves the segmentation and the later analysis and classification of the image. In 2013, Duric et al. [8] performed a comparison between ultrasound tomographies and digital mammographies showing that both methods are positively associated with the identification of the amount of dense tissue. They performed the study by comparing the volume-averaged sound speed of the breast in ultrasound tomography and mammographic percent density in mammographies.

Segmentation involves splitting a digital image into non-overlapping groups of pixels to make it easy to interpret [9]. This process is useful to find objects and boundaries, and it can be used in many applications such as object detection, recognition, machine vision, and medical images [10]. A perfect segmentation should have uniform and homogeneous characteristics, boundaries should be simple, interiors should not have small holes if possible, and adjacent regions should have different characteristics [11]. Since the previous characteristics are hard to achieve, there are techniques that try to find the best segmentation for digital images, each of them with advantages and disadvantages [10].

According to Fu et al. [12], such techniques can be conveniently summarized into three types: edge detection, region extraction, and clustering segmentation.

Edge detection is a technique based on distinguishing regions with the highest change in grey level. It also detects discontinuities in depth and surface, changes in material properties, and variations in scene illumination and brightness.

Several studies report the progression and implementation of edge detection. Dollár and Zitnick [13] uses a segmentation technique with a structured learning framework applied to random decision forests. Malik et al. [14] describe the Canny edge detection method to identify lines in a Finger Knuckle Print, a biometric identifier adopted in recent years.

Qi et al. [15] worked infrared images and algorithms to transform edges into curves and produce intersections for breast cancer detection. Later on, Mencattini et al. [16] proposed an algorithm to detect micro-calcifications and masses. It also detects the edges using an enhancement procedure after a transformation process.

Song et al. [17] used Fully Convolutional Networks to segment and detect corners in aerial images of buildings. Their work achieved good performance in this task, outperforming several algorithms in their corresponding comparison.

The region extraction approach splits an image into regions using merging or dividing techniques. Fan et al. [18] proposed an image segmentation method using edges as first elements, incorporating pixels into each region at each iteration. A similar technique was used by Yan et al. in 2003 [19] using local entropy, getting good performance with noisy images. The use of a region extraction algorithm is convenient because a region comprises more pixels and achieves better performances with noisy images, providing more information than edge detection techniques [18], which face issues with noisy images. Combining both region extraction and edge detection provides more detail information about the image [18].

Clustering or feature thresholding consider choosing a threshold value that maps all the pixels into different clusters. These threshold values can be grey level, gradients or percentages. Yao et al. [20] used an algorithm based on clustering to segment fish images. Salem [21] reported a similar technique to find white cells using a k-means algorithm. And Patel and Sinha [22] used a clustering algorithm based on adaptive k-means, diversifying the parameters to boost image segmentation's performance. In general, clustering is easy to implement, simple, and the results are easy to interpret.

On one hand, the extraction of regions seeks to identify the region of interest for further analysis. On the other hand, clustering algorithms seek to group regions with similar characteristics regardless of whether they represent something relevant.

The clustering approach for segmentation is the protagonist in this work.

There are recent works on the detection of breast cancer using image processing and deep learning to achieve good results in both branches. In 2018, Mambou et al. [23] performed a comparative study of several algorithms for breast cancer detection using Infrared Thermal Imaging, obtaining relevant results in this field.

One of the classic methods for image thresholding is the Otsu method [24], which uses clustering to generate binary images. This method is still one of the most referenced thresholding methods because of its simplicity and good performance. This method is sensitive if the region of interest (ROI) or the background is bigger than the other region since it will classify the other pixels incorrectly. The Otsu method finds the optimal threshold by maximizing or minimizing variances between or within each class, respectively [25]. This method has been successfully used for image thresholding in many applications [26–28].

In this paper, we explore two algorithms based on populations, the novel bat algorithm (NBA) and the genetic algorithm (GA). In addition, we explore the simulated annealing algorithm (SA) based on a single trajectory. The GA is based on an evolutionary system and the NBA is based on swarm optimization.

There are a lot of segmentation techniques, but there is no method to determine which is the most useful algorithm for the segmentation of specific digital images. In this paper, we focus on meta-heuristics applied to the segmentation of digital images, specifically, mammography images.

This paper is organized as follows. The review of recent works in mammography image segmentation is presented in Section 2, with an explanation of the algorithms under test, while giving insights into their functioning. In Section 3, the proposed methodology is described. Section 4 provides the results and the analysis of the experiments. Finally, in Section 5 we present the discussion and conclusions of this work and future work in this area.

2. Meta-Heuristics and Applications for Image Segmentation

Meta-heuristics are algorithms that mimic fauna behavior or biological systems to solve computational problems such as optimization, classification, or segmentation [29]. Some algorithms used to solve optimization problems are not effective when the problem is nondeterministic since this requires many computational resources. Therefore, using meta-heuristics is one of the best choices because of their stochastic behavior [30]. These algorithms use randomly generated values to conduct a local search while exploring a larger space of solution.

In this paper, we present the use of some meta-heuristics applied to segmentation. We explain these meta-heuristics below.

2.1. Simulated Annealing

Simulated annealing (SA) [31] is a searching algorithm mainly designed for global optimization problems [32].

The simulated annealing (SA) algorithm is an optimization method inspired by the tempering of metals used since 5000 B.C. and belongs to a class of local search algorithms (LSA) commonly called threshold algorithms (TA).

To understand the simulated annealing method, it is necessary to understand that this technique is used in the industry to obtain more resistant or more crystalline properties to improve the qualities of materials. This same principle is adapted to a computational algorithm that works by mimicking this process.

The process consists of "melting" the material (heating it to a very high temperature). In this situation, the atoms gain a "random" distribution within the material structure, and the system energy is maximal. Then, the temperature is reduced in stages, allowing the atoms to remain in equilibrium in each of these stages (that is, the atoms reach an optimal configuration for that temperature). At the end of the process, the atoms form a highly regular crystalline structure, thus, the material achieves maximum strength, and the system energy is minimal.

The experiments report that if the process sharply reduces the temperature or if there is not enough time in each stage, the structure of the material is not optimal.

This algorithm presents three main phases: heating the material to a predetermined temperature, maintaining the temperature that allows the molecules to accommodate in states of minimum energy, and then a slow cooling of the material to allow an increase in the size of the crystals and a reduction of their defects. In each iteration, the algorithm tests the neighbors to find better solutions, considering new solutions with a probability of using that solution even if it is not a better solution.

Simulated annealing (SA) is a simple trajectory method that starts with a certain state, S. Through a particular process, it generates a neighbor state, S', to the current environment. If the energy, or evaluation of the state S', is better than the one of S, the element S changes to S'.

If the evaluation of S' is worse than that of S, S' is chosen instead of S, with some probability depending on the differences of the evaluations of both states and the current temperature T. The probability of choosing a worse state instead of the current state allows a local optimum to be left in order to reach the global optimum. In a minimization process, the probability of choosing a worse state is calculated by:

$$p = e^{-1 * \frac{f'(x) - f(x)}{T}} \qquad (1)$$

In this work, given a state S, we obtain a neighboring state S' of the following form:

$$S'(i) = S(i) \pm rand(0, 255) \qquad (2)$$

The simulated annealing algorithm has several stages. Each stage corresponds to a lower temperature than the previous stage (this refers to the monotony: after each stage, the temperature goes down and the system cools). Therefore, a criterion of temperature change is required ("how much time" is waited at each stage to result in the system achieving its "thermal equilibrium").

If the temperature lowers sufficiently slowly (the temperature parameter and the generation of enough transitions) at each temperature, it can achieve the optimum configuration.

Recently, Manwar et al. [32] used an algorithm similar to simulated annealing to reduce the nonlinearity of Galvo scanners. In that study, they evaluated the algorithm in different frequencies to synthesize the signal. Their method showed better results compared to other methods for the compensation of Galvo scanners.

Similarly, in 2018 Fayyaz et al. [33], used simulated annealing (SA) to optimize an amplitude modulator showing the efficiency and effectiveness of SA when finding the optimum wavefront shape for focusing light.

2.2. Genetic Algorithms

In natural systems, different individuals of a population compete for resources to survive and reproduce. Those individuals with a better biological adaptation to their environment will preserve and enhanced their genetic material and pass that information to future generations, contrary to what happens with less adapted individuals, who will generate a smaller number of descendants, and the possibility of transmitting their genetic material information to the next generations is much smaller.

From an artificial systems perspective, computational thinking maps the adaptation coming from natural evolution processes into algorithms used for optimization problems. These heuristics-based algorithms try to imitate, in a certain way, the natural evolutionary process to get the best results for a specific problem. The evolutionary algorithms use a population to find solutions and identify the best one to solve an optimization problem. Among them are the genetic algorithms (GAs) that are computational models capable of adapting or recreating themselves based on mutation, crossing and selection phenomena [34], biological metaphors principles of the natural evolution of the species proposed by Darwin in 1859 [35]. Research on GAs began to develop in the 1950s, when some biologists tried to simulate genetic systems on a computer, such as A. Fraser [36], who did something very similar to GAs using chains and phenotypes, trying to simulate them on the SILLIAC computer. Years later in 1975, John H. Holland [37] was the first to gather and develop the critical mass of ideas from systems theory, mathematics, and computational science to get the principles of evolution to search for the optimal results of a particular problem. The principles developed by Holland led to the birth of the GAs and their use in the theory of intelligent adaptive systems.

As stated before, GAs algorithms work with a population of individuals included within a specific solution of a fitness function for the optimization problem. Each of these individuals has a value of the function being tested; this value represents the adaptation of the individual to the environment. The better the adaptation of an individual to the problem, the more likely they are to transmit their genes to the next generations. The solutions that obtain the best results for the solution of the aim function will be those that are preserved during the process of optimization for future generations of individuals. Therefore, this value of the function will determine if an individual is a good candidate or not and will guide the search for good solutions.

With individuals chromosomally coded, the algorithm can carry out the evolution of the solutions by following the steps described in Figure 1 [38]:

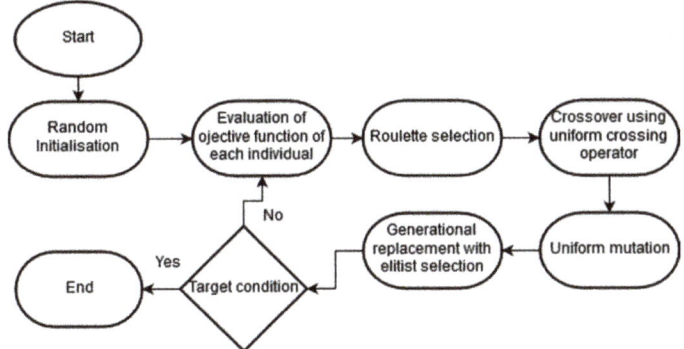

Figure 1. Flowchart of genetic algorithm.

As a summary, the genetic algorithm starts with the randomness generation of population at the beginning and uses the crossing (exchange of genetic material between two individuals), mutation (alteration of the internal elements of each individual), and the selection operators (operators designed to select the most suitable individuals for each problem) to select the best solutions in each iteration.

2.3. Novel Bat Algorithm

The bat algorithm (BA) is an algorithm proposed by Yang and Hossein in 2012 [39] as an algorithm used to solve optimization problems. Later, Meng et al. [40] transformed the algorithm to include the Doppler effect, giving this version the name of novel bat algorithm (NBA). This algorithm uses bats' behaviors and the echolocation ability to identify objects in a determinate space. The bats use echolocation to sense distance and change velocity and frequency to search for solutions for the fitness function that is being optimized.

One of the novel features of the NBA algorithm is the consideration and modelling of the compensation of the Doppler effect on the pulses emitted and the environmental noise. Another aspect to highlight is the inclusion of the quantum behavior of bats, which allows these virtual bats to find food in different habitats to find the optimal overall solution to the problem.

There are different variants of bat-inspired algorithms based on modelling some echolocation characteristics of bats. Therefore, the authors of the BA proposal in [41] provide a set of rules for optimization algorithms based on the behavior of bats, as described in [40]:

- To measure distance, bats use echolocation, which also allows them to differentiate between their prey and obstacles.
- Bats fly randomly with a velocity q_i to a position z_i with a fixed frequency f_{min}, varying the wavelength λ and volume V_0 to look for their prey. Depending on the proximity of the target, they can automatically adjust the wavelength of their emitted pulses (frequency), and adjust the pulse emission rate $l \in [0, 1]$.

- Although the volume may vary in various ways in reality, it is important to consider that the noise varies from a positive number V_0 to a constant minimum value V_{min}.

To this brief set of rules, we must add a couple more, which present a direct relationship with the new features proposed by the NBA algorithm [40]:

- All bats can search for food in different habitats.
- All bats can compensate for the Doppler effect by echo.

Finally, $z_{i,j}^G$ ($i \in \{1, \ldots, n\}$, $j \in \{1, \ldots, d\}$) characterizes the position and $q_{i,j}^G$ the velocities in an iteration G, of n bats looking for food in a d-dimensional space.

Updating the positions, velocities and other parameters of bats' behavior takes G number of iterations. Each one will perform the following steps explained in Figure 2:

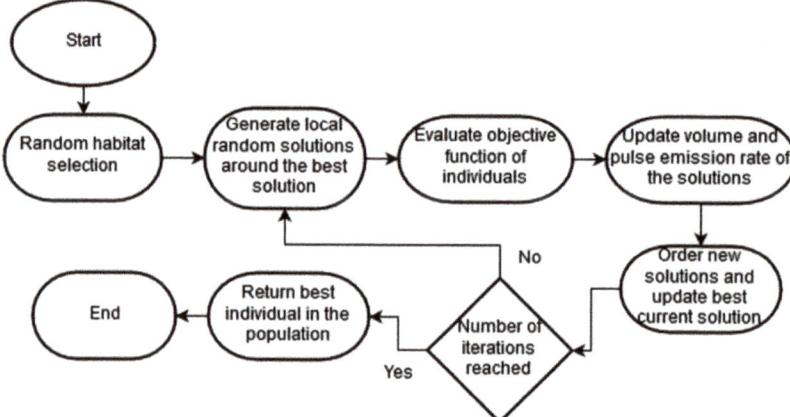

Figure 2. Flowchart of the novel bat algorithm.

As seen, this algorithm can self-adapt throughout the optimization process, because of the compensation it makes for the Doppler effect in echoes and the adjustments it makes to the frequency according to the proximity of individuals to the solution. It is also worth mentioning the addition of habitat selection methods to BA, since these two types of behavior of bats, mechanical and quantum, allow the algorithm to have a better convergence and diversity of solutions.

This algorithm was used for the segmentation of mammography images by González-Patiño et al. in 2016 [42], showing lower segmentation errors compared to the Otsu method.

The NBA and GA algorithms are population-based algorithms so they perform a more extensive search throughout the search space, which provides an advantage compared to SA which is a single-trajectory algorithm.

3. Proposed Method

In this work, we propose a new method for the segmentation of mammographic images through meta-heuristics. This process is relevant because most meta-heuristics are designed only as optimization algorithms.

This methodology consists of three stages:

1. In the first stage, a segmentation algorithm processes each image to simplify their analysis.
2. In the second stage, a group of radiological experts defines the characteristics that are relevant to the diagnosis. We define 11 characteristics of clinical and general data, six characteristics of lesion intensity descriptors, 13 characteristics of texture descriptors of the lesion, and eight descriptors

of the shape and location of the lesion. With these 38 descriptors, a data bank is formed for subsequent classification.
3. Once the database is formed, a classification algorithm is used to get a pre-diagnosis of the corresponding lesion.

We explain each stage in the following subsections.

3.1. Image Segmentation as an Optimization Problem

The proposal of the present investigation for phase 1 of the method consists of the modification of the meta-heuristic algorithms previously mentioned in Section 2, and its application to the segmentation of mammography images.

Performing the characterization of the mammographic image through the use of meta-heuristic algorithms outlines the problem of segmenting the image as an optimization problem. For this purpose, the Dunn index definition is considered as the objective function in all cases.

We use the Dunn index [43] as the fitness function for the meta-heuristic algorithms, so that the higher the Dunn index, the better the segmentation. Thus, in the three meta-heuristic algorithms, this index is the objective function to be maximized, and it is defined in Equation (3). We calculate this index as the minimum outer-cluster distance over the maximum inter-cluster distance:

$$f = \frac{Minimum\ distance\ between\ pixels\ of\ a\ different\ group}{Maximum\ distance\ between\ pixels\ of\ the\ same\ group} \quad (3)$$

The optimization function allows you to calculate the dispersion between the pixels of different groups and pixels of the same group. It is sought to maximize this function since we prefer for the numerator to be a very large value and the denominator to be a tiny value.

The distance calculated in Equation (3) is the Minkowski distance of order $p = 1$. This distance for the order p is defined Equation (4) with $p \geq 1$.

$$(\sum_{i=1}^{n} |x_i - y_i|^p)^{\frac{1}{p}} \quad (4)$$

The Minkowski distance for order 1 is reduced to a difference between pixel values between 0 and 255.

Three-dimensional vectors for the coding of the solutions are used; the components are defined in the interval [0,255], and represent a grey level. Three components are defined since it is desired to segment the image into three regions: background, breast area, and lesion. An example of this coding is shown in Figure 3. Figure 3 represents an individual with three components, each of them representing a grey level.

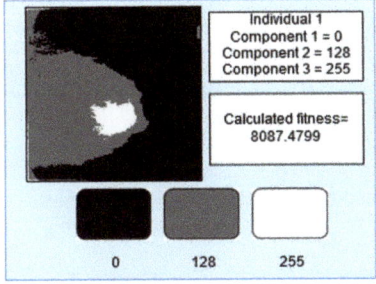

Figure 3. Example of segmentation using the solution of Individual 1.

This coding allows us to consider each solution as a candidate segmentation. Considering the solution of Figure 3, where the components represent a region of the image, we can obtain segmentation, as shown in the same figure, for Individual 1. For each pixel in the image, we calculate the distance to each component value, and we will assign the pixel to the component with the lowest distance.

According to Zhang et al. [44], the evaluation of the segmentation uses the average squared color error (F, Equation (5)), which penalizes over-segmentations, and its improved version (F', Equation (6)), which penalizes segmentations with a much greater number of small regions.

$$F = \sqrt{N} \sum_{j=1}^{N} \frac{ej^2}{\sqrt{Sj}} \qquad (5)$$

$$F' = \frac{1}{1000 * SI} \sqrt{\sum_{b=1}^{MaxArea} N(b)^{1+\frac{1}{b}}} \sum_{j=1}^{N} \frac{ej^2}{\sqrt{Sj}} \qquad (6)$$

$$ej^2(Rj) = \sum_{p \in Rj} \left(Cx(p) - \hat{C}x(Rj) \right)^2 \qquad (7)$$

$$\hat{C}x(Rj) = \frac{\sum_{p \in Rj} Cx(p)}{Sj} \qquad (8)$$

where N represents the number of regions in the image, Sj is the quantity of pixels in the region j, SI is the area of the image, $N(b)$ is the number of regions of the segmented image that have exactly b units of area, ej^2 is the squared colour error of region j defined in Equation (7). $Cx(p)$ is the value of component x for pixel p and $\hat{C}x(Rj)$ is the average value of component x in the region j is defined in Equation (8).

In Figure 4, you can observe an example of the use of the Dunn index to identify the best segmentation. We observe three individuals with three components, where each component represents a grey level used to generate the segmented image. The calculated fitness, shown in Figure 4, is the value calculated using the Dunn index, which measures how good the segmentation is according to the distance between clusters.

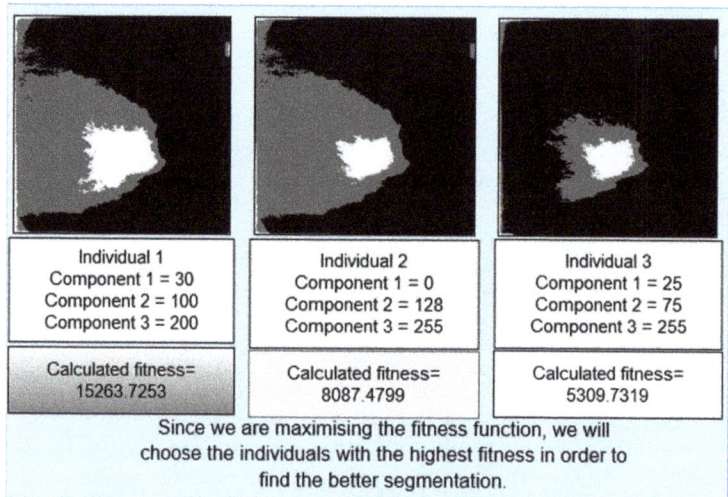

Figure 4. Example of individual values and calculation of fitness function.

In this paper we compare four segmentation algorithms, the Otsu method [24] and three customized meta-heuristic algorithms [31,40,45], to perform the segmentation task.

3.2. Extraction of Characteristics

For the diagnosis stage, a group of radiologists defined the characteristics. The characteristics are extracted from the region of interest of the segmented image and each definition is presented by Moura [46] and it explains the formulas for each calculation. This manuscript will present some descriptors in Table 1.

Table 1. Characteristics extracted from the images.

Group	Description	Features
Clinical and general data	General information of the patient and mammography	Age, breast density, image view, mammography nodule, mammography calcification, etc.
Intensity descriptors	Computed directly from the grey-levels of the pixels inside the lesion's contour	Mean, standard deviation, skewness, kurtosis, minimum and maximum.
Texture descriptors	Computed from the grey-level co-occurrence matrix related to the bounding box of lesion's contour	Energy, contrast, correlation, entropy, etc.
Shape and location descriptors	Descriptors of the lesion's contour	Area, perimeter, center of mass, circularity, elongation, form, solidity and extent.

3.3. Classification of the Lesion

For this step, a new artificial immune system (AIS) is proposed, which presents a competitive performance for breast cancer classification.

The proposed algorithm uses two different responses (adaptive immune response and innate immune response), which can be seen analogously as the training phase and classification phase in pattern recognition. The innate immune response will use the antibodies generated in the previous response to classifying each pattern according to the most similar antibody, while the adaptive immune response is based on grouping the patterns into random groups without replacement for each of the classes.

Later, a mean pattern (antigen) will be calculated from the groups formed above. The class of the grouped patterns will be assigned to each generated pattern. The mean calculates the performance of the aforementioned set of patterns, using the antibodies as models to classify.

The algorithm performs the adjustment of the antibodies by making the antibodies approach their most similar pattern and moving them away from different class patterns.

Subsequently, it generates clones for each antibody, and it controls this increase, obtaining an average of the clones generated by each antibody.

In the adaptive immune response, the algorithm will generate artificial antibodies that will be used as structures that will remember recognized patterns (antigens) and will have the ability to identify new antigens.

Once it completes the adaptive immune response (training phase), it will use the final antibodies generated for the classification of the other patterns (innate immune response). This classification will be performed by finding the closest antibody to each pattern, so the assigned class of each pattern will be one of the closest antibodies.

As a summary:

The adaptive immune response consists of five phases:

1. Detection of antigenic macromolecules: Groups are generated where patterns are randomly grouped according to their class.
2. Activation of B lymphocytes: Representative antibodies are generated for a set of patterns by the average in each group.
3. Immune response regulation: The antibodies are adjusted, approaching them with similar antibodies and moving away from different antibodies.
4. Development of adaptive immunity: Cloning of the antibodies is performed. Surplus antibodies are also eliminated in an elitist way.

5. Resolution of threat: Antibodies are stored in the immune memory to be used in the innate immune response.

The Innate immune response consists of one phase:

1. Resolution of threat: For each pattern, its similarity will be compared with the antibodies stored in immune memory. The antibody class of the antibody with greater similarity will be assigned.

The results applied in mammographic images are presented in the following section.

4. Experimental Results

Experimental tests were performed by segmenting images with three proposed segmentation algorithms. These results were compared to the images produced by the Otsu method.

Subsequently, the errors obtained by each algorithm and the best algorithm to segment are presented.

In the same way, the classification of six datasets using three algorithms are carried out (the presented model and two classic classification algorithms).

Performances are calculated and the algorithm with the best performance for each dataset is obtained.

4.1. Selection of the Best Segmentation Algorithm

To select the best segmentation algorithm, three meta-heuristic algorithms and the Otsu method were used to analyze mammographies from the Breast Cancer Digital Repository [46], which come from a Portuguese breast cancer database of real patients, which comprises 362 segmentations. The Faculty of Medicine of the University of Porto, Portugal provided this database made by expert radiologists, and they proved by biopsies the classifications provided (benign or malign).

Otsu parameters were used as a default and implemented in Matlab™. The parameters for the NBA were: three bats, three regions, 10 iterations, and other parameters as proposed by Meng et al. [40] were used. These parameters showed a lower error for segmentation according to the study of González-Patiño et al. in 2016 [42]. For the genetic algorithm, 50 individuals were used, with three regions, 10 iterations, a mutation rate of 0.05 and a crossover rate of 0.7. Finally, for simulated annealing an initial temperature of 100,000 and a cooling rate of 0.05 were used.

Errors for each image were calculated and we present the mean of the results in Table 2. The following images show two of the experiments carried out using images from mammographies. The first row shows the original mammography image, the region of interest and the region of interest overlapped in the original image. The second row shows the results applying the Otsu method, and the binarized segmentation obtained by the novel bat algorithm and the genetic algorithm. The final row shows the segmented image produced by the simulated annealing algorithm applied to segmentation.

Table 2. Mean of the errors calculated for the segmentation of the 362 mammographies.

Algorithm	F mean	F' mean
Otsu method	15.324×10^6	12.880×10^{-3}
Simulated Annealing	14.889×10^6	12.519×10^{-3}
Novel Bat Algorithm	10.758×10^6	9.030×10^{-3}
Genetic Algorithm	9.888×10^6	8.313×10^{-3}

Figure 5 shows a similar image (in shape) produced by Otsu and GA, while the NBA method segmented a different region. The NBA segmented a smaller region, which seems to be more accurate in relation to the original segmented image. The four algorithms got similar regions since the mammography has a very defined region of interest, however this is not the case in all images.

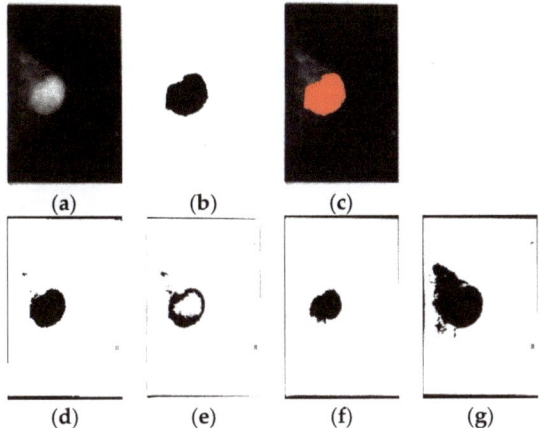

Figure 5. (a) Original image, (b) Region of interest (ROI) segmented by an expert radiologist, (c) Overlap of the ROI over the original image, (d) Image segmented by the Otsu method, (e) Image segmented by the novel bat algorithm (NBA), (f) Image segmented by the genetic algorithm (GA), (g) Image segmented by simulated annealing (SA).

The behavior observed in the meta-heuristics results from the simplicity of the algorithms; given that GA is a more complex algorithm than the simulated annealing and the NBA.

Figures 5 and 6 reveals that the images segmented by GA have a smaller area compared to the images segmented by the Otsu method or NBA. This is relevant because the region obtained by GA is more similar to the original region obtained by the expert radiologist. It is necessary to take into consideration that the image acquired by some algorithms includes noisy pixels that do not correspond to the desired image, which is why in the results it is shown that GA got a better segmentation when contemplating all the corresponding pixels. In both cases, the algorithm which produced the most similar region to the original segmented image by the expert was genetic algorithms. Table 2 and Figure 7 show the mean of the errors for all the 362 mammography images of the dataset.

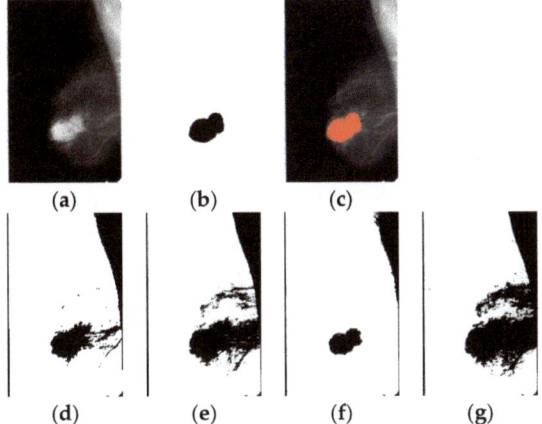

Figure 6. (a) Original image, (b) ROI segmented by an expert radiologist, (c) Overlap of the ROI over the original image, (d) Image segmented by the Otsu method, (e) Image segmented by NBA, (f) Image segmented by GA, (g) Image segmented by SA.

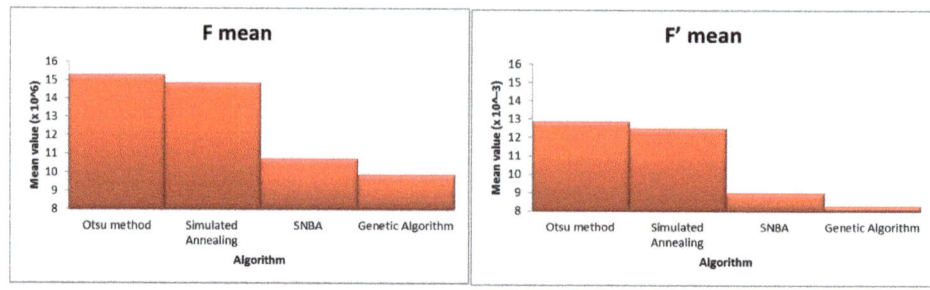

Figure 7. F and F' Mean error values of each algorithm.

Table 2 shows a descending error of the algorithms, which showed that Otsu has the highest error while NBA and GA have lower errors, with GA being the algorithm with the lowest error.

Even if the Otsu method is a frequently used method for image thresholding, it can be observed that the Otsu method's error is 1.42 times higher than that of the NBA and 1.54 times higher than that of the GA. The error acquired by the simulated annealing algorithm was close to but lower than the Otsu method. In addition, the NBA's error is 1.08 times higher than the GA's; showing that the GA has the lowest mean error for the segmentation of breast cancer mammographies. The GA showed a lower error because of its capability to explore a higher number of candidate solutions, which resulted in a better segmentation compared to the NBA, which also explores a large number of candidate solutions.

Run times for each algorithm were calculated, and the mean run time is shown in Table 3.

Table 3. Mean time for each algorithm.

Algorithm	Mean Time (sec)
Otsu method	0.051
Simulated annealing	130.406
Novel bat algorithm	53.830
Genetic algorithm	170.830

According to the mean times, Otsu method had the lowest mean run time, in contrast to its higher error in the segmentation. The genetic algorithm had the lowest segmentation error, but it had the highest mean run time.

4.2. Classification of the Lesions

The proposed classification algorithm with six cancer-related datasets was tested, and the results are presented with the classification performances in the following section, including the comparison with two classic algorithms utilized in experiments and published in the literature.

The datasets used were:

1. Breast Cancer Digital Repository (BCDR) [46]. The Faculty of Medicine of the University of Porto, Portugal provided this dataset, and it is used to explore the computer-based detection and diagnosis methods.
2. Breast Cancer Wisconsin (Original) Data Set (BCWO) [47,48]. The University of Wisconsin, USA provided this dataset. This dataset contains data collected from clinical cases by Dr. William Wolberg.
3. Breast Cancer Wisconsin (Prognostic) Data Set (BCWP) [48]. Dr. Wolberg provided this dataset, and it contains the follow-up of breast cancer cases.
4. Lung Cancer Data Set (LCDS) [48]. Hong and Young used this dataset to apply the k-nearest neighbor method, showing a biased result. It contains data on lung cancer patients.

5. Mammographic Mass Data Set (MMDS) [49]. Matthias Elter donated this dataset to the University of California, Irvine (UCI) repository in October 2007. It contains data of mammographic mass lesions from Breast Imaging Report Database Systems.
6. Haberman's Survival Data Set (HSDS) [48]. Tjen-Sien Lim donated this dataset in 1999. It contains data related to the survival of patients who underwent breast cancer surgery.

Similarly, a brief description of the classical algorithms used for classification is included. Support Vector Machines (SVM) [50] are algorithms used for classification and regression. These algorithms build a model that represents the patterns of the training set. Their main objective is to find a hyperplane that separates two classes. These algorithms work ideally for two classes; however, it is possible to use strategies that allow them to be used to build a model able to separate two classes.

Repeated incremental pruning to produce error reduction (RIPPER) [51] is an algorithm based on association rules with reduced error pruning; this technique is frequently used in decision tree algorithms. The generation of rules is performed by applying pruning operators to reduce the error. The algorithm ends when the error is increased after a pruning operation.

Regarding the classification process, we present the results of classifier accuracy comparing the two algorithms with the artificial immune system in Table 4.

Table 4. Classification performances.

Dataset	SVM	RIPPER	AIS
BCDR	**80.111**	75.414	64.360
BCWO	96.996	95.565	**97.420**
BCWP	76.263	77.778	**79.290**
LCDS	56.250	50.000	**68.750**
MMDS	79.501	**82.934**	78.980
HSDS	72.876	73.856	**76.471**

We repeated the classification process 10 times and averaged the performance of each algorithm. The classification accuracy is calculated as the number of correctly classifying elements among the total number of elements. The best accuracy for each dataset is bolded in Table 4.

As it can be observed, the artificial immune system (AIS) proposed got the best performance in four out of six datasets, which is interesting considering that the other algorithms are widely used in the literature.

5. Discussion and Conclusions

The meta-heuristic-based segmentations showed a lower error compared to the Otsu method, which is relevant since the Otsu method is one of the classic and still used methods for thresholding. These algorithms showed better performances than the Otsu method, even when using them without testing different configurations.

When contrasting the meta-heuristic with the highest error (SA) and the Otsu method, it can be observed that the error presented in simulated annealing is lower than that of Otsu, which represents an important fact since Otsu is a well-known and useful method for thresholding.

According to the runtime test, it was observed that the Otsu method had the lowest run time. However, implementing this method with the Matlab™ software resulted in an optimized and reviewed version of the algorithm. Implementing the meta-heuristic algorithms was a dummy one, with no use of parallel programming.

Many of the meta-heuristics have numerous parameters, which increases the performances of the algorithm depending on the values for each parameter. In this work, we change no parameters for such algorithms, and we did not prove other configurations, but in future work, this could be improved.

The classification model presented (AIS) showed a good performance in most of the datasets where it was tested. This is relevant since it is a new intelligent computing algorithm that opens a research gap in bio-inspired algorithms.

The proposed method can be used as a guide for the radiologists when there is a high demand for processing mammograms and it can also be used as a second opinion in critical cases.

In future work, we can implement preprocessing for enhancing the contrast of the images, to compare the use of preprocessing before segmenting using the meta-heuristics. This could help the segmentation algorithms to perform a better segmentation. This could help the segmentation algorithms to perform a better segmentation.

Concerning the classification algorithm, it is convenient to explore the use of mutation strategies, the global optima, and to improve the search for the desired solution.

Author Contributions: Conceptualization, A.-J.A.-C., D.G.-P., Y.V.-R., and F.K.; Methodology, A.-J.A.-C., D.G.-P., Y.V.-R., and F.K.; Software, Y.V.-R., D.G.-P.; Validation, Y.V.-R., D.G.-P.; Formal Analysis, A.-J.A.-C., D.G.-P., Y.V.-R.; Investigation, A.-J.A.-C., D.G.-P., Y.V.-R., and F.K.; Resources, ; Data Curation, A.-J.A.-C. and Y.V.-R.; Writing—Original Draft Preparation, D.G.-P., and Y.V.-R.; Writing—Review and Editing, A.-J.A.-C., D.G.-P., and Y.V.-R.; Visualisation, A.-J.A.-C. and Y.V.-R.; Supervision, A.-J.A.-C.; Project Administration, A.-J.A.-C. and D.G.-P.

Funding: The authors of the present paper would like to thank the following institutions for their economic support to develop this work: Comisión de Operación y Fomento de Actividades Académicas del Instituto Politécnico Nacional (COFAA-IPN), Centro de Investigación en Computación (Centro de Investigación en Computación-IPN), Centro de Innovación y Desarrollo Tecnológico en Cómputo (CIDETEC), Escuela Superior de Ingeniería Mecánica y Eléctrica (ESIME), and Consejo Nacional de Ciencia y Tecnología (CONACYT).

Acknowledgments: The authors would like to thank the Instituto Politécnico Nacional (Secretaría Académica, Comisión de Operación y Fomento de Actividades Académicas, Secretaría de Investigación y Posgrado, Centro de Investigación en Computación, and Centro de Innovación y Desarrollo Tecnológico en Cómputo), the Consejo Nacional de Ciencia y Tecnología (Conacyt), and Sistema Nacional de Investigadores in México for their economic support to develop this work. In addition, the authors gratefully acknowledge the Electrical and Computer Engineering department of the University of Waterloo, Ontario in Canada for their support to develop this work.

Conflicts of Interest: The authors declare no conflict of interest.

References

1. Siegel, R.L.; Miller, K.D.; Jemal, A. Cancer statistics, 2016. *CA: Cancer J. Clin.* **2016**, *66*, 7–30. [CrossRef]
2. Peng, L.; Chen, W.; Zhou, W.; Li, F.; Yang, J.; Zhang, J. An immune-inspired semi-supervised algorithm for breast cancer diagnosis. *Comput. Methods Programs Biomed.* **2016**, *134*, 259–265. [PubMed]
3. Oeffinger, K.C.; Fontham, E.T.H.; Etzioni, R.; Herzig, A.; Michaelson, J.S.; Shih, Y.-C.T.; Walter, L.C.; Church, T.R.; Flowers, C.R.; LaMonte, S.J.; et al. Breast cancer screening for women at average risk: 2015 guideline update from the American Cancer Society. *JAMA* **2015**, *314*, 1599–1614. [CrossRef] [PubMed]
4. Kalager, M.; Zelen, M.; Langmark, F.; Adami, H.-O. Effect of screening mammography on breast-cancer mortality in Norway. *N. Engl. J. Med.* **2010**, *363*, 1203–1210. [CrossRef] [PubMed]
5. Nelson, H.D.; Tyne, K.; Naik, A.; Bougatsos, C.; Chan, B.K.; Humphrey, L. Screening for breast cancer: An update for the US Preventive Services Task Force. *Ann. Intern. Med.* **2009**, *151*, 727–737. [CrossRef] [PubMed]
6. Nyström, L.; Bjurstam, N.; Jonsson, H.; Zackrisson, S.; Frisell, J. Reduced breast cancer mortality after 20+ years of follow-up in the Swedish randomized controlled mammography trials in Malmö, Stockholm, and Göteborg. *J. Med. Screen.* **2017**, *24*, 34–42. [CrossRef]
7. Yu, S.; Wu, S.; Zhuang, L.; Wei, X.; Sak, M.; Neb, D.; Hu, J.; Xie, Y. Efficient segmentation of a breast in B-mode ultrasound tomography using three-dimensional GrabCut (GC3D). *Sensors* **2017**, *17*, 1827. [CrossRef]
8. Duric, N.; Boyd, N.; Littrup, P.; Sak, M.; Myc, L.; Li, C.; West, E.; Minkin, S.; Martin, L.; Yaffe, M.; et al. Breast density measurements with ultrasound tomography: A comparison with film and digital mammography. *Med. Phys.* **2013**, *40*, 13501. [CrossRef]
9. Pham, D.L.; Xu, C.; Prince, J.L. Current methods in medical image segmentation 1. *Annu. Rev. Biomed. Eng.* **2000**, *2*, 315–337. [CrossRef]
10. Iglesias, J.E.; Sabuncu, M.R. Multi-atlas segmentation of biomedical images: A survey. *Med. Image Anal.* **2015**, *24*, 205–219.

11. Munot, K.; Mehta, N.; Mishra, S.; Chaturvedi, R.N. a Review on Image Segmentation Techniques With an Application Perspective. *Int. J. Adv. Res. Comput. Sci.* **2017**, *8*, 846–850.
12. Russ, J.C.; Matey, J.R.; Mallinckrodt, A.J.; McKay, S. The image processing handbook. *Comput. Phys.* **1994**, *8*, 177–178. [CrossRef]
13. Dollár, P.; Zitnick, C.L. Fast edge detection using structured forests. *IEEE Trans. Pattern Anal. Mach. Intell.* **2015**, *37*, 1558–1570. [CrossRef] [PubMed]
14. Malik, J.; Dahiya, R.; Girdhar, D.; Sainarayanan, G. Finger knuckle print authentication using Canny edge detection method. *Int. J. Signal Imaging Syst. Eng.* **2016**, *9*, 333–341. [CrossRef]
15. Qi, H.; Snyder, W.E.; Head, J.F.; Elliott, R.L. Detecting breast cancer from infrared images by asymmetry analysis. In Proceedings of the 22nd Annual International Conference of the IEEE Engineering in Medicine and Biology Society, Chicago, IL, USA, 23–28 July 2000; Volume 2, pp. 1227–1228.
16. Mencattini, A.; Salmeri, M.; Lojacono, R.; Frigerio, M.; Caselli, F. Mammographic images enhancement and denoising for breast cancer detection using dyadic wavelet processing. *IEEE Trans. Instrum. Meas.* **2008**, *57*, 1422–1430. [CrossRef]
17. Song, W.; Zhong, B.; Sun, X. Building Corner Detection in Aerial Images with Fully Convolutional Networks. *Sensors* **2019**, *19*, 1915. [CrossRef]
18. Fan, J.; Yau, D.K.Y.; Elmagarmid, A.K.; Aref, W.G. Automatic image segmentation by integrating color-edge extraction and seeded region growing. *IEEE Trans. Image Process.* **2001**, *10*, 1454–1466.
19. Yan, C.; Sang, N.; Zhang, T. Local entropy-based transition region extraction and thresholding. *Pattern Recognit. Lett.* **2003**, *24*, 2935–2941. [CrossRef]
20. Yao, H.; Duan, Q.; Li, D.; Wang, J. An improved K-means clustering algorithm for fish image segmentation. *Math. Comput. Model.* **2013**, *58*, 790–798. [CrossRef]
21. Salem, N.M. Segmentation of white blood cells from microscopic images using K-means clustering. In Proceedings of the 2014 31st National Radio Science Conference (NRSC), Cairo, Egypt, 28–30 April 2014; pp. 371–376.
22. Patel, B.C.; Sinha, G.R. An adaptive k-means clustering algorithm for breast image segmentation. *Int. J. Comput. Appl.* **2010**, *10*, 35–38. [CrossRef]
23. Mambou, S.; Maresova, P.; Krejcar, O.; Selamat, A.; Kuca, K. Breast cancer detection using infrared thermal imaging and a deep learning model. *Sensors* **2018**, *18*, 2799. [CrossRef] [PubMed]
24. Otsu, N. A threshold selection method from gray-level histograms. *Automatica* **1975**, *11*, 23–27. [CrossRef]
25. Liu, D.; Yu, J. Otsu method and K-means. In Proceedings of the 2009 Ninth International Conference on Hybrid Intelligent Systems, Shenyang, China, 12–14 August 2009; Volume 1, pp. 344–349.
26. Lu, C.; Zhu, P.; Cao, Y. The segmentation algorithm of improvement a two-dimensional Otsu and application research. In Proceedings of the 2010 2nd International Conference on Software Technology and Engineering (ICSTE), San Juan, PR, USA, 3–5 October 2010; Volume 1, pp. V1-76–V1-79.
27. Yan, W.G.; Wang, C.J.; Guo, J. One extended OTSU flame image recognition method using RGBL and stripe segmentation. *Appl. Mech. Mater.* **2012**, *121*, 2141–2145. [CrossRef]
28. Chen, Q.; Zhao, L.; Lu, J.; Kuang, G.; Wang, N.; Jiang, Y. Modified two-dimensional Otsu image segmentation algorithm and fast realisation. *IET Image Process.* **2012**, *6*, 426–433. [CrossRef]
29. Binitha, S.; Sathya, S.S. A survey of bio inspired optimization algorithms. *Int. J. Soft Comput. Eng.* **2012**, *2*, 137–151.
30. Mendoza, J.E.; Rousseau, L.-M.; Villegas, J.G. A hybrid metaheuristic for the vehicle routing problem with stochastic demand and duration constraints. *J. Heuristics* **2016**, *22*, 539–566. [CrossRef]
31. Khachaturyan, A.G.; Semenovskaya, S.V.; Vainstein, B. A statistical-thermodynamic approach to determination of structure amplitude phases. *Sov. Phys. Crystallogr.* **1979**, *24*, 519–524.
32. Manwar, R.; Zafar, M.; Podoleanu, A.; Avanaki, M. An Application of Simulated Annealing in Compensation of Nonlinearity of Scanners. *Appl. Sci.* **2019**, *9*, 1655. [CrossRef]
33. Fayyaz, Z.; Mohammadian, N.; Salimi, F.; Fatima, A.; Tabar, M.R.R.; Avanaki, M.R.N. Simulated annealing optimization in wavefront shaping controlled transmission. *Appl. Opt.* **2018**, *57*, 6233–6242. [CrossRef]
34. Banzhaf, W.; Nordin, P.; Keller, R.E.; Francone, F.D. *Genetic Programming: An Introduction*; Morgan Kaufmann San Francisco: San Francisco, CA, USA, 1998; Volume 1.
35. Darwin, C. *On the Origin of Species by Means of Natural Selection, or the Preservation of Favoured Races in the Struggle for Life*; Murray: Glasgow, UK, 1859.

36. Fraser, A.S. Simulation of Genetic Systems by Automatic Digital Computers II. Effects of Linkage on Rates of Advance Under Selection. *Aust. J. Biol. Sci.* **1957**, *10*, 484–491. [CrossRef]
37. Holland, J.H. *Adaptation in Natural and Artificial Systems: An introductory Analysis with Applications to Biology, Control and Artificial Intelligence*; University of Michigan Press: Ann Arbor, MI, USA, 1975.
38. Sastry, K.; Goldberg, D.; Kendall, G. Genetic Algorithms. In *Search Methodologies: Introductory Tutorials in Optimization and Decision Support Techniques*; Burke, E.K., Kendall, G., Eds.; Springer: Boston, MA, USA, 2005; pp. 97–125. ISBN 978-0-387-28356-2.
39. Yang, X.-S.; Hossein Gandomi, A. Bat algorithm: A novel approach for global engineering optimization. *Eng. Comput.* **2012**, *29*, 464–483. [CrossRef]
40. Meng, X.-B.; Gao, X.Z.; Liu, Y.; Zhang, H. A novel bat algorithm with habitat selection and Doppler effect in echoes for optimization. *Expert Syst. Appl.* **2015**, *42*, 6350–6364. [CrossRef]
41. Yang, X.-S. A New Metaheuristic Bat-Inspired Algorithm. In *Nature Inspired Cooperative Strategies for Optimization (NICSO 2010)*; González, J.R., Pelta, D.A., Cruz, C., Terrazas, G., Krasnogor, N., Eds.; Springer: Berlin/Heidelberg, Germany, 2010; pp. 65–74. ISBN 978-3-642-12538-6.
42. González-Patiño, D.; Villuendas-Rey, Y.; Argüelles-Cruz, A.J. Mammogram image segmentation using bioinspired novel bat swarm clustering. *Res. Comput. Sci.* **2016**, *118*, 87–96.
43. Dunn, J.C. A fuzzy relative of the ISODATA process and its use in detecting compact well-separated clusters. *J. Cybern.* **1973**, *3*, 32–57. [CrossRef]
44. Zhang, H.; Fritts, J.E.; Goldman, S.A. Image segmentation evaluation: A survey of unsupervised methods. *Comput. Vis. Image Underst.* **2008**, *110*, 260–280. [CrossRef]
45. Davis, L. *Handbook of Genetic Algorithms*; CumInCAD: New York, NY, USA, 1991.
46. Moura, D.C.; López, M.A.G. An evaluation of image descriptors combined with clinical data for breast cancer diagnosis. *Int. J. Comput. Assist. Radiol. Surg.* **2013**, *8*, 561–574. [CrossRef]
47. Mangasarian, O.L.; Wolberg, W.H. *Cancer Diagnosis via Linear Programming*; Department of Computer Sciences, University of Wisconsin-Madison: Madison, WI, USA, 1990; pp. 1–18.
48. Lichman, M. *UCI Machine Learning Repository 2013*; School of Information and Computer, University of California: Irvine, CA, USA, 2013.
49. Elter, M.; Schulz-Wendtland, R.; Wittenberg, T. The prediction of breast cancer biopsy outcomes using two CAD approaches that both emphasize an intelligible decision process. *Med. Phys.* **2007**, *34*, 4164–4172. [CrossRef]
50. Platt, J.C. Fast training of support vector machines using sequential minimal optimization. *Adv. Kernel Methods* **1999**, *1*, 185–208.
51. Cohen, W.W. Fast effective rule induction. In Proceedings of the Twelfth International Conference on Machine Learning, Tahoe City, CA, USA, 9–12 July 1995; pp. 115–123.

 © 2019 by the authors. Licensee MDPI, Basel, Switzerland. This article is an open access article distributed under the terms and conditions of the Creative Commons Attribution (CC BY) license (http://creativecommons.org/licenses/by/4.0/).

Article

Quantitative Analysis of Benign and Malignant Tumors in Histopathology: Predicting Prostate Cancer Grading Using SVM

Subrata Bhattacharjee [1], Hyeon-Gyun Park [1], Cho-Hee Kim [2], Deekshitha Prakash [1], Nuwan Madusanka [1], Jae-Hong So [2], Nam-Hoon Cho [3] and Heung-Kook Choi [1,*]

1. Department of Computer Engineering, u-AHRC, Inje University, Gimhae 50834, Korea
2. Department of Digital Anti-Aging Healthcare, Inje University, Gimhae 50834, Korea
3. Department of Pathology, Yonsei University Hospital, Seoul 03722, Korea
* Correspondence: cschk@inje.ac.kr; Tel.: +82-010-6733-3437

Received: 12 June 2019; Accepted: 22 July 2019; Published: 24 July 2019

Abstract: An adenocarcinoma is a type of malignant cancerous tissue that forms from a glandular structure in epithelial tissue. Analyzed stained microscopic biopsy images were used to perform image manipulation and extract significant features for support vector machine (SVM) classification, to predict the Gleason grading of prostate cancer (PCa) based on the morphological features of the cell nucleus and lumen. Histopathology biopsy tissue images were used and categorized into four Gleason grade groups, namely Grade 3, Grade 4, Grade 5, and benign. The first three grades are considered malignant. K-means and watershed algorithms were used for color-based segmentation and separation of overlapping cell nuclei, respectively. In total, 400 images, divided equally among the four groups, were collected for SVM classification. To classify the proposed morphological features, SVM classification based on binary learning was performed using linear and Gaussian classifiers. The prediction model yielded an accuracy of 88.7% for malignant vs. benign, 85.0% for Grade 3 vs. Grade 4, 5, and 92.5% for Grade 4 vs. Grade 5. The SVM, based on biopsy-derived image features, consistently and accurately classified the Gleason grading of prostate cancer. All results are comparatively better than those reported in the literature.

Keywords: prostate cancer; histopathology; microscopic; tissue image; segmentation; morphological; quantitative; classification; SVM

1. Introduction

Prostate adenocarcinoma, a type of prostate cancer, is the second most commonly diagnosed cancer. In the United States, the incidence of prostate cancer ranks first among all malignant tumors in men. The Gleason score is currently the most common grading system of prostate adenocarcinoma and is widely used to assess the prognosis of men with prostate cancer using samples from a prostate biopsy. There are some diagnostic protocols for cancer grading, for which microscopic evaluation of tissue specimens is required. For this, the samples need to be appropriately stained using Hematoxylin and Eosin (H&E) compounds. The cancer grade is assessed by a pathologist based on the morphological features of lumen and cell nucleus observed in the tissue. Cancer diagnosis and grading based on digital pathology have become increasingly complex due to the increase in cancer occurrence and specific treatment options for patients [1].

In South Korea, the incidence of prostate cancer is increasing significantly. Prostate cancer (PCa) is the fifth most common cancer among males in Korea and the expected cancer deaths in 2018 were 82,155 [2]. The detection of prostate cancer has always been a major issue for pathologists and medical practitioners, for both diagnosis and treatment. Usually, the cancer detection process in histopathology consists of categorizing stained microscopic biopsy images into malignant and benign.

The Gleason grade grouping system defines Gleason scores ≤ 6 as grade 1, score $3 + 4 = 7$ as grade 2, score $4 + 3 = 7$ as grade 3, score $4 + 4$, $3 + 5$ or $5 + 3 = 8$ as grade 4, and score $4 + 5$, $5 + 4$ or $5 + 5 = 9$ or 10 as grade 5. The Gleason score is obtained by adding the primary (most common) and secondary (second most common) scores from H&E stained tissue microscopic images. This system was developed by Dr. Donald F Gleason, who was a Pathologist in Minnesota, and members of the Veterans Administration Cooperative Urological Research Group (VACURG) [3]. This system was tested on a large number of patients, including long-term follow-ups and is considered an outstanding success.

In recent years, an excellent and important addition to microscopy and digital imaging has been developed for microscopes that are used to convert stained tissue slides into whole slide digital images. This allows for more efficient computer-based viewing and analysis of histopathology. Early diagnosis and treatment are required, to avoid the enlargement of cancer cells in the prostate gland and control the spreading of more aggressive tumors to other parts of the body.

The digital pathology field has grown dramatically over recent years, largely due to technological advancements in image processing and machine learning algorithms, and increases in computational power. As part of this field, many methods have been proposed for automatic histopathological image analysis and classification. In this paper, color segmentation, based on k-means clustering method, is proposed for microscopic biopsy tissue image processing, and the watershed algorithm has been implemented to separate touching cell nuclei in tissue images.

This approach can be implemented in different ways; however, the marker selection approach has been carried out in this study to control over-segmentation. Diagnosing prostate cancer from a biopsy tissue image under a microscope is difficult for the pathologists and doctors. Therefore, machine learning and deep learning techniques are developed for computerized classification and cancer grading. In this study, a machine learning classification method is proposed in order to classify Gleason grade groups of prostate cancer. From a perspective of computer engineering, since the regular procedure of diagnosing prostate cancer and grading is difficult and time consuming; therefore, automated computerized methods are in high demand and are essential for medical image analysis.

2. Literature Review

Tabesh et al. [4] extracted features that describe color, texture, and morphology from 367 and 268 H&E image patches, which were acquired from tissue microarray (TMA) datasets. These features were used for support vector machine (SVM) classification. They achieved an accuracy of 96.7% and 81% for predicting benign vs. malignant and low-grade vs. high-grade classifications, respectively, using 5-fold cross-validation.

Doyle et al. [5] proposed a cascade approach to the multi-class grading problem. They used cascade binary classification to maximize inter- and intra-class accuracy rather than the conventional one-shot classification and one-versus-all approaches to multi-class classification. In the proposed cascade approach, each division is classified separately and independently.

Nir et al. [6] proposed some novel features based on intra- and inter-nuclei properties for classification. They trained their classifier on 333 tissue microarray (TMA) cores annotated by six pathologists for different Gleason grades and used SVM classification to achieve an accuracy of 88.5% and 73.8% for cancer detection (benign vs. malignant) and low vs. high grade (Grade 3 vs. Grade 4, 5), respectively.

Doyle et al. [7] extracted nearly 600 image texture features to perform pixel-wise Bayesian classification at each image scale to obtain the corresponding likelihood scene. The authors achieved an accuracy of 88.0% for distinguishing between benign and malignant samples.

Rundo et al. [8] proposed Fuzzy C-Means (FCM) clustering algorithm for prostate multispectral MRI morphologic data processing and segmentation. The authors used co-registered T1w and T2w

MR image series and achieved an average dice similarity coefficient 90.77 ± 7.75, with respect to 81.90 ± 6.49 and 82.55 ± 4.93 by processing T2w and T1w imaging alone, respectively.

Jiao et al. [9] used combined deep learning and SVM methods for breast masses classification. The methods were applied to the Digital Database for Screening Mammography (DDSM) dataset and achieved high accuracy under two objective evaluation measures. The authors used nearly 600 images, out of these, 50% were benign and 50% were malignant. The classification accuracy achieved in this paper was 96.7% for distinguishing between benign and malignant samples.

Hu et al. [10] presented a novel mass detection system for digital mammograms, which integrated a visual saliency model with deep learning techniques. The authors used combined deep learning and SVM methods for image and feature classification, respectively. They achieved an average accuracy of 91.5% in mass detection between cancer and benign datasets.

Naik et al. [11] presented a method for automated histopathology images. They have demonstrated the utility of glandular and nuclear segmentation algorithm in accurate extraction of various morphological and nuclear features for automated grading of prostate cancer, breast cancer, and distinguishing between cancerous and benign breast histology specimen. The authors used a SVM classifier for classification of prostate images containing 16 Gleason grade 3 images, 11 grade 4 images, and 17 benign epithelial images of biopsy tissue. They achieved an accuracy of 95.19% for grade 3 vs. grade 4, 86.35% for grade 3 vs. benign, and 92.90% for grade 4 vs. benign.

Nguyen et al. [12] introduced a novel approach to grade prostate malignancy using digitized histopathological specimens of the prostate tissue. They have extracted tissue structural features from the gland morphology and co-occurrence texture features from 82 regions of interest (ROI) with 620 × 550 pixels to classify a tissue pattern into three major categories: benign, grade 3 carcinoma, and grade 4 carcinoma. The authors proposed a hierarchical (binary) classification scheme and obtained 85.6% accuracy in classifying an input tissue pattern into one of the three classes.

Albashish et al. [13] proposed some texture features, namely Haralick, Histogram of Oriented Gradient (HOG), and run-length matrix, which have been extracted from nuclei and lumen images individually. They used a total of 149 images with 4140 × 3096 pixels, and the dataset was randomly divided into 50% for training and 50% for testing. An ensemble machine learning classification system was proposed, and achieved an accuracy of 88.9% for Grade 3 vs. Grade 4, 92.4% for benign vs. Grade 4, and 97.85% for benign vs. Grade 3. These accuracies were averaged over 50 simulation runs and statistical significance.

Diamond et al. [14] used morphological and texture features to classify the sub-region of 100 × 100 pixels and subjected each to image-processing techniques. They classified a tissue image into either stroma or prostatic carcinoma. In addition, the authors used lumen area to discriminate benign tissue from the other two classes. As a result, 79.3% of sub-regions were correctly classified.

Ding et al. [15] introduced an automated image analysis framework capable of efficiently segmenting microglial cells from histology images and analyzing their morphology. Their experiments show that the proposed framework is accurate and scalable for large datasets. They extracted three types of features for SVM classification, namely Mono-fractal, Multi-fractal, and Gabor features.

Yang et al. [16] used image processing and machine learning algorithms to analyze the smear images captured by the developed image-based cytometer. A low-cost, portable image-based cytometer was built for image acquisition from Giemsa stained blood smear. The authors selected 50 images manually for the training set, out of these, 25 images were parasites and 25 images were non-parasites. The selected images were then segmented separately to extract the features for Support Vector Machine (SVM) classification, and they used linear kernel classifier to train and test these features.

3. Materials and Methods

3.1. Tissue Image Dataset

The histopathology images that were congregated to create our dataset are sub-images of benign and malignant samples. These sub-images were cropped from the whole-slide microscopic tissue images stained with H&E, shown in Figure 1. The data were collected from Severance Hospital of Yonsei University and the grading of these data was histologically confirmed by a pathologist. The whole slide size in Figure 1a–d is 33,584 × 70,352 pixels. The patch image magnification is 40× for Figure 1e–h and the image size is 512 × 512 pixels. We selected 400 sub-images for feature extraction and SVM classification. These were divided into four groups, namely Grade 3, Grade 4, Grade 5, and Benign.

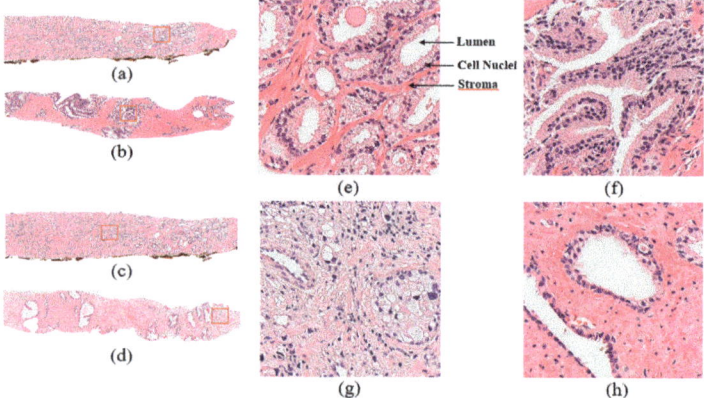

Figure 1. Microscopic biopsy images stained with Hematoxylin and Eosin (H&E) compound; (**a–d**) whole slide tissue images of Grade 3, Grade 4, Grade 5, and Benign; and (**e–h**) the regions of interest (ROIs) taken from whole-slide images (**a**), (**b**), (**c**), (**d**) respectively. The dark blue is the cell nucleus, pink is the stroma, and white is the lumen.

Figure 1 shows the sub-images that were used to detect cell nuclei and classify prostate cancer. It is a very challenging task to classify different Gleason grades because images usually contain many clusters and overlapping objects. Figure 2 shows the entire proposed process for predicting cancer gradings based on microscopic images. The pipeline model includes original biopsy image, region of interest (ROI) segmentation, watershed segmentation, features extraction, classification, and analysis results [16].

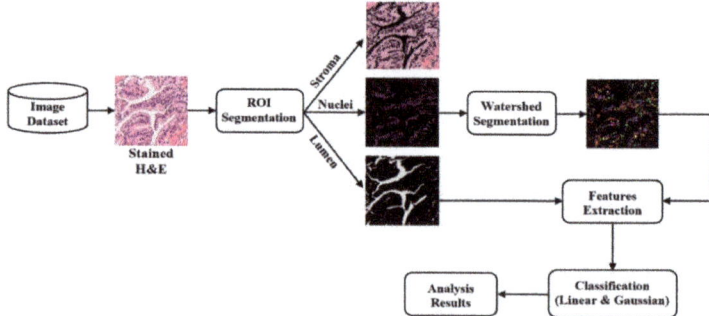

Figure 2. Proposed pipeline model for predicting cancer grading from microscopic biopsy images.

3.2. ROI Segmentation

Image segmentation plays an important role in medical image processing systems. The nuclei and lumen of prostate cancer are the most important components of histopathological images [17]. To identify cell nuclei and lumen from images and carry out systematic processing, a K-means clustering algorithm was applied using MATLAB R2018a (The MathWorks, Natick, MA, USA) [18], where image pixels were partitioned into three clusters (thus, k = 3). The segmented components from the tissue images are: stroma, lumen, and the cell nucleus. However, nucleus and lumen components were selected for feature extraction and SVM classification, as shown in Figure 3 [19].

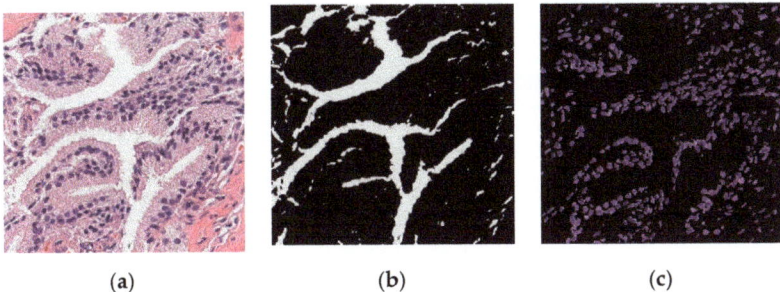

Figure 3. Image segmentation using K-means algorithm: (**a**) original tissue image; (**b**) lumen segmentation; and (**c**) nucleus segmentation.

According to our visual results, the K-means based method is best suited for microscopic biopsy images. K-means segmentation has been applied here to separate the nucleus and lumen tissue components from microscopic biopsy images. The K-means algorithm uses iterative modification to produce a final result. The following algorithm iterates between two steps:

1. Data assignment step:

$$\underset{c_k \in C}{\operatorname{argmin}} \, dist(c_k, x)^2 \tag{1}$$

2. Centroid update step:

$$c_k = \frac{1}{|s_k|} \sum_{x_k \in s_k} x_k \tag{2}$$

The K-means algorithm is composed of the following steps:

1. Specify k, number of cluster to be generated.
2. Select k random points as cluster centers.
3. Assign each instance to its closest cluster center using the Euclidean distance.
4. Calculate the centroid mean for each cluster and use it as a new cluster center.
5. Reassign all the instances to the closest cluster center.
6. Iterate until there is no change in the cluster center.

3.3. Watershed Segmentation

The watershed transform is an image processing technique that can be applied to a binary image for object segmentation. In the segmented images of nucleus tissue components, we observed that there were many overlapping cell nuclei. We separated these connected objects by applying the watershed segmentation algorithm [20,21]. This method was used to extract nucleus-based morphological features for SVM classification. We validated this algorithm experimentally and found that it performs better than other cell nuclei separation algorithms. It is one of the well-known methods for separating overlapping objects [22].

Algorithm for Watershed Segmentation

According to the algorithm, $g(x, y)$ and M_i is the image pixel value and the regional minima, respectively. The iteration steps of the algorithm are as follow:

$$T[n] = \{(x, y) \mid g(x, y) < n\} \tag{3}$$

$$n = min + 1 \text{ to } n = max + 1 \tag{4}$$

$$C_n(M_n) = C(M_i) \cap T[n] \tag{5}$$

where $T[n]$ is the set of coordinates of a point in $g(x, y)$, n is the flooding stage, and $C_n(M_i)$ is the set of coordinates of points in the catchment basin.

$$C_n(M_n) = 1, \text{ at } (x, y); \text{if } (x, y) \in C(M_i) \text{ and } (x, y) \in T[n] \tag{6}$$

$$C_n(M_n) = 0, otherwise \tag{7}$$

We computed the results of the above two equations and viewed the resulting binary image.

$$C[n] = \bigcup_{i=1}^{R} C(M_i) \tag{8}$$

$$C[max+1] = \bigcup_{i=1}^{R} C(M_i) \tag{9}$$

where $C[n]$ is the union of the flood catchment basin portions at stage set n, $C[max+1]$ is the union of all catchment basins. As per Equations (8) and (9), $C[n]$ is the subset of $T[n]$ and $C[n-1]$ is the subset of $C[n]$. Hence, each connected component of $C[n-1]$ is the connected in exactly one connected component of $T[n]$.

We used the following steps to separate overlapping nuclei:

1. Converted 24-bit/pixel RGB color image to binary using adaptive thresholding method.
2. Removed the noise from the binary image.
3. Applied the Euclidean distance transform to a binary image to generate a distance map.
4. Used a Gaussian filter to smooth the distance map.
5. Applied inverse distance transform after smoothing the distance map.
6. Identified local minima using markers on the inverse distance transform image.
7. Finally, applied watershed segmentation based on local minima points, iterating until all overlapping objects were segmented.

We used the described watershed segmentation algorithm to separate the overlapping cell nuclei. This has been used previously for nucleus counting and to extract features for classification [23]. Figure 4 shows the necessary steps for watershed segmentation, including segmenting the nuclei image, converting to a binary image, applying the Euclidean distance transform, and labeling the watershed image using color mapping.

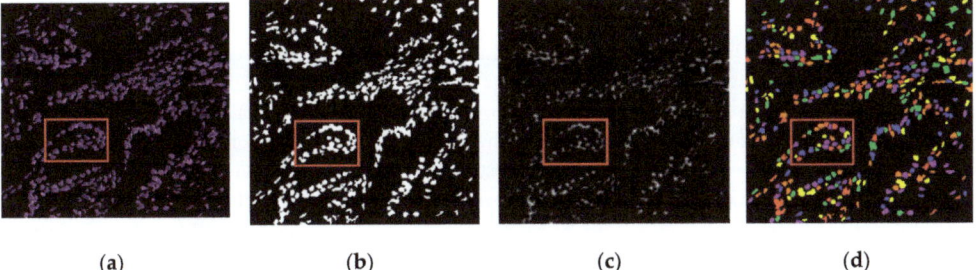

Figure 4. Overview of watershed segmentation: (**a**) original segmented image of nucleus tissue components; (**b**) noise-removed binary image; (**c**) Euclidean distance transform on binary image; and (**d**) result of the watershed algorithm and labelled nuclei using color mapping.

However, at the beginning of the watershed segmentation, there were some errors leading to over-segmentation, which caused some objects to be divided into several parts, as shown in Figure 5a. To show an example of over-segmentation, we used a cropped image that was taken from the region marked with a red box in Figure 4. First, to control over-segmentation, we used an approach called the marker-selection watershed transform to improve the segmentation results [24]. This approach determines markers for each region of interest and transforms the distance map image in such a way that the region markers are the only local minima of the resulting image. Second, after the Euclidean distance transform, we applied a Gaussian filter to smooth the distance map and then applied internal markers to the smoothed inverse results of the distance transform, as shown in Figure 5b. Third, the watershed algorithm was applied to the marker selection image, as shown in Figure 5c. Finally, the resulting image appeared after removing the noise and watershed lines, and the centroid of each nucleus was labelled, as shown in Figure 5d.

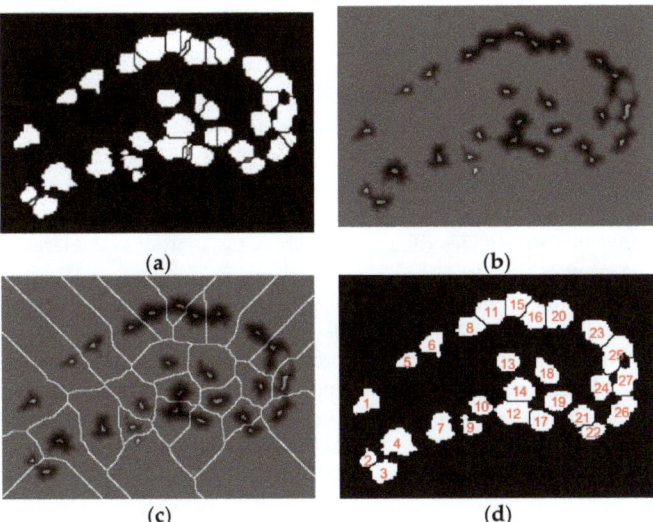

Figure 5. Improvement of over-segmentation: (**a**) over-segmented objects; (**b**) markers applied to the inverse results of the distance transform; (**c**) applied watershed algorithm on images (b); and (**d**) the resulting image after removing the noise and watershed line, and the centroid of the nucleus has been labelled.

3.4. Feature Extraction

Feature extraction is a very important step in the analysis of prostate cancer and prediction of cancer grades from microscopic biopsy images. The shape and morphological features of prostate cancer are described in References [25,26]. Although different features have been considered for prostate cancer grading and classification, morphological and texture feature extraction is the most common. Training and testing were performed based on the selected data, which were extracted from tissue images. In total, 19 features were extracted from the cell nucleus and lumen and, among these, 14 significant features were selected for SVM classification. The morphological features of the cell nucleus and lumen considered in this paper are: area, perimeter, major axis length, minor axis length, circularity, diameter, nucleus to nucleus distance, nucleus to nucleus minimum distance, eccentricity, and compactness. After watershed segmentation was performed on the nucleus images, cellular level features were extracted to detect and grade prostate cancer using the SVM classification method [27,28]. We used both region- and contour-based methods on the segmented nucleus and lumen images to gather data about the morphological features. To compare all of the extracted features and find the significant features, we used Fisher's coefficient and analysis of variance (ANOVA) to identify the most significant features [29,30]. Table 1 shows descriptions of the significant features of the cell nucleus and lumen. According to the statistical test, all of these features are highly statistically significant ($p < 0.001$).

Table 1. Proposed features for support vector machine (SVM) binary classification to classify Gleason grading of prostate cancer.

Feature Type	Feature Description
Nucleus features	Area, perimeter, major axis length, minor axis length, circularity, diameter, compactness, nucleus to nucleus average distance, nucleus to nucleus minimum distance
Lumen features	Area, perimeter, major axis length, minor axis length, eccentricity

3.5. Support Vector Machine (SVM) Classification

In this paper, we used SVM classification of morphological features for cell nucleus and lumen to predict the Gleason grading of prostate cancer. Classification of the various Gleason grade groups from microscopic biopsy images is a very challenging task [31,32]. The classification accuracy depends on different classifiers and their kernel types. An SVM is a supervised learning technique, but it can be applied to both classification and regression problems [33,34]. SVMs can generate optimal hyperplane in an iterative manner that maximizes the margin, where the margin is the largest distance to the nearest training data point of any class.

For classification purposes, we experimented with a few classifiers, such as logistic regression (LR), linear discriminant analysis (LDA), and SVMs. We selected SVMs for this analysis because they achieved better accuracy. Supervised learning approaches generally proceed as follows: prepare the data set for training and testing; choose an appropriate algorithm; select features to fit the model; train the model; use the trained model for prediction. In SVM classification, linear and Gaussian kernel are used to classify samples as benign and malignant and discriminate between Grade 3 vs. Grade 4, 5 and Grade 4 vs. Grade 5 of the Gleason grade groups [35].

We used 2-fold cross-validation to train the model and compared the performance of the different classification models. Later, we adjusted the K-fold cross-validation manually to improve the accuracy [36,37]. The linear kernel, K, maps the original data with the kernel function,

$$K(x) = (x \cdot x' + c) \tag{10}$$

where x is the data and c is a constant.

In SVM classification, the gaussian kernel function, used for binary classification was expressed by:

$$K(x, x') = exp(-\gamma \|x - x'\|^2), \ \gamma = \frac{1}{2\sigma^2} \quad (11)$$

where x, x' is the feature vector, $\|x - x'\|^2$ is the Euclidean distance between two feature vectors, γ is a hyper-parameter, which changes the smoothness of the kernel function, and σ is a free parameter.

To classify Gleason grade groups, we used the proposed binary classification approach, which divides the multi-category classification into multiple two-category groupings. Each division in Figure 6 represents a separate and independent classification, amounting to three binary divisions. In the first sequence, all of the samples in the dataset were classified as "malignant" vs. "benign". Within the cancer group, we separated the dataset between Grade 3 vs. Grade 4+5, and Grade 4 vs. Grade 5, and further classified these using different SVM models [38–40].

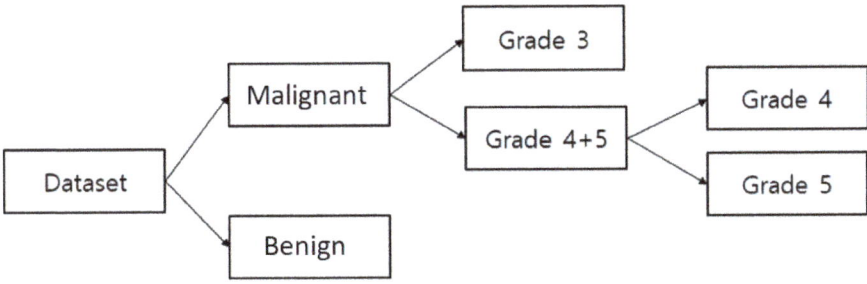

Figure 6. Proposed binary method for support vector machine (SVM) classification. Three different classifiers have been used here for binary classification and each group is classified independently and separately.

4. Results and Discussion

Quantitative analysis was performed on each cancerous image based on the four prostate cancer tissue groups (Grade 3, Grade 4, Grade 5, and Benign). We implemented the proposed method using MATLAB R2018a. We performed data analysis to analyze the components of the nuclei, which were segmented from prostate tissue images.

In this paper, 400 images were used in total. Of these, 240 were used for training and 160 were used for testing. The number of images considered for each group was 100, and these were classified as malignant vs. benign, Grade 3 vs. Grade 4+5, and Grade 4 vs. Grade 5. Each image was 24-bits/pixel with a size of 512 × 512 pixels. All of the possible results are shown in Tables 2–4, where we show the confusion matrices of SVM binary classification for training and testing separately.

Table 2. Confusion matrix of SVM binary classification—Malignant vs. Benign.

	Training: 99.2%				Testing: 88.7%		
Train	Malignant	Benign	Data	Test	Malignant	Benign	Data
Malignant	60	0	60	Malignant	34	6	40
Benign	1	59	60	Benign	3	37	40

Table 3. Confusion matrix of SVM binary classification—Grade 3 vs. Grade 4, 5.

	Training: 91.7%				Testing: 85.0%		
Train	Grade 3	Grade 4+5	Data	Test	Grade 3	Grade 4+5	Data
Grade 3	55	5	60	Grade 3	36	4	40
Grade 4+5	5	55	60	Grade 4+5	8	32	40

Table 4. Confusion matrix of SVM binary classification—Grade 4 vs. Grade 5.

	Training: 95.0%				Testing: 92.5%		
Train	Grade 4	Grade 5	Data	Test	Grade 4	Grade 5	Data
Grade 4	54	6	60	Grade 4	36	4	40
Grade 5	0	60	60	Grade 5	2	38	40

Tables 2–4 show the confusion matrices used to evaluate the performance of machine learning algorithms and the classifiers on a set of train and test data. We have shown these confusion matrix tables to get a better idea about the errors of a classification model. Each one of these tables is divided into two parts to show the correctly classified and misclassified data with respect to the training and testing process respectively.

In Table 5, we used four types of performance metrics, namely, accuracy, sensitivity, specificity, and Matthews's correlation coefficient (MCC). These metrics were calculated using our confusion matrices, i.e., true positive (TP), true negative (TN), false positive (FP), and false negative (FN). We multiplied the accuracy by 100% to normalize it with respect to the other measurements. The four types of performance metrics used in Table 5 are explained as follow,

1. Accuracy is measure of the proportion of correctly classified samples.

$$Accuracy = \frac{TP+TN}{TP+TN+FP+FN} \times 100 \tag{12}$$

2. Sensitivity is a measure of the proportion of positive correctly classified samples.

$$Sensitivity = \frac{TP}{TP+FN} \times 100 \tag{13}$$

3. Specificity is a measure of the proportion of negative correctly classified samples.

$$Specificity = \frac{TN}{TN+FP} \times 100 \tag{14}$$

4. Matthew's correlation coefficient (MCC) is the eminence of binary class classification. It is a correlation coefficient between target and predictions.

$$MCC = \frac{TP \times TN - FP \times FN}{\sqrt{((TP+FN)(TP+FP)(TN+FN)(TN+FP))}} \times 100 \tag{15}$$

Table 5 shows the classification results of the proposed method for three different groups. The SVM binary classification accuracy, sensitivity, specificity, and MCC for malignant vs. benign are 88.7%, 91.8%, 86.0%, and 70.2%, respectively. For Grade 3 vs. Grade 4+5, the classification accuracy, sensitivity, specificity, and MCC are 85.0%, 81.8%, 88.8%, and 70.3, respectively. For Grade 4 vs. Grade 5, the classification accuracy, sensitivity, specificity, and MCC are 92.5%, 94.7%, 95.0%, and 85.1, respectively.

Table 5. Evaluation results and performance metrics for three binary divisions using SVM.

Groups	Accuracy (%)	Sensitivity (%)	Specificity (%)	MCC (%)
Malignant vs. Benign	88.7	91.8	86.0	70.2
Grade 3 vs. Grade 4, 5	85.0	81.8	88.8	70.3
Grade 4 vs. Grade 5	92.5	94.7	95.0	85.1

For the purpose of validation, we also performed prostate cancer grading classification using multilayer perceptron (MLP) technique in Weka, shown in Table 6. MLP is a class of feed-forward artificial neural network, which consists of at least three layers of node: an input layer, hidden layer, and an output layer. Each node is a neuron except input nodes and uses a non-linear activation function. MLP utilizes a supervised learning technique like SVM. From the results shown in Tables 5 and 6, we can see that the proposed SVM binary classification works significantly better than MLP, and the highest accuracy obtained was 92.5%, for Grade 4 vs. Grade 5. First, classification was performed to detect cancer in all of the samples in the dataset. The second and third classification was performed within the cancer group for low- and high-grade cancer detection. In Figure 7, the bar graph shows the comparison results for the three different binary divisions that are used for SVM classification.

Table 6. Evaluation results for three binary divisions using the multilayer perceptron (MLP) classification technique.

Groups	Training Accuracy (%)	Testing Accuracy (%)
Malignant vs. Benign	99.0	81.0
Grade 3 vs. Grade 4, 5	98.0	75%
Grade 4 vs. Grade 5	97.5	76.25

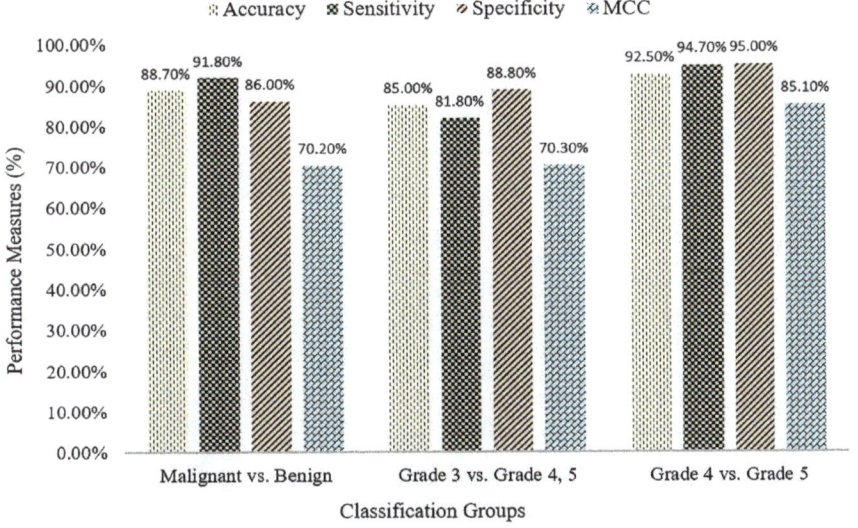

Figure 7. Comparison graph of support vector machine (SVM) classification accuracy among three binary divisions. The classification accuracies of the three groups are very close to each other, and the highest accuracy obtained was 92.50%, for grade 4 vs. grade 5. Matthew's correlation coefficient (MCC) indicates the quality of binary classification among the three classification groups.

To predict, automatically, prostate cancer gradings, we used machine learning and deep learning algorithms such as SVM and MLP, respectively. To do so, we first applied image segmentation as a preprocessing step. Secondly, we converted the images from RGB to binary to carry out watershed segmentation. Thirdly, we calculated a set of morphological features based on the segmented nucleus and lumen tissue images. Finally, the SVM and MLP classification was performed based on the significant features selected.

We can see that the results of the comparison between SVM classification accuracy in Table 7 and Figure 8 vary between one-shot and binary classifiers. When we classified our data using multi-class or one-shot classifiers, the classification accuracies for benign, Grade 3, Grade 4, and Grade 5 are 60%, 55%, 85%, and 50%, respectively. Using the proposed binary classification approach, the accuracies for the same groups are 92.5%, 90.0%, 90.0%, and 95.0%, respectively. Comparing both classifiers simultaneously, we can see that the results obtained using the binary classifier are better than those obtained using multi-class or one-shot classifier. Table 8 shows the comparison results of MLP classifier between one-shot and binary classification. After comparing the results between SVM and MLP classification methods, we can say that the proposed method, SVM, achieved better results than MLP. In one-shot classification, the entire dataset is classified into four groups simultaneously. In this case, the errors in one class affect the performance of the others, negatively impacting the classification accuracy. Thus, the model cannot make correct predictions. Whereas, in binary classification, the entire dataset is separated into three groups and each group is classified separately and independently. In this case, the errors in one class do not affect the performance of the other class.

Table 7. Support vector machine (SVM) classifier, comparison between one-shot and binary classification.

One-Shot Classification		Binary Classification	
Groups	Accuracy (%)	Groups	Accuracy (%)
Benign	60.0	Benign	92.5
Grade 3	55.0	Grade 3	90.0
Grade 4	85.0	Grade 4	90.0
Grade 5	50.0	Grade 5	95.0
Total	65.5	Total	92.0

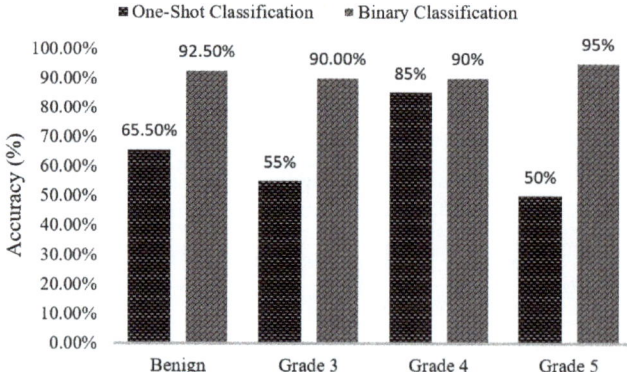

Figure 8. Comparison between support vector machine (SVM) classifiers among the four Gleason grade groups. In the case of one-shot classification, the classifier could not accurately distinguish among the four groups. In the case of binary classification, the classifier was almost always accurate, with little variation.

In Table 9, we compare the accuracy of different standard classification methods with our proposed method. The classification accuracy achieved for the class low vs. high grade using the proposed method is higher than other methods described in the literature. On cancer diagnosis, when classified Malignant vs. Benign, our result is better than Nir et al. (2018) and Doyle et al. (2006), but not higher compared to Tabesh et al. (2017), because they used different types of features that are extracted from the tissue image, namely color channel histogram, fractal dimension, fractal code, wavelet, and MAGIC. The authors of Reference [4] computed the features of epithelial nuclei objects in the tissue image, whereas, our method computed the features of all nuclei objects existing in the biopsy prostate tissue image.

Table 8. Multilayer perception (MLP) classifier, comparison between one-shot and binary classification.

One-Shot Classification		Binary Classification	
Groups	Accuracy (%)	Groups	Accuracy (%)
Benign	37.5	Benign	87.5
Grade 3	67.5	Grade 3	90.0
Grade 4	45.0	Grade 4	75.0
Grade 5	70.0	Grade 5	77.5
Total	55.5	Total	82.5

Table 9. Comparison between the proposed method and other standard methods for the classification of prostate cancer gradings.

Authors	Classification Methods	Classes	Accuracy
Tabesh et al. (2007) [4]	kNN	Malignant vs. Benign	96.7%
		Low vs. High Grade	81.0%
Doyle et al. (2012) [5]	Decision Tree (DT)	Grade 3	77.0%
		Grade 4	76.0%
		Grade 5	95.0%
Nir et al. (2018) [6]	SVM	Malignant vs. Benign	88.5%
		Low vs. High Grade	73.8%
Doyle et al. (2006) [7]	Bayesian	Malignant vs. Benign	88.0%
Rundo et al. [8]	Fuzzy C-Means	Multispectral (Tw1 & Tw2)	90.77%
Naik et al. [11]	SVM	Grade 3 vs. Grade 4	95.19%
		Benign vs. Grade 3	86.35%
		Benign vs. Grade 4	92.90%
Albashish et al. (2017) [12]	SVM	Grade 3 vs. Grade 4	88.9%
		Benign vs. Grade 3	97.9%
		Benign vs. Grade 4	92.4%
Nguyen et al. (2012) [13]	SVM	Benign, Grade 3, and Grade 4 carcinoma	85.6%
Proposed	SVM	Malignant vs. Benign	88.7%
		Low vs. High Grade	85.0%
		Grade 4 vs. Grade 5	92.5%
		Grade 3	90.0%
		Grade 4	90.0%
		Grade 5	95.0%

5. Conclusions

In this study, we have developed a computerized grading system for digitized histopathology images using supervised learning methods. The segmentation process for biopsy tissue image was performed using the k-means algorithm and touching cells were separated using the watershed algorithm. Morphological features were selected for prostate cancer grading and diagnosis. Gaussian and linear kernels were used for the classification of prostate histopathological images. Using these kernels, we observed some improvements in the results, and gradually increased the performance of the model used for training and testing. The parameters of the kernel play a vital role in the classification process, and the best combination of C and γ was selected for better classification accuracy. Satisfactory classification results were obtained using the extracted morphological features, and these features were extracted from the sub-images, viewable in 40× magnification. The quantitative analysis described here is remarkably flexible in terms of implementation. The SVM binary classification method presented in this paper is used to classify malignant vs. benign, Grade 3 vs. Grade 4+5, and Grade 4 vs. Grade 5. Our results are satisfactory and comparable with those reported in the literature and produced quantitative measures based on the features extracted from microscopic biopsy tissue images. In order to justify our proposed method, SVM, we also carried out features classification using MLP. One-shot and binary classification results were compared to show the differences in two classifications accuracies. In future studies, we will improve our classification accuracy using the combinations of multiple features. Deep learning and machine learning techniques will be used for comparative analysis, where, image classification will be performed using the convolutional neural network (CNN) and feature classification will be performed using support vector machine (SVM), respectively.

Author Contributions: Conceptualization, S.B., H.-G.P. and N.M.; Formal analysis, S.B., D.P., J.-H.S. and N.-H.C.; Methodology, S.B.; Project administration, C.-H.K.; Resources, H.-G.P., C.-H.K. and N.-H.C.; Supervision, H.-K.C.; Validation, S.B.; Visualization, N.M., J.-H.S., N.-H.C. and H.-K.C.; Writing—original draft, S.B.; Writing—review & editing, N.M.

Funding: This research was funded by the Ministry of Trade, Industry, and Energy (MOTIE), Korea, grant number (R&D, P0002072).

Acknowledgments: This research was financially supported by the Ministry of Trade, Industry, and Energy (MOTIE), Korea, under the "Regional Specialized Industry Development Program (R&D, P0002072)" supervised by the Korea Institute for Advancement of Technology (KIAT).

Ethical Approval: All subjects provided written informed consent for their participation in the study, which was approved by the Institutional Ethics Committee at College of Medicine, Yonsei University, Korea (IRB no. 1-2018-0044).

Conflicts of Interest: The authors declare that they have no conflicts of interest.

References

1. Braunhut, B.L.; Punnen, S.; Kryvenko, O.N. Updates on Grading and Staging of Prostate Cancer. *Surg. Pathol. Clin.* **2018**, *11*, 759–774. [CrossRef] [PubMed]
2. Chung, M.S.; Shim, M.; Cho, J.S.; Bang, W.; Kim, S.I.; Cho, S.Y.; Rha, K.H.; Hong, S.J.; Koo, K.C.; Lee, K.S.; et al. Pathological Characteristics of Prostate Cancer in Men Aged < 50 Years Treated with Radical Prostatectomy: A Multi-Centre Study in Korea. *J. Korean Med. Sci.* **2019**, *34*, 1–10.
3. Gleason, D.F. Histologic grading of prostate cancer: A perspective. *Hum. Pathol.* **1992**, *23*, 273–279. [CrossRef]
4. Tabesh, A.; Teverovskiy, M.; Pang, H.Y.; Kumar, V.P.; Verbel, D.; Kotsianti, A.; Saidi, O. Multifeature Prostate Cancer Diagnosis and Gleason Grading of Histological Images. *IEEE Trans. Med. Imaging* **2007**, *26*, 1366–1378. [CrossRef] [PubMed]
5. Doyle, S.; Feldman, M.D.; Shih, N.; Tomaszewski, J.; Madabhushi, A. Cascaded Discrimination of Normal, Abnormal, and Confounder Classes in Histopathology: Gleason Grading of Prostate Cancer. *BMC Bioinform.* **2012**, *13*, 282. [CrossRef] [PubMed]
6. Nir, G.; Hor, S.; Karimi, D.; Fazli, L.; Skinnider, B.F.; Tavassoli, P.; Turbin, D.; Villamil, C.F.; Wang, G.; Wilson, R.S.; et al. Automatic grading of prostate cancer in digitized histopathology images: Learning from multiple experts. *Med. Image Anal.* **2018**, *50*, 167–180. [CrossRef] [PubMed]

7. Doyle, S.; Madabhushi, A.; Feldman, M.; Tomaszeweski, J. A Boosting Cascade for Automated Detection of Prostate Cancer from Digitized Histology. *Comput. Vis.–ECCV 2012* **2006**, *4191*, 504–511.
8. Rundo, L.; Militello, C.; Russo, G.; Garufi, A.; Vitabile, S.; Gilardi, M.C.; Mauri, G. Automated Prostate Gland Segmentation Based on an Unsupervised Fuzzy C-Means Clustering Technique Using Multispectral T1w and T2w MR Imaging. *Information* **2017**, *8*, 49. [CrossRef]
9. Jiao, Z.; Gao, X.; Wang, Y.; Li, J. A deep feature based framework for breast masses classification. *Neurocomputing* **2016**, *197*, 221–231. [CrossRef]
10. Hu, Y.; Li, J.; Jiao, Z. Mammographic Mass Detection Based on Saliency with Deep Features. *Int. Conf.* **2016**, 292–297.
11. Naik, S.; Doyle, S.; Agner, S.; Madabhushi, A.; Feldman, M.; Tomaszewski, J. Automated gland and nuclei segmentation for grading of prostate and breast cancer histopathology. In Proceedings of the 2008 5th IEEE International Symposium on Biomedical Imaging: From Nano to Macro, Paris, France, 14–17 May 2008; pp. 284–287.
12. Albashish, D.; Sahran, S.; Abdullah, A.; Abd Shukor, N.; Md Pauzi, H.S. Lumen-Nuclei Ensemble Machine Learning System for Diagnosing Prostate Cancer in Histopathology Images. Pertanika. *J. Sci. Technol.* **2017**, *25*, 39–48.
13. Nguyen, K.; Sabata, B.; Jain, A.K. Prostate cancer grading: Gland segmentation and structural features. *Pattern Recognit. Lett.* **2012**, *33*, 951–961. [CrossRef]
14. Diamond, J.; Anderson, N.H.; Bartels, P.H.; Montironi, R.; Hamilton, P.W. The use of morphological characteristics and texture analysis in the identification of tissue composition in prostatic neoplasia. *Hum. Pathol.* **2004**, *35*, 1121–1131. [CrossRef] [PubMed]
15. Ding, Y.; Pardon, M.C.; Agostini, A.; Faas, H.; Duan, J.; Ward, W.O.C.; Easton, F.; Auer, D.; Bai, L. Novel Methods for Microglia Segmentation, Feature Extraction, and Classification. *IEEE/ACM Trans. Comput. Boil. Bioinform.* **2017**, *14*, 1366–1377. [CrossRef] [PubMed]
16. Yang, D.; Subramanian, G.; Duan, J.; Gao, S.; Bai, L.; Chandramohanadas, R.; Ai, Y. A Portable Image-Based Cytometer for Rapid Malaria Detection and Quantification. *PLoS ONE* **2017**, *12*, 1–18. [CrossRef] [PubMed]
17. Irshad, H.; Veillard, A.; Roux, L.; Racoceanu, D. Methods for Nuclei Detection, Segmentation, and Classification in Digital Histopathology: A Review—Current Status and Future Potential. *IEEE Rev. Biomed. Eng.* **2014**, *7*, 97–114. [CrossRef] [PubMed]
18. Majid, M.A.; Huneiti, Z.A.; Balachandran, W.; Balarabe, Y. Matlab as a Teaching and Learning Tool for Mathematics: A Literature Review. *Int. J. Arts Sci.* **2013**, *6*, 23–44.
19. Wählby, C.; Lindblad, J.; Vondrus, M.; Bengtsson, E.; Björkesten, L.; Wä Hlby, C.; Bjö Rkesten, L. Algorithms for Cytoplasm Segmentation of Fluorescence Labelled Cells. *Anal. Cell. Pathol.* **2002**, *24*, 101–111. [CrossRef] [PubMed]
20. Choi, H.J.; Choi, H.K. Grading of renal cell carcinoma by 3D morphological analysis of cell nuclei. *Comput. Boil. Med.* **2007**, *37*, 1334–1341. [CrossRef]
21. Mouelhi, A.; Sayadi, M.; Fnaiech, F.; Mrad, K.; Ben Romdhane, K. Automatic image segmentation of nuclear stained breast tissue sections using color active contour model and an improved watershed method. *Biomed. Signal. Process. Control* **2013**, *8*, 421–436. [CrossRef]
22. Shiels, C.; Adams, N.M.; Islam, S.A.; Stephens, D.A.; Freemont, P.S. Quantitative Analysis of Cell Nucleus Organisation. *PLoS Comput. Boil.* **2007**, *3*, e138. [CrossRef] [PubMed]
23. Kumar, R.; Srivastava, R.; Srivastava, S. Detection and Classification of Cancer from Microscopic Biopsy Images Using Clinically Significant and Biologically Interpretable Features. *J. Med. Eng.* **2015**, *2015*, 1–14. [CrossRef] [PubMed]
24. Choi, H.K.; Jarkrans, T.; Bengtsson, E.; Vasko, J.; Wester, K.; Malmström, P.U.; Busch, C. Image Analysis Based Grading of Bladder Carcinoma. Comparison of Object, Texture and Graph Based Methods and Their Reproducibility. *Anal. Cell. Pathol.* **1997**, *15*, 1–18. [CrossRef] [PubMed]
25. Peng, Y.; Jiang, Y.; Yang, C.; Brown, J.B.; Antic, T.; Sethi, I.; Schmid-Tannwald, C.; Giger, M.L.; Eggener, S.E.; Oto, A. Quantitative Analysis of Multiparametric Prostate MR Images: Differentiation between Prostate Cancer and Normal Tissue and Correlation with Gleason Score—A Computer-aided Diagnosis Development Study. *Radiology* **2013**, *267*, 787–796. [CrossRef] [PubMed]

26. Doyle, S.; Hwang, M.; Shah, K.; Madabhushi, A.; Feldman, M.; Tomaszeweski, J. Automated Grading Of Prostate Cancer Using Architectural And Textural Image Features. In Proceedings of the 2007 4th IEEE International Symposium on Biomedical Imaging: From Nano to Macro, Arlington, VA, USA, 12–15 April 2007; pp. 1284–1287.
27. Loukas, C.; Kostopoulos, S.; Tanoglidi, A.; Glotsos, D.; Sfikas, C.; Cavouras, D. Breast Cancer Characterization Based on Image Classification of Tissue Sections Visualized under Low Magnification. *Comput. Math. Methods Med.* **2013**, *2013*, 1–7. [CrossRef]
28. Emiliozzi, P.; Maymone, S.; Paterno, A.; Scarpone, P.; Amini, M.; Proietti, G.; Cordahi, M.; Pansadoro, V. Increased Accuracy of Biopsy Gleason Score Obtained By Extended Needle Biopsy. *J. Urol.* **2004**, *172*, 2224–2226. [CrossRef] [PubMed]
29. Wei, L.; Yang, Y.; Nishikawa, R.M. Microcalcification classification assisted by content-based image retrieval for breast cancer diagnosis. *Pattern Recognit.* **2009**, *42*, 1126–1132. [CrossRef]
30. Mazo, C.; Alegre, E.; Trujillo, M. Classification of cardiovascular tissues using LBP based descriptors and a cascade SVM. *Comput. Methods Programs Biomed.* **2017**, *147*, 1–10. [CrossRef]
31. Ribeiro, M.G.; Neves, L.A.; Nascimento, M.Z.D.; Roberto, G.F.; Martins, A.S.; Tosta, T.A.A. Classification of colorectal cancer based on the association of multidimensional and multiresolution features. *Expert Syst. Appl.* **2019**, *120*, 262–278. [CrossRef]
32. Huang, P.W.; Lee, C.H. Automatic Classification for Pathological Prostate Images Based on Fractal Analysis. *IEEE Trans. Med. Imaging* **2009**, *28*, 1037–1050. [CrossRef]
33. Sahran, S.; Albashish, D.; Abdullah, A.; Shukor, N.A.; Pauzi, S.H.M. Absolute cosine-based SVM-RFE feature selection method for prostate histopathological grading. *Artif. Intell. Med.* **2018**, *87*, 78–90. [CrossRef] [PubMed]
34. Molina, J.F.G.; Zheng, L.; Sertdemir, M.; Dinter, D.J.; Schönberg, S.; Rädle, M. Incremental Learning with SVM for Multimodal Classification of Prostatic Adenocarcinoma. *PLoS ONE* **2014**, *9*, e93600.
35. Cortes, C.; Vapnik, V. Support-Vector Networks. *Mach. Learn.* **1995**, *20*, 273–297. [CrossRef]
36. Fondon, I.; Sarmiento, A.; García, A.I.; Silvestre, M.; Eloy, C.; Polónia, A.; Aguiar, P. Automatic classification of tissue malignancy for breast carcinoma diagnosis. *Comput. Boil. Med.* **2018**, *96*, 41–51. [CrossRef] [PubMed]
37. Liang, C.; Bian, Z.; Lv, W.; Chen, S.; Zeng, D.; Ma, J. A computer-aided diagnosis scheme of breast lesion classification using GLGLM and shape features: Combined-view and multi-classifiers. *Phys. Medica* **2018**, *55*, 61–72. [CrossRef] [PubMed]
38. Li, J.; Weng, Z.; Xu, H.; Zhang, Z.; Miao, H.; Chen, W.; Liu, Z.; Zhang, X.; Wang, M.; Xu, X.; et al. Support Vector Machines (SVM) classification of prostate cancer Gleason score in central gland using multiparametric magnetic resonance images: A cross-validated study. *Eur. J. Radiol.* **2018**, *98*, 61–67. [CrossRef] [PubMed]
39. Hai, J.; Tan, H.; Chen, J.; Wu, M.; Qiao, K.; Xu, J.; Zeng, L.; Gao, F.; Shi, D.; Yan, B. Multi-level features combined end-to-end learning for automated pathological grading of breast cancer on digital mammograms. *Comput. Med. Imaging Graph.* **2019**, *71*, 58–66. [CrossRef] [PubMed]
40. Doyle, S.; Feldman, M.; Tomaszewski, J.; Madabhushi, A. A Boosted Bayesian Multiresolution Classifier for Prostate Cancer Detection from Digitized Needle Biopsies. *IEEE Trans. Biomed. Eng.* **2012**, *59*, 1205–1218. [CrossRef]

© 2019 by the authors. Licensee MDPI, Basel, Switzerland. This article is an open access article distributed under the terms and conditions of the Creative Commons Attribution (CC BY) license (http://creativecommons.org/licenses/by/4.0/).

Article

Applications of Capacitive Imaging in Human Skin Texture and Hair Analysis

Christos Bontozoglou [1,*,†] and Perry Xiao [1,2,*,†]

1 School of Engineering, London South Bank University, 103 Borough Road, London SE1 0AA, UK
2 Biox Systems Ltd., Technopark Building, 90 London Road, London SE1 6LN, UK
* Correspondence: bontozoc@lsbu.ac.uk (C.B.); perry.xiao@lsbu.ac.uk (P.X.)
† These authors contributed equally to this work.

Received: 1 August 2019; Accepted: 24 December 2019; Published: 29 December 2019

Abstract: This article focuses on the extraction of information from human skin and scalp hair for evaluation of a subject's condition in the cosmetic and pharmaceutical industries. It uses capacitive images from existing hand-held research equipment and it applies image processing algorithms to expand their possible applications. The literature review introduces the readers into the field of skin research, and it highlights pieces of information that can be extracted by in vivo skin and ex vivo hair measurements. Then, the selected scientific equipment is presented, and Maxwell-based electrostatic simulations are employed to evaluate the measurement apparatus. Image analysis algorithms are suggested for (a) the detection of polygons on the human skin texture, (b) the estimation of wrinkles length and (c) the observation of hair water sorption capabilities by capacitive imaging systems. Finally, experiments are conducted to evaluate the performance of the presented algorithms and the results are compared with the literature. The results indicate that capacitive imaging systems can be used for skin age classification, detection and tracking of skin artifacts (e.g., wrinkles, moles or scars) and calculation of water content in hair samples.

Keywords: texture; skin microrelief; water sorption; aging; hair

---

1. Introduction

The electrical properties of skin and hair alongside their texture and anatomy provide information about a person's health, efficiency of drug delivery and effects from application of cosmetic products. For these reasons, scientists from a variety of fields use non-invasive instruments to extract such information and achieve experimental results that strengthen their research. As part of this introduction, selected publications are illustrating the above points right after a summary of human skin structure and hair anatomy. Then, the use of capacitive imaging in various research fields is demonstrated by summarizing selected research in the literature.

The skin is the largest human organ in terms of surface area and its thickness varies from 5 μm to 1 mm (or more) in different areas of the body [1]. It protects internal organs from environmental influences, and it regulates the water body loss. As illustrated in Figure 1 (left), the skin is separated into three main layers: epidermis, dermis and subcutis. Among other functions, the epidermis provides chemical and diffusion protection, the dermis protects from external mechanical forces and the subcutis connects the skin to the underlying tissue [1]. The outermost sublayer, target of non-invasive instruments, is called stratum coreum and consists mainly of dead keratinocytes. The textural information of this layer, or skin microrelief, is affected by the internal body health, age and living habits as well as by environmental influences [2]. The visible part of hair (Figure 1 right) is shaped as a three-layered shaft of dead protein filament: the cuticle, cortex and medulla [3]. The cuticle is a thin surrounding layer of roof tiles-like structures with about five degrees inclination from the hair shaft core. The cortex is the thickest layer of the hair, it consists of tightly packed keratin cells and it is responsible for the hair color as well as most of the water holding capabilities. Last, the medulla is a soft unstructured keratin that forms the core in the thicker human hair (e.g., scalp hair) [4].

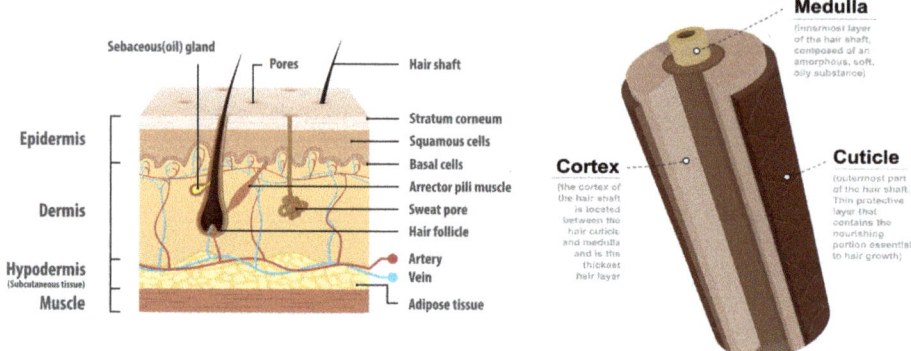

Figure 1. Human skin [5] and hair anatomy [6].

In the literature, a variety of methods have been used to extract information from the human skin surface. Zhang et al. [7] used a capacitive imaging system to measure solvent concentration and penetration in human skin to demonstrate how such a system can support skin clinical trial studies. In the same work, skin damage by intense washing, tape stripping and SLS irritation was characterized. In the area of skin aging analysis, Corcuff et al. [8] studied the skin furrow response during arm extension based on image analysis of negative skin replicas. They provided evidence that younger people can buffer skin strain between primary and secondary line orientations while elderly subjects tend to have furrows only in one orientation that rotate during extension. In more recent years, Zahouani et al. [9] achieved better classification between primary and secondary skin lines by measuring their depth using three-dimensional confocal microscopy. Experimental results on 120 Caucasian women confirmed that secondary lines fade with age while the depth of primary lines increases. A different approach of evaluating skin aging was employed by [10,11], where the surface area of polygons shaped between skin furrow was measured and it was found to associate well with subjects' chronological age.

As illustrated for the area of skin research, hair samples are also analyzed by scientists to detect health and cosmetic conditions. Wosu et al. [12] reviewed 39 studies that associate hair cortisol concentrations to stress psychiatric symptoms and disorders. This approach was found to be more accurate in detecting long-term stress because cortisol concentration measurements from blood, urine and saliva samples are influenced by living habits and environment conditions. Furthermore, Kristensen et al. [13] used hair samples from 266 women to determine that hair dyeing and frequent washing does not affect cortisol measurements. In a different scientific field, Boll et al. [14] employed ATR FT-IR spectroscopy to differentiate between dyed and undyed hair. Such information can be used in forensic hair analysis, given their static classification was found to be 98.1 ± 3.0% accurate in detecting whether a sample is dyed or not, but also in identifying the brand and the color of the dye. In the field of cosmetic science, Barba et al. [15] used thermogravimetric methodology to measure the water content of hair and to assess hair damage from bleaching and straightening. They found that bleached hair shown 3.9% reduced water holding capabilities while the straighten hair 9.5%.

In this work, a capacitive imaging system is used to extract information from skin and hair samples. To the best of our knowledge, Lévêque and Querleux [16] first used this technology for human skin characterization and surface hydration mapping in 2003. They used a fingerprint system to measure the distribution of skin surface capacitance in different body sites, to detect main microrelief orientation and density as well as to support the system's usability in the field of skin research by performing side-by-side experiments with Corneometer CM812. Later on, Batisse et al. [17] demonstrated advantages of such technology over other skin hydration apparatuses by pointing out the importance of visually observing the contact of a capacitive sensor with the skin during lips moisturization and volar forearm inflammation experiments. Since then, capacitive imaging systems have been used in order to examine various skin conditions e.g., mapping of psoriasis and acne lesions, assessment of sun exposure, skin aging, damage or irritation [18–22]. Furthermore, it is worth mentioning successful attempts to improve the capacitive imaging technology. Bevilacqua and Gherardi [23] achieved depth profiling up to 50 μm by fusing the image with pressure information monitored by a subsystem attached on the back of the measurement apparatus. Also, Huang et al. developed a wearable capacitive imaging system using an "ultrathin, stretchable sheets with arrays of embedded impedance sensors for precise measurement and spatially multiplexed mapping" [24].

In the following sections, the measurement principle of capacitive imagining for non-invasive skin and hair measurements is analyzed. Then, image processing algorithms are suggested to extract quantitative values from skin and hair samples. Finally, three experiments are conducted to evaluate both the imaging system but also the integrity of the selected algorithms.

2. Materials and Methods

2.1. Measurement Apparatus

To capture hydration information in pharmaceutical and cosmetic industries non-invasive capacitance sensing systems are employed. In the field of skin research, the most widely known single pixel capacitive system is the Corneometer CM825 (Courage+Khazaka, Cologne, Germany) [25]. This instrument has a 7 × 7 mm sensor in interdigital electrode geometry with 250 μm spatial wavelength and 50 μm electrode width (Figure 2). Its sensor is covered with a 20 μm thick layer of glass for galvanic contact protection [26]. The capacitive imaging sensors are based on the same technology with Corneometer and they are widely known as fingerprint sensors. Their sensing surface consists of a two-dimensional array of miniaturized capacitive pixels that capture the hydration map of the sample. In this work the Epsilon E100 (Biox Systems Ltd., London, UK) [27] is used. It has a sensing surface of 12.8 × 15 mm filled with 300 × 256 pixels and a 2 μm coat of silicon dioxide for galvanic contact protection. The accompanying software exports hydration information in a format of color-coordinated images. In those images, brighter pixels denote higher while darker pixels denote lower hydration levels.

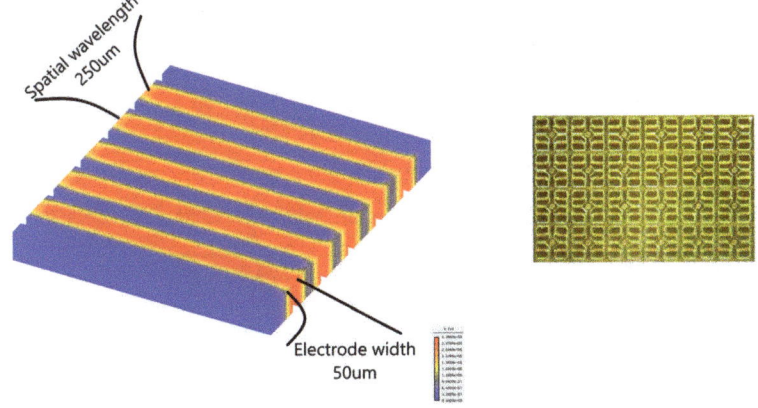

Figure 2. **Left**, 3D representation of scalar potential for Corneometer's electrodes as produced using Agros2D [28]. **Right**, representation of the fingerprint sensor pixels embedded in Epsilon E100 [27].

2.2. Measurement Depth Simulation

In capacitive measurements, the image resolution and sensitivity are important as well as the measurement depth. For in vivo skin measurements, the penetration depth of the electric field should not exceed the thickness of stratum corneum, which consists of dead and keratinized cells [1]. Otherwise, the pixel readouts will saturate because of the higher conductance in deeper skin layers. By contrast, in hair measurements the depth should be sufficient to reach the medulla to capture information from all layers in the hair shaft. In this work, the measurement depth is defined as the distance from the sensor surface for which the electric potential drops by 97%. To the best of our knowledge, this comes in agreement with experimental and theoretical results achieved by Huanyu Cheng et al. [29], where the system capacity is used as a reference instead of the electric potential. Of course, in either approach, the system response changes with the hydration level of the sample, so the penetration depth of the electric field varies accordingly. In order to evaluate the performance of Epsilon E100, the penetration depth for different insulating materials is simulated using Maxwell's equations for both E100 and CM825 and the results are compared.

A property of electromagnetics is that Maxwell's equations are linear. As a result, we can refer to an arbitrary distribution of charges by simple addition of the relevant contributions. In the case of interdigital sensors, such as Corneometer, for the charge distribution residing on a strip (x_1, x_2), the electric potential at an arbitrary point (x, y) is given by the expression in Equation (1). The simulation defined 6 driving and 5 sensing rectangular electrodes in the dimensions provided by CM825 literature. The electric potential was calculated for points on a line segment from the surface of the middle driving electrode to 60 µm perpendicular displacement.

$$\Phi(x,y) = \lambda k_e (\sinh^{-1} \frac{x_2 - x}{|y|} - \sinh^{-1} \frac{x_1 - x}{|y|}) \tag{1}$$

where:

Φ = the electric potential
λ = the surface charge density and
k_e = the dielectric of the insulating material under examination $(k e_0)$

In the case of annular or disk sensors, such as Epsilon E100, Equation (2) calculates the electric potential in any given point perpendicular to the center of the electrode. For the purpose of this simulation, one driving electrode disk and one sensing electrode ring were used. The electric potential was measured from the center of the sensor surface to 60 µm distance on a perpendicular line segment.

$$\Phi(z) = \frac{\lambda}{2k_e}(\sqrt{R_{out}^2 + z^2} - \sqrt{R_{in}^2 + z^2}) \tag{2}$$

where:

R_{out} = the outer electrode radius
R_{in} = the inner electrode radius
z = the perpendicular point distance from the center of the sensor surface

The hydration level of the samples is simulated by changing the dielectric permittivity parameter (k_e). The latter ranges from 1 to 80.1 in 20 °C, where higher value denotes more hydrated samples. For both instruments, the dielectric permittivity of the protective layers is assumed 3.9 and the simulation is repeated for samples with dielectric permittivity of 80, 7 and 3.9. The simulation results in Figure 3 show the Corneometer CM825 measurement depth ranges from 10 to 40 µm, while for Epsilon E100 from 5 to 22 µm. Assuming that stratum corneum thickness is between 10 µm and 40 µm, the results validate that both instruments are appropriate for human skin measurements with Corneometer with greater penetration depth. On hair measurements, assuming sample thickness 70–150 µm, both instruments have insufficient penetration depth but still reach fraction of the cortex layer.

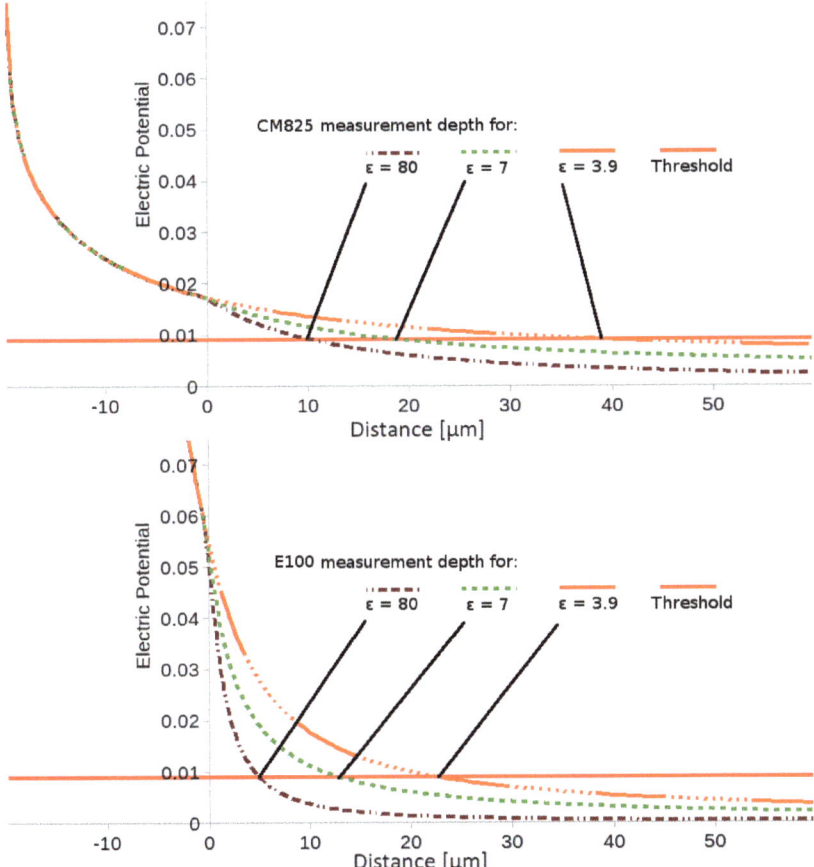

Figure 3. Maxwell-based simulation results to compare measurement depth between Corneometer CM825 and Epsilon E100. The end of the protective layers is overlapping with the y-axis (x = 0). Therefore, electric potential readouts at the left of the y-axis (x < 0) are inside the protective layer, while readouts at the right of the y-axis (x > 0) are taken from within the sample under examination. Also, the electric potential on the electrode surface is configured to 0.3 V, but the upper part is cropped for better visualization of the results. The horizontal line at 0.09 V represents the 3% depth threshold.

2.3. Analysis of Capacitive Image

In this section, image processing algorithms are applied on skin and hair capacitive images to demonstrate how texture and color information can be used in cosmetic and pharmaceutical industries. More specifically, algorithms are presented: (a) to detect the average surface area of skin polygons and associate the results with the chronological age of the subject, (b) to estimate the length of skin furrows and track progression over the course of time and (c) to calculate the water content and loss rate in human hair.

2.3.1. Skin Polygons Detection

The human skin texture is highly inhomogeneous, and it worsens with age because the elasticity fibers network is weakening and the strain is not buffered efficiently [8,9]. Therefore, in order to detect the skin polygons across a wide range of age groups, a segmentation algorithm is required that focuses only on pixels in contact with the skin, it suppresses wrinkles until they become sizeless and it does not need predefined seeds. According to [30], segmentation algorithms may be separated into two categories, the graph-based and gradient-based segmentation. In graph-based segmentation, weights are calculated for each pair of neighboring pixels and superpixels are identified by minimizing a cost function. As it is illustrated in Figure 4, graph-based segmentation algorithms are not fit for this application because they do not suppress skin furrow.

Figure 4. Example application of graph-based segmentation algorithm on capacitive image from volar wrist area. Left the original and right the segmented frame. For this demonstration, algorithm developed by Felzenszwalb and Huttenlocher is used [31].

To the best of our knowledge, Bevilacqua A. and Gherardi A. [18] first apply the gradient-based segmentation algorithm by Vincent and Soille [32] on skin capacitive images. In order to cope with pixel noise and skin hydration inhomogeneity, the frames are pre-processed with a normalization filter. This filter controls the over-/under-segmentation of the sample but given the vast skin texture variety between subjects and sites a global configuration for absolute polygons' count is not feasible. Thus, the reliability of this approach is experimentally strengthened by calculating the correlation between polygons density and subject age and comparing the results with these reported in the literature using manual polygons' count [33]. An example output of this segmentation algorithm is shown in Figure 5.

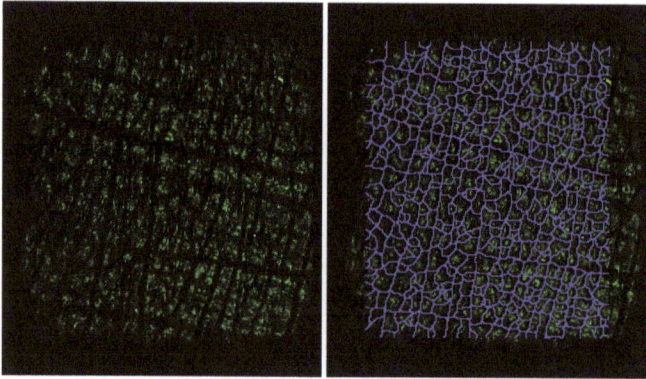

Figure 5. Example of gradient-based segmentation algorithm applied on capacitive image from the palm thenar. Right, the original and left the segmented frame. The lines that overlap with skin furrows represent the sizeless boundaries between segments. Reproduced with permission from [33], John Wiley and Sons, 2018.

2.3.2. Feature Length Estimation

Lévêque and Querleux [16] applied the Gray Level Co-Occurrence Matrix (GLCM) on capacitive images to calculate the angle between primary and secondary skin lines in the area of volar forearm and estimate the age of the subject. While this measurement apparatus has shown poor performance in calculating the skin anisotropy index compared to 3D confocal microscopy due to lack of depth profiling, it does not restrict the application on 2D feature extraction [33]. The GLCM is calculated by Equation (3) [34].

$$P(i,j,d,\theta) = \sum_{x=0}^{n}\sum_{y=0}^{m} \delta_{iI_1}\delta_{jI_2} \qquad (3)$$

where:

i, j = Greyscale level
δ_{kI} = Kronecker delta
I_1 = I(x,y)
I_2 = I(x + dcosθ, y + dsinθ)

In this study, the greyscale levels of the image are reduced from 256 to three (i.e., below, target and over), with the drawback of reducing classification capabilities [35], to strengthen the correlation between pixels within the target level. The length of the target feature d_{length} towards any discrete orientation θ_{target} is estimated by the displacement value d for which ∇P approaches zero ($d_{length} = d_{\nabla P \to 0}$). The feature length is converted from pixels to SI units using the sensor DPI from the instrument's specification (i.e., 50 μm). Of course, selecting a region of interest to eliminate effects from neighboring areas and using repositioning algorithms between frames to track the same wrinkle over the course of time are required [7].

2.3.3. Hair Water Content

In previous work [7], the skin water content percentage and solvent concentration are calculated from capacitive images using Equation (4). In that study, a normalized cross-correlation algorithm detects the same skin area across measurements taken before and after the solvent application on the skin, which allows measurement of the solvent concentration in skin more accurately. In order to apply the same equation on hair samples, the repositioning algorithm is replaced with a range threshold [36]. In this way, only the pixels in contact with hair samples are isolated increasing the accuracy of water percentage calculation. Figure 6 demonstrates how this range threshold excludes pixels with bad contact on a single frame of hair sample.

$$\text{Water Content}[\%] = 100 \frac{\epsilon_m - \epsilon_{dry}}{\epsilon_{water} - \epsilon_{dry}} \qquad (4)$$

where ϵ_m, ϵ_{dry} and ϵ_{water} the dielectric permittivity of the sample, the dry sample and this of deionized water correspondingly.

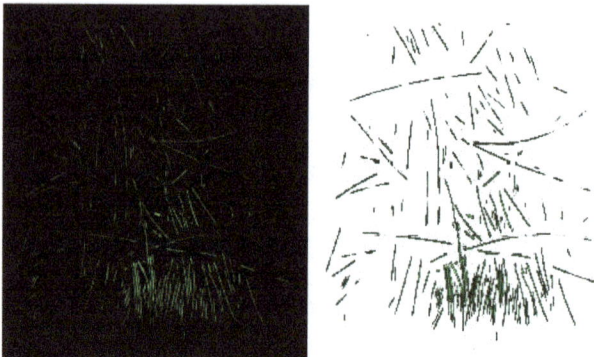

Figure 6. Example of threshold application on hair capacitive image. On the left the original frame, where acclimatized hair on bright green and air/bad contact on darker green shades. Right, the software output after a range threshold is applied to exclude pixels with bad contact before calculating water content.

3. Results

In this section, three experiments are conducted to examine the performance of the presented algorithms on capacitive images. The first experiment evaluates Vincent and Soille segmentation algorithm to automatically count the skin polygons, i.e., the skin areas shaped between wrinkles. For this purpose, capacitive images were recorded from 12 volunteers aging from 12 to 74 years old. The samples were taken from the middle volar forearm area while the arm was in resting position to reduce strain. Then, the segmentation algorithm was applied using Epsilon E100 software and the average number of polygons per square millimeter was correlated against the subjects' age. The results in Table 1 demonstrate that the average number of polygons per surface area decreases with age. The calculated correlation (−0.71) comes in agreement with previous studies in the literature [10,11].

Table 1. Experimental results for skin polygons per surface area in middle volar forearm across 12 volunteers in different age groups using capacitive images. The correlation of the average number of polygons per mm^2 against subjects age is calculated to −0.71. Reproduced with permission from [33], John Wiley and Sons, 2018.

Age	12	16	26	28	31	48	48	50	60	60	64	74	R^2	p-Value
Polygons/mm^2	12.1	10.7	11.6	10.9	10.3	10.4	8.8	7.9	8.0	7.7	6.0	8.8	−0.71	<0.0006

The second experiment consists of a short comparison between C-Cube, a calibrated digital spectroscope (Pixience, Toulouse, France) [37], and Epsilon E100 in feature length estimation. The same skin area of volar forearm was captured with both instruments and three furrows were randomly selected (Figure 7). C-Cube software provides the length measurement as a default feature by drawing the linear segment of interest on the captured frame (Figure 7 left). Epsilon E100 does not provide such feature, so the region of interest was cropped, and our length estimation algorithm was applied (Figure 7 right). The results in Table 2 suggest that if there are no neighboring artifacts in capacitive images, such systems can calculate the length of a furrow with good accuracy.

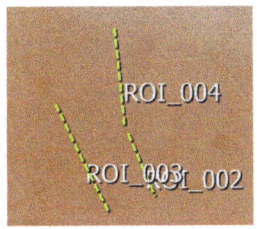
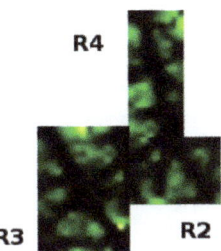

Figure 7. Area of volar forearm captured with C-Cube spectroscope (**left**) and Epsilon E100 capacitive imaging system (**right**). R2-4 the three randomly selected furrows for the comparative experiment.

Table 2. Results of comparative study between C-Cube spectroscope and Epsilon E100 to examine accuracy of wrinkle length estimation using GLCM. The length of three wrinkles in the area of volar forearm was compared and the correlation between the two measurement methods is calculated to 0.9.

Apparatus	R2	R3	R4	R^2
C-Cube [mm]	0.6	0.9	1.1	-
Epsilon E100 [mm]	0.7	0.8	1.0	0.9

In the final experiment, the ability of capacitive imaging sensors to measure hair water content and desorption rate are examined. For this purpose, scalp hair samples from three volunteers were washed in deionized water and dried before left to acclimatize overnight in three different humidity chambers. Saturated salt solutions adjusted the relative humidity levels while both temperature and relative humidity were logged every 10 s using SHT35 by Sensirion [38]. The selected salts are potassium nitrate (85% RH), sodium chloride (75% RH) and magnesium nitrate (67% RH). After acclimatization, the samples were moved in 21 °C & 35% RH conditions, side by side, and they were held against the sensor surface with a plastic plug provided by the manufacturer. The system was capturing video frames until the water loss rates reached a flat state or until the video exceeded 5.5 h. In order to target only the pixels in contact with hair, a range filter from 3.5 to 80 was applied on each frame. The selection of these limits is based on previous work with Epsilon E100 [39]. Five video instances from the same hair sample per acclimatization chamber are shown in Figure 8.

Two observations are made from the experimental results in Figure 9 and their summary in Table 3. First, the hair water content right after acclimatization correlates well with the relative humidity of the chamber. Second, the hair samples from younger subjects tend to hold water for a longer period of time. The latter comes in agreement with Xiao P. et al. [40], stating that lower diffusion rates are observed in younger subjects meaning better water holding capabilities. Note that in many occasions the sample never reached the expected baseline. In those cases, the lowest water content readout was used in the calculations for Table 3.

Table 3. Results of scalp hair water loss experiment using Epsilon E100. The left side of the table shows that the water content % correlates well with the %RH in the acclimatization chamber. The right side of the table shows how long it takes for the sample to lose 75% of its initial water content.

	Water Content [%]			75% Water Loss [s]		
%RH	Age 30	Age 33	Age 50	Age 30	Age 33	Age 50
85	6.0	3.9	4.1	2625	2810	2377
75	4.7	3.3	3.7	1728	2135	1100
67	3.8	3.0	3.4	1490	1604	914
R^2	0.99	0.97	0.99	-	-	-

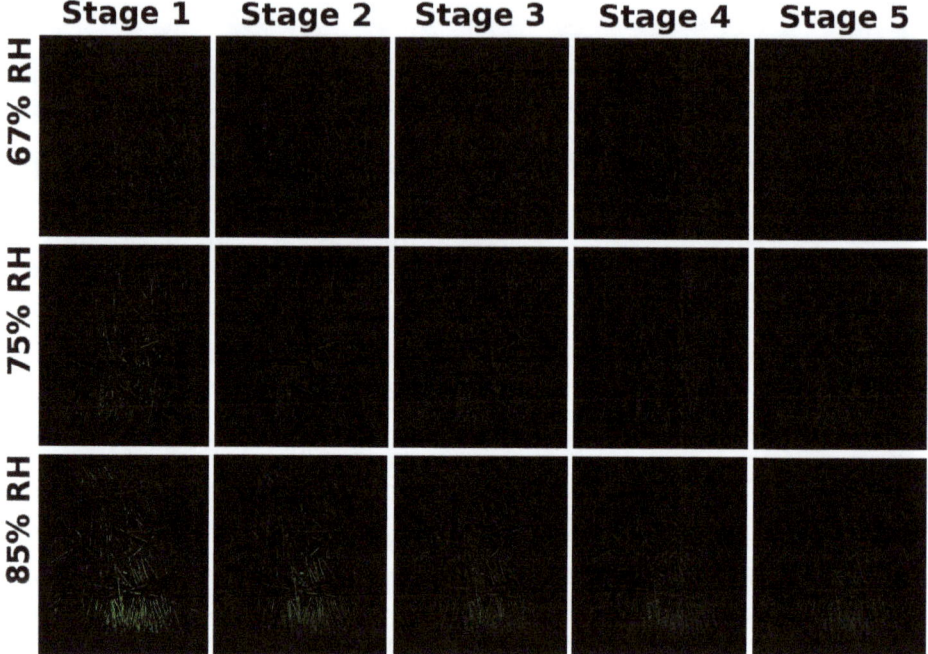

Figure 8. Video snapshots from water desorption in hair samples. The first row shows five frames from hair capacitive images over time after acclimatization in 67%RH. Rows two and three show the same sample after acclimatization in 75% and 85%RH chambers correspondingly. The contrast is modified to highlight hair samples.

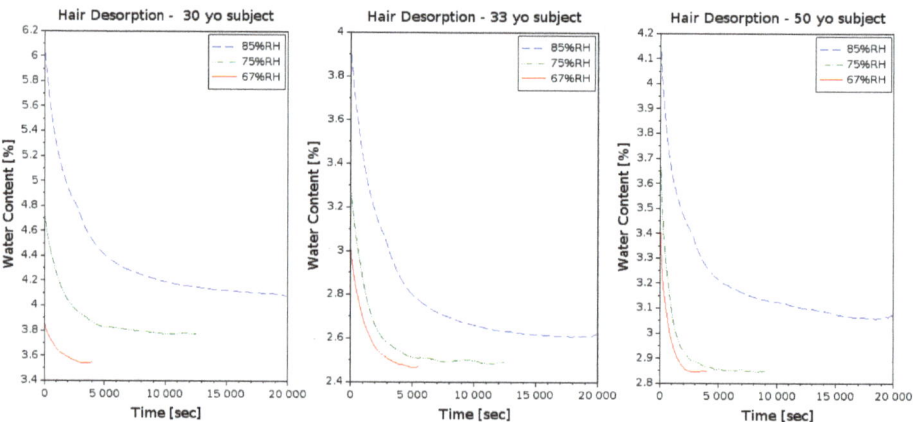

Figure 9. Hair water content desorption curves using samples from three volunteers and for three different humidity acclimatization chambers.

4. Discussion

In this study, we achieved to summarize the importance of skin and hair analysis in a variety of scientific fields, we introduced and analyzed the apparatus of capacitive imaging systems using a Maxwell-based simulation, we suggested algorithms for information extraction using such equipment and we conducted experiments to evaluate the overall system performance.

The simulation compared the penetration depth of the electric field between Corneometer C825 and Epsilon E100. The results show that both instruments have satisfactory penetration depth for skin measurements, with CM825 reaching twice the measurement depth. This implies that if stratum corneum thickness is less than 40 μm, errors should be found in a side-by-side comparative study. Such results have not been found in the literature and might be of interest to achieve in future work for validation of our simulation. Another conclusion we draw by this simulation is that both instruments have insufficient measurement depth for hair analysis. Their electric field reaches the shaft cortex and it will give some reasonable readouts, but it could not represent the absolute hair water content.

Our experiments focused on the evaluation of capacitive imaging systems in cosmetic and skin research studies. The first experiment demonstrated how capacitive images can be used to extract skin texture information. While this could be performed with any calibrated spectroscope, expanding applications of existing laboratory equipment is one of our goals. The suggested algorithm successfully detected the skin polygons and measured their average surface area. The results associate well with subjects' age, giving -0.71 correlation with <0.0006 statistical significance. Furthermore, the achieved correlation value agrees with [10,11], where the correlation between age and polygons density were calculated to -0.64 and -0.65 in the dorsal hand and volar forearm correspondingly.

The second experiment used the GLCM to estimate the length of wrinkles. In order to determine the reliability of this method, we compared our results with calibrated spectroscopy for a small group of samples. The experiment was not extended further because the need to bring the capacitive sensor in contact with the sample results to skin deformation. This is enough to twist the frame and make repositioning algorithms to fail identifying the same wrinkle. Nevertheless, the same logic could be applied on objects with greater surface area (e.g., moles or scars) and track changes in their dimensions over time.

Our last experiment focused on measuring water desorption rate from scalp hair samples. The experimental results shown that the measurement apparatus is capable of differentiating desorption rates from young and elder subjects. This means that the comparative interpretation of the results between samples are in agreement with the literature [40], indicating that such systems can be used in hair analysis studies. Unfortunately, the observed desorption rates are lower than the ones reported in similar studies using different measurement methods (e.g., DVS or thermogravimetric) and for many of our samples the expected baseline was not reached. More specifically, Xiao et al. [40] found that it takes only 58 min for soaked hair samples to return to their baseline hydration using DVS.

To conclude, we believe that capacitive imaging sensors can be used for skin texture analysis and human skin age classification. We also believe that evidence is found for capacitive imaging application on hair water loss studies. This will require a sensor with greater penetration depth and a better sample-holding mechanism.

Author Contributions: Conceptualization, C.B. and P.X.; Methodology, C.B. and P.X.; Software, C.B.; Validation, C.B. and P.X.; Formal Analysis, C.B.; Investigation, C.B.; Resources, P.X.; Data Curation, C.B.; Writing—Original Draft Preparation, C.B.; Writing—Review & Editing, C.B. and P.X.; Visualization, C.B.; Supervision, P.X.; Project Administration, P.X.; Funding Acquisition, P.X. All authors have read and agreed to the published version of the manuscript.

Funding: This research received no external funding.

Conflicts of Interest: The authors declare no conflict of interest.

References

1. Johnsen, G.K. Skin Electrical Properties and Physical Aspects of Hydration of Keratinized Tissues. Ph.D. Thesis, University of Oslo, Oslo, Norway, 2010.
2. Isik, B.; Gurel, M.S.; Erdemir, A.T.; Kesmezacar, O. Development of skin aging scale by using dermoscopy. *Skin Res. Technol.* **2013**, *19*, 69–74. [CrossRef] [PubMed]
3. Bontozoglou, C.; Zhang, X.; Patel, A.; Lane, M.E.; Xiao, P. In Vivo Human Hair Hydration Measurements by Using Opto-Thermal Radiometry. *Int. J. Thermophys.* **2019**, *40*, 22. [CrossRef]

4. Wang, X.J.; Dhond, R.P.; Sorin, W.V.; Nelson, J.S.; Newton, S.A.; Milner, T.E. Characterization of human scalp hairs by optical low-coherence reflectometry. *Opt. Lett.* **1995**, *20*, 524. [CrossRef] [PubMed]
5. iStock.com/Paladjai. Vol.2 Structure of the Skin Info Graphics Illustration Vector on White Background. Beauty Concept. Stock Illustration. Available online: https://www.istockphoto.com/ (accessed on 17 December 2019).
6. iStock.com/Iv__design. Science of Hair. Anatomical Training Poster. Hair Structure. Detailed Medical Vector Illustration Stock Illustration. Available online: https://www.istockphoto.com/ (accessed on 17 December 2019).
7. Zhang, X.; Bontozoglou, C.; Chirikhina, E.; Lane, M.E.; Xiao, P. Capacitive Imaging for Skin Characterizations and Solvent Penetration Measurements. *Cosmetics* **2018**, *5*, 52. [CrossRef]
8. Corcuff, P.; de Lacharrière, O.; Lévêque, J.L. Extension-induced changes in the microrelief of the human volar forearm: Variations with age. *J. Gerontol.* **1991**, *46*, M223–M227. [CrossRef] [PubMed]
9. Zahouani, H.; Djaghloul, M.; Vargiolu, R.; Mezghani, S.; Mansori, M.E.L. Contribution of human skin topography to the characterization of dynamic skin tension during senescence: Morpho-mechanical approach. *J. Phys. Conf. Ser.* **2014**, *483*, 012012. [CrossRef]
10. Gao, Q.; Yu, J.; Wang, F.; Ge, T.; Hu, L.; Liu, Y. Automatic measurement of skin textures of the dorsal hand in evaluating skin aging. *Skin Res. Technol.* **2013**, *19*, 145–151. [CrossRef]
11. Trojahn, C.; Dobos, G.; Schario, M.; Ludriksone, L.; Blume-Peytavi, U.; Kottner, J. Relation between skin micro-topography, roughness, and skin age. *Skin Res. Technol.* **2015**, *21*, 69–75. [CrossRef]
12. Wosu, A.C.; Valdimarsdóttir, U.; Shields, A.E.; Williams, D.R.; Williams, M.A. Correlates of cortisol in human hair: Implications for epidemiologic studies on health effects of chronic stress. *Ann. Epidemiol.* **2013**, *23*, 797–811.e2. [CrossRef]
13. Kristensen, S.K.; Larsen, S.C.; Olsen, N.J.; Fahrenkrug, J.; Heitmann, B.L. Hair dyeing, hair washing and hair cortisol concentrations among women from the healthy start study. *Psychoneuroendocrinology* **2017**, *77*, 182–185. [CrossRef]
14. Boll, M.S.; Doty, K.C.; Wickenheiser, R.; Lednev, I.K. Differentiation of hair using ATR FT-IR spectroscopy: A statistical classification of dyed and non-dyed hairs. *Forensic Chem.* **2017**, *6*, 1–9. [CrossRef]
15. Barba, C.; Méndez, S.; Martí, M.; Parra, J.L.; Coderch, L. Water content of hair and nails. *Thermochim. Acta* **2009**, *494*, 136–140. [CrossRef]
16. Leveque, J.L.; Querleux, B. SkinChipR, a new tool for investigating the skin surface in vivo. *Skin Res. Technol.* **2003**, *9*, 343–347. [CrossRef] [PubMed]
17. Batisse, D.; Giron, F.; Lévêque, J.L. Capacitance imaging of the skin surface. *Skin Res. Technol.* **2006**, *12*, 99–104. [CrossRef] [PubMed]
18. Bevilacqua, A.; Gherardi, A. Age-related skin analysis by capacitance images. In Proceedings of the 17th International Conference on Pattern Recognition, ICPR 2004, Cambridge, UK, 26–26 August 2004.
19. Xhauflaire-Uhoda, E.; Piérard, G.E. Skin capacitance imaging of acne lesions. *Skin Res. Technol.* **2007**, *13*, 9–12. [CrossRef]
20. Bazin, R.; Laquieze, S.; Rosillo, A.; Lévêque, J.L. Photoaging of the chest analyzed by capacitance imaging. *Skin Res. Technol.* **2010**, *16*, 23–29. [CrossRef]
21. Xhauflaire-Uhoda, E.; Mayeux, G.; Quatresooz, P.; Scheen, A.; Piérard, G.E. Facing up to the imperceptible perspiration. Modulatory influences by diabetic neuropathy, physical exercise and antiperspirant. *Skin Res. Technol.* **2011**, *17*, 487–493. [CrossRef]
22. Pan, W.; Zhang, X.; Lane, M.; Xiao, P. The occlusion effects in capacitive contact imaging for in vivo skin damage assessments. *Int. J. Cosmet. Sci.* **2015**, *37*, 395–400. [CrossRef]
23. Bevilacqua, A.; Gherardi, A. Characterization of a capacitive imaging system for skin surface analysis. In Proceedings of the 2008 First Workshops on Image Processing Theory, Tools and Applications, Sousse, Tunisia, 23–26 November 2008; IEEE: Piscataway, NJ, USA, 2008; pp. 1–7. [CrossRef]
24. Huang, X.; Cheng, H.; Chen, K.; Zhang, Y.; Zhang, Y.; Liu, Y.; Zhu, C.; Chi Ouyang, S.; Kong, G.W.; Yu, C.; et al. Epidermal Impedance Sensing Sheets for Precision Hydration Assessment and Spatial Mapping. *IEEE Trans. Biomed. Eng.* **2013**, *60*, 2848–2857. [CrossRef]
25. Bauer, H. Courage + Khazaka Electronic, Köln-Corneometer CM 825 (E). Available online: https://www.courage-khazaka.de/en/scientific-products/all-products/16-wissenschaftliche-produkte/alle-produkte/183-corneometer-e (accessed on 1 July 2019).

26. Barel, A.O.; Clarys, P. Skin Capacitance. In *Non Invasive Diagnostic Techniques in Clinical Dermatology*; Springer-Verlag: Berlin/Heidelberg, Germany, 2014; pp. 357–366.
27. Biox Epsilon Model E100 Specifications. Available online: https://www.biox.biz/Products/Epsilon/E100PSpecs.php (accessed on 1 July 2019).
28. Agros2D. Available online: http://www.agros2d.org/ (accessed on 18 December 2019).
29. Cheng, H.; Zhang, Y.; Huang, X.; Rogers, J.A.; Huang, Y. Analysis of a concentric coplanar capacitor for epidermal hydration sensing. *Sens. Actuators A Phys.* **2013**, *203*, 149–153. [CrossRef]
30. Achanta, R.; Shaji, A.; Smith, K.; Lucchi, A.; Fua, P.; Süsstrunk, S. SLIC superpixels compared to state-of-the-art superpixel methods. *IEEE Trans. Pattern Anal. Mach. Intell.* **2012**, *34*, 2274–2282. [CrossRef] [PubMed]
31. Felzenszwalb, P.F.; Huttenlocher, D.P. Efficient Graph-Based Image Segmentation. *Int. J. Comput. Vis.* **2004**, *59*, 167–181. [CrossRef]
32. Vincent, L.; Soille, P. Watersheds in digital spaces: An efficient algorithm based on immersion simulations. *IEEE Trans. Pattern Anal. Mach. Intell.* **1991**, *13*, 583–598. [CrossRef]
33. Bontozoglou, C.; Zhang, X.; Xiao, P. Micro-relief analysis with skin capacitive imaging. *Skin Res. Technol.* **2019**, *25*, 165–170. [CrossRef]
34. Bianconi, F.; Chirikhina, E.; Smeraldi, F.; Bontozoglou, C.; Xiao, P. Personal identification based on skin texture features from the forearm and multi-modal imaging. *Skin Res. Technol.* **2017**, *23*, 392–398. [CrossRef]
35. Clausi, D.A. An analysis of co-occurrence texture statistics as a function of grey level quantization. *Can. J. Remote Sens.* **2002**, *28*, 45–62. [CrossRef]
36. Pan, W.; Zhang, X.; Chirikhina, E.; Bontozoglou, C.; Xiao, P. Measurement of Skin Hydration with a Permittivity Contact Imaging System. In Proceedings of the IFSCC Conference Zurich 2015, Zurich, Switzerland, 21–23 September 2015.
37. Pixience. The C-Cube: The New Standard for Digital Dermoscopy. Available online: https://www.pixience.com/products/presentation-2/?lang=en (accessed on 28 July 2019).
38. SHT3x (RH/T)—Digital Humidity Sensor | Sensirion. Available online: https://www.sensirion.com/en/environmental-sensors/humidity-sensors/digital-humidity-sensors-for-various-applications/ (accessed on 31 May 2019).
39. Xiao, P.; Bontozoglou, C. Capacitive contact imaging for in-vivo hair and nail water content measurements. *H & PC* **2015**, *10*, 62–65.
40. Xiao P.; Ciortea LI.; Bontozoglou C.; Imhof, R.E. Hair Water Content & Water Holding Capacity Measurements. In Proceedings of the 7th International Conference on Applied Hair Science, Red Bank, NJ, USA, 8–9 June 2016.

© 2019 by the authors. Licensee MDPI, Basel, Switzerland. This article is an open access article distributed under the terms and conditions of the Creative Commons Attribution (CC BY) license (http://creativecommons.org/licenses/by/4.0/).

Article

Use of Texture Feature Maps for the Refinement of Information Derived from Digital Intraoral Radiographs of Lytic and Sclerotic Lesions

Rafał Obuchowicz [1,*], Karolina Nurzynska [2], Barbara Obuchowicz [3], Andrzej Urbanik [1] and Adam Piórkowski [4]

1. Department of Diagnostic Imaging, Jagiellonian University Medical College, 19 Kopernika Street, 31-501 Cracow, Poland
2. Institute of Informatics, Faculty of Automatic Control, Electronics, and Computer Science, Silesian University of Technology, Akademicka 16, 44-100 Gliwice, Poland
3. Department of Conservative Dentistry with Endodontics, Jagiellonian University Collegium Medicum, Montelupich. 4, 31-155 Cracow, Poland
4. Department of Biocybernetics and Biomedical Engineering, AGH University of Science and Technology, Mickiewicza 30, 30-059 Cracow, Poland
* Correspondence: rafalobuchowicz@su.krakow.pl; Tel.: +48-12-424-7494

Received: 17 May 2019; Accepted: 19 July 2019; Published: 24 July 2019

Abstract: The aim of this study was to examine whether additional digital intraoral radiography (DIR) image preprocessing based on textural description methods improves the recognition and differentiation of periapical lesions. (1) DIR image analysis protocols incorporating clustering with the k-means approach (CLU), texture features derived from co-occurrence matrices, first-order features (FOF), gray-tone difference matrices, run-length matrices (RLM), and local binary patterns, were used to transform DIR images derived from 161 input images into textural feature maps. These maps were used to determine the capacity of the DIR representation technique to yield information about the shape of a structure, its pattern, and adequate tissue contrast. The effectiveness of the textural feature maps with regard to detection of lesions was revealed by two radiologists independently with consecutive interrater agreement. (2) High sensitivity and specificity in the recognition of radiological features of lytic lesions, i.e., radiodensity, border definition, and tissue contrast, was accomplished by CLU, FOF energy, and RLM. Detection of sclerotic lesions was refined with the use of RLM. FOF texture contributed substantially to the high sensitivity of diagnosis of sclerotic lesions. (3) Specific DIR texture-based methods markedly increased the sensitivity of the DIR technique. Therefore, application of textural feature mapping constitutes a promising diagnostic tool for improving recognition of dimension and possibly internal structure of the periapical lesions.

Keywords: digital intraoral radiography; image preprocessing; periapical lesions; texture analysis

1. Introduction

The importance of assessment of periapical lesions in clinical decision-making is well known. Osteolytic lesions form as periapical lesions in response to inflammatory infiltrates and are often associated with morbidity of the root canal pulp [1–4]. Recognition of osteolytic changes provides important information about the viability of a tooth, which influences decision-making during the treatment process. The relevant anatomical structures themselves are often small, which hinders acquisition of adequate anatomical outline. Moreover, the relative complexity of the region is increased by the presence of superimposing structures that result in "anatomical noise". All of these factors contribute to the difficulty in recognition of bone resorption on radiographic images, which impedes the accuracy of diagnosis using digital intraoral radiography (DIR) images and may result in periapical lesions going undetected or detection is inadequate [5–7].

All of the above-mentioned factors contribute to the relatively low sensitivity of lesion detection of 70% reportedly associated with DIR images, which is markedly less than that of cone-beam computed tomography (CBCT) [8]. While visualization of lesions on CBCT is superior to that on DIR, the radiation dose the patient is exposed to via CBCT is considerably higher than that associated with conventional radiography. Therefore, use of CBCT is questionable, especially as a follow-up modality. In the current work, we present how effective DIR image analyses are with the use of image post-processing in order to refine the acquired information.

Texture feature analysis was first used to evaluate the structure of osteoporotic bone [9–11], where fractal dimension and 13 Haralick features were used for osteoporosis classification on mandibular X-ray images [9]. Similar techniques were applied to analyze periapical bone loss [12–15], where the Gray-Level Co-occurrence Matrix and Fractal Brownian Motion Model were used for bone-loss area detection [15] and localization [12]. Numerous features formed the basis for segmentation [13] and bone loss degree measurement [14]. Other applications of texture analysis were used for periapical bone healing [16–18], where radiological assessment of treatment effectiveness of guided bone regeneration was measured. The most similar research to the presented is considered in [19,20], where the authors tried to detect the type of cyst using the Gray-Level Co-occurrence Matrix and its related properties. By using textural analysis to enhance bone representation derived from DIR images, trabecular structure may be depicted more informatively, and the shapes of different anatomical structures may be determined more accurately. Such techniques may also facilitate more precise determination of changes in the periapical bone region.

The aim of the present study was to examine the applicability of different texture analysis techniques to radiographic dental images for the refinement and possible differentiation of periapical lesions.

2. Materials and Methods

2.1. Ethics Approval and Consent to Participate

The study protocol was designed in accordance with the guidelines of the Declaration of Helsinki and the Good Clinical Practice Declaration Statement. Particular care was taken to ensure the safety of personal data, and all images were anonymized before processing. Written informed consent for the publication of clinical details and anonymized clinical images was obtained from the scientific committee and management department of the dental clinic. The usual requirement for informed consent from patients was waived in view of the retrospective nature of the research.

2.2. Experiment Overview

The experiment described in this work consisted of three main parts:

1. Data collection—the medical-dental data describing the lesions was gathered. In all the cases, it was assured that a quality of obtained dental images was adequate. The data was carefully scrutinized for technically improper images to be rejected. Only 12-bit images recorded in digital imaging and communications in medicine (DICOM) format were accepted. Personal information in images was concealed by special software.
2. Texture feature map computation—the digitized radiographs were of various quality, therefore some standardization was necessary. Sometimes, improving the contrast by image processing methods (e.g., histogram stretching or equalization - HISTEQ) was sufficient, yet in the presented problem it was not satisfactory. Therefore, for each image a set of texture feature maps were prepared. Those maps may also be characterized with low contrast, hence again (as in the preprocessing step) the standard methods for its improvement were applied in the post-processing stage.

3. Results evaluation—finally the data was revised by experienced radiologists whose statements were the basis for the assessment of results.

A detailed description of these three parts of the experiment is given below and is presented in Figure 1.

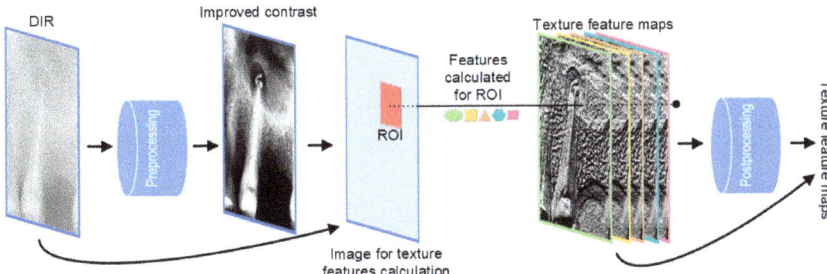

Figure 1. Image processing system schema. DIR, digital intraoral radiography. ROI, region of interest.

2.3. Dataset Description

Sixty-five anonymized DIR images from patients who attended the dental clinic from 2015 to 2017 were used in the study. Sixty-five dental DIR images, consisting of 35 images showing lytic lesions and 30 showing sclerotic lesions, were subjected to analysis. Radiographic material of patients of both sexes aged 26–57 years was used in the study. The images were selected from the institutional picture archiving and communication system (PACS). The selection criteria were acceptable image quality and suspicion of the presence of a periapical lesion on DIR.

Periapical radiographs were obtained using a dental X-ray system (Carestream Trophy with RVG 5200, Kodak, Rochester, NY, USA). Digital images were acquired at 70 kVp and 7 mA with a mean exposure time of 0.05 s, image dimensions of 1200 × 1600 pixels, and a pixel size of 0.018 mm. Digital images were saved in 16-bit digital imaging and communications in medicine (DICOM) format in the local PACS.

2.4. Texture Feature Map Computation

The original DIR images were analyzed using OsiriX (Pixmeo) on a Mac OS-based platform. The DIR images underwent texture preprocessing in the MATLAB environment (MathWorks, Natick, MA, USA) on Windows. The clustering of image colors was implemented using the clustering with a k-means approach (CLU). The co-occurrence matrices (COM) [21], first-order features (FOF), gray-tone difference matrices (GTDM) [22], run-length matrices (RLM) [23], and local binary patterns (LBP) [24,25] were applied. The details of the texture methods are described in [26]. Most of the mentioned methods (COM, FOF, GTDM, and RLM) in the original version compute several features to describe the whole image content. However, such an approach would not be useful for diagnostic purposes, so a new image (called a "texture feature map") reflecting the feature values calculated in a small region of interest was computed. As a consequence, several texture feature maps (depending on the number of features designed for each texture operator) were generated using this technique. In the current study, each texture feature map was generated using the "moving window" approach where, for each pixel, the new feature value was calculated on the basis of data collected in a square window (with sides of an odd number of pixels in length), to ensure that the considered pixel is in the center. In the current study, a 21 × 21 pixel square was used. This size achieves a consensus between computational overhead resulting in image processing time (which grows exponentially with the size and statistical stability of the results, where 441 elements used to fill a histogram of 256 bins is sufficient to achieve statistically reliable results) and image quality. For the LBP texture operator, the radius, R, was in a range from 3 to 15 pixels and there were 8 samples taken in the circular neighborhood in the described experiments. In the case of clustering, from 10 up to 50 clusters were considered. The RLM calculated the matrix for images quantized to 32 and 64 colors, while the matrix stored the maximal length of 10 elements in a run. The square neighborhood applied to calculate the GTDM matrix was evaluated for sides 3, 7, and 11, yet the smallest one returned the best result. Some examples of texture feature maps achieved with the techniques described here are presented in Figure 2.

2.5. Pre- and Post-Processing of Images

The DICOM images store shade information using 12 bits, the aim of which is to save as much detailed information in the scanned data as possible, while most algorithms for image processing are used to work with gray-scale images coding the information on 8 bits. Therefore, in order to process the data, the depth of the color was reduced, and the images were converted to 8-bit color coding. This operation makes it possible to compute the texture feature maps with the standard approach to texture processing and has been proven to remove some of the noise [27].

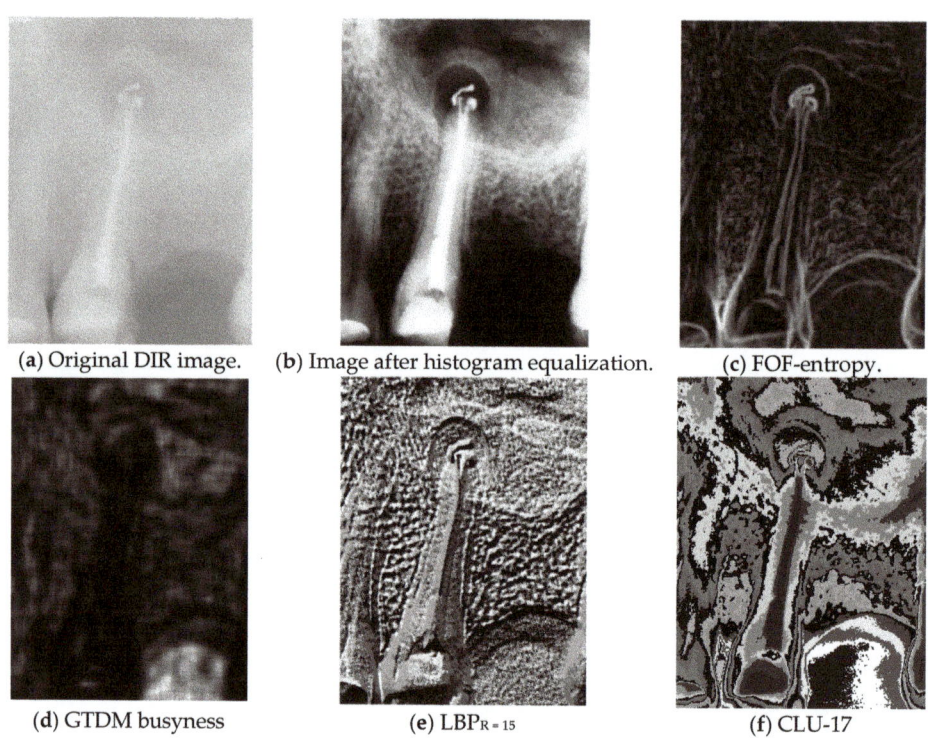

(a) Original DIR image. (b) Image after histogram equalization. (c) FOF-entropy.

(d) GTDM busyness (e) $LBP_{R=15}$ (f) CLU-17

Figure 2. Various aspects of image preprocessing. (a) Original data. (b) The same image after histogram equalization shows improved contrast but does not depict the structure clearly in the tooth region. (c–f) Examples of feature maps derived from a digital intraoral radiographic image via various texture analysis methods. CLU, clustering with k-means approach; DIR, digital intraoral radiography; FOF, first-order features; GTDM, gray-tone difference matrices; LBP, local binary patterns.

When an image is of low quality, particularly when it lacks sharpness and contrast, the histogram equalization operation can be applied to ensure that the whole color range is used, thereby rendering objects more easily visible, as shown in Figure 2b. However, this approach is prone to failure when used on images that contain both very dark and very light objects, as is often the case with plain radiographic images. Hence, there is a need for more sophisticated methods to depict the data in a more informative manner. Nevertheless, application of the histogram equalization (HEQ) method in the preprocessing stage (e.g., before the texture feature map is computed) was tested, and when followed by the RLM technique, proved to be useful because the texture feature map quality improved significantly.

On the other hand, the contrast of some textural feature maps was low and did not present the content clearly. For those, histogram stretching after the final result was applied which aims in scaling the pixel values in order to assure use of the full range of 8-bit color coding (0–255). This transformation does not change the image content but makes it easier for a radiologist to evaluate. There were also some cases where application of histogram equalization gave a better effect.

2.6. Experiment Methodology

The native DICOM images and texture feature maps obtained were analyzed on a 4K retina monitor by two radiologists with 10 and 30 years of experience in analysis of classical bone radiograms including dental. Standard DIR images were assessed first. The pictures were then examined separately using the techniques described above, i.e., CLU, COM, FOF, GTDM, RLM, and LBP. The prepared feature images were evaluated for radiodensity, border definition, and tissue contrast. Figure 3 presents the analyzed regions. The aforementioned parameters were estimated separately to evaluate assumptive improvement of visualization and the subsequently increased accuracy in detection of periapical lesions. Radiodensity analysis was used to evaluate bone density changes, border definition presented edge definition of the changes, and gray-scale contrast meant tissue contrast was used for evaluation of the lesion character. All features were summarized in the evaluation chart. Results were encoded in 1/0 code where 0 meant lack of the recognition of the feature and 1 meant its visual confirmation. In order to comprehensively evaluate the texture feature map usability, two main clinical issues were analyzed—sclerotic lesions and lytic lesions. Interrater reliability was at the level of 98% (on the basis of concordance correlation) where doubtful cases were established on the basis of interrater consensus.

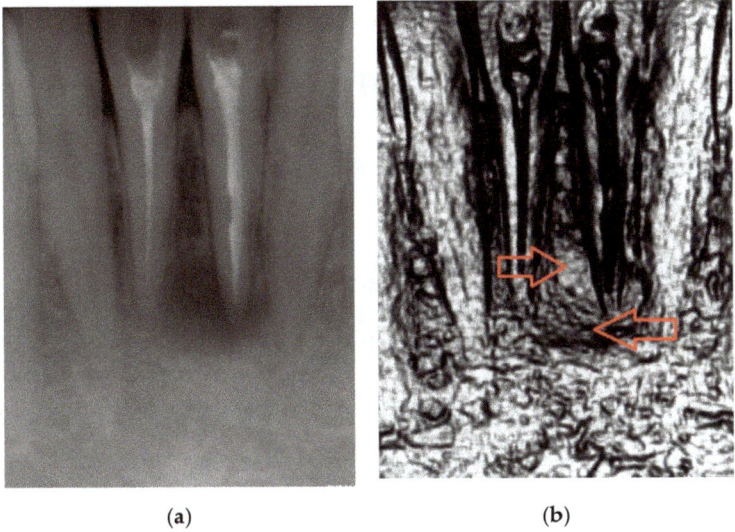

(a) (b)

Figure 3. *Cont.*

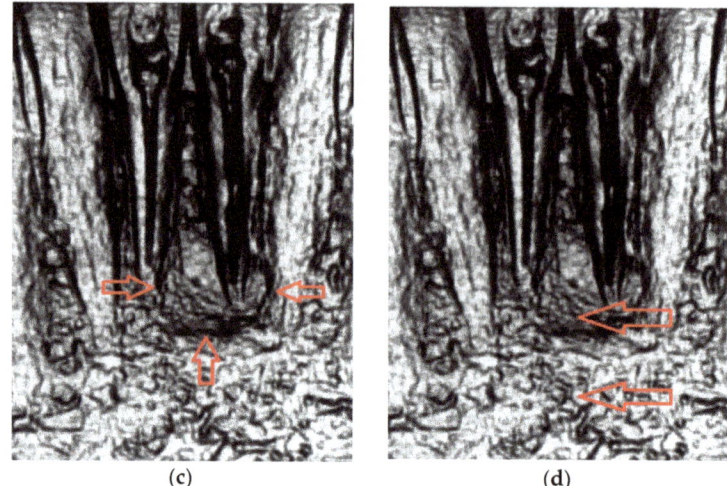

Figure 3. Example of assessment of a periapical lesion. (**a**) Entry digital intraoral radiographic image. (**b**) Differentiation of radiodensity in the lesion (different shapes are pointed out by the arrows). (**c**) Border definition (pointed out by small arrows). (**d**) Tissue contrast between lesion and the neighborhood (shown by the arrows).

3. Results

Figure 4 shows the performance of each image processing approach with respect to radiological changes. The original DIR images are presented in the top row, next to the same images after histogram equalization transformation. As shown, despite the images gaining better contrast, the visibility of changes did not improve substantially. In the next row in Figure 4, the region where changes existed is enclosed within a red line. In the following rows, the texture feature maps computed for CLU, FOF energy, GTDM busyness, (HEQ) RLM, short run high gray level run emphasis, and LBP are given. Recognition of the borders of the lesions and their internal structure was markedly improved in comparison with the initial DIR image. In the FOF group, the contours of lytic changes were represented effectively. LBP and CLU revealed previously hidden information about the internal structure of the lesions and tissue contrast. (HEQ) RLM was found to improve visualization of lytic and sclerotic lesions markedly.

The delineation of sclerotic lesions and internal pattern recognition were achieved with CLU, RLM, and LBP texture feature maps.

The potential utility of each method was calculated based on data derived from experts, and changes are expressed as sensitivity and specificity for the different groups of texture feature maps that are gathered in the bar plots presented in Figure 5. All samples presenting lesions and marked as such by experts take place for true positive (TP) cases. When there was a change unnoticed by the expert, there was a true negative (TN) result. Then, when the expert noticed the change in the DIR data without lesions, a false positive (FP) result was recorded. False negative (FN) results corresponded to the situation in which the data presented any changes, and the expert confirmed it. Consequently, the formulas for sensitivity and specificity are as follows:

$$Sensitivity = \frac{TP}{TP + FN},$$

$$Specificity = \frac{TN}{TN + FP}.$$

A very low percentage value meant that the transformation did not improve visibility of a lesion, while a high percentage value indicated improved visibility and a marked increase in potential lesion recognition after texture map utilization in comparison to DIR. Performance of each image processing approach with respect to radiological changes is presented on separate graphs in Figure 5, illustrating the sensitivity and specificity of the proposed methods. Recognition and differentiation of the lytic lesions after use of the texture feature maps showed the highest sensitivity for the (HEQ) RLM texture feature map, with scores of 94%, 89%, and 94% for radiodensity, border definition, and tissue contrast, respectively; specificity for recognition of these parameters was 86%, 89%, and 43%. The next best performing texture feature map was CLU, with a sensitivity of 83%, 77%, and 80%, and a specificity of 74%, 97%, and 51% for radiodensity, border definition, and tissue contrast, respectively. FOF texture feature maps showed relatively low sensitivity for lytic lesions (60%, 69%, and 51%) but high specificity (94%, 91%, and 69%) for recognition of the three chosen radiological features.

For the sclerotic lesions, the (HEQ) RLM texture feature map was again found to have the best performance, with a sensitivity of 97%, 80%, and 97% for recognition of radiodensity, border definition, and tissue contrast, respectively. The specificity of the following texture feature maps was lower for recognition of radiodensity changes (47%) but better for border definition (90%) and tissue contrast differentiation (53%). FOF texture feature maps performed well in detection of sclerotic lesions, with recognition of radiodensity, border definition, and tissue contrast in 73%, 70%, and 57% of cases, respectively, with high specificity for the chosen features of 60%, 83%, and 73%. No important refinement of recognition of sclerotic lesions was observed for CLU texture feature maps, which had low sensitivity of values of 60%, 60%, and 3% for the three radiological features.

The highest sensitivity for detection of sclerotic lesions was shown for the (HEQ) RLM texture feature maps in terms of radiodensity differentiation, border definition, and tissue contrast recognition, but its specificity was not higher than that of the CLU and FOF texture feature maps. FOF texture feature maps showed good sensitivity for detection of sclerotic lesions and had better specificity than the CLU and RLM texture feature maps.

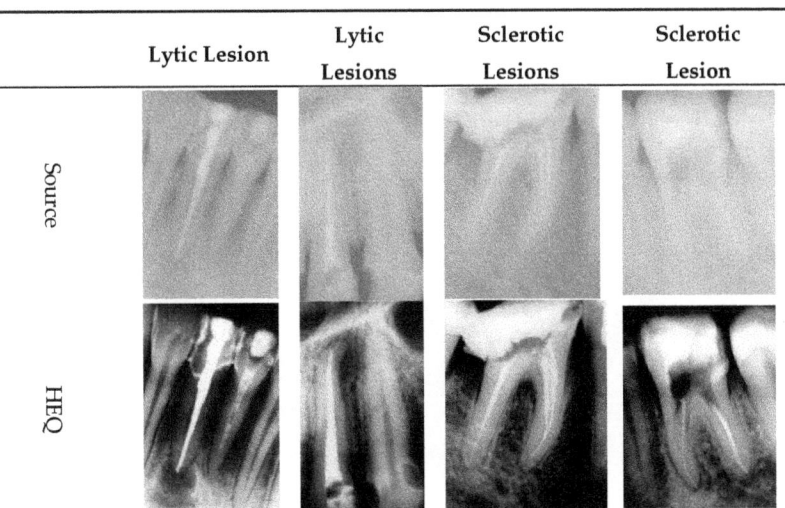

Figure 4. *Cont.*

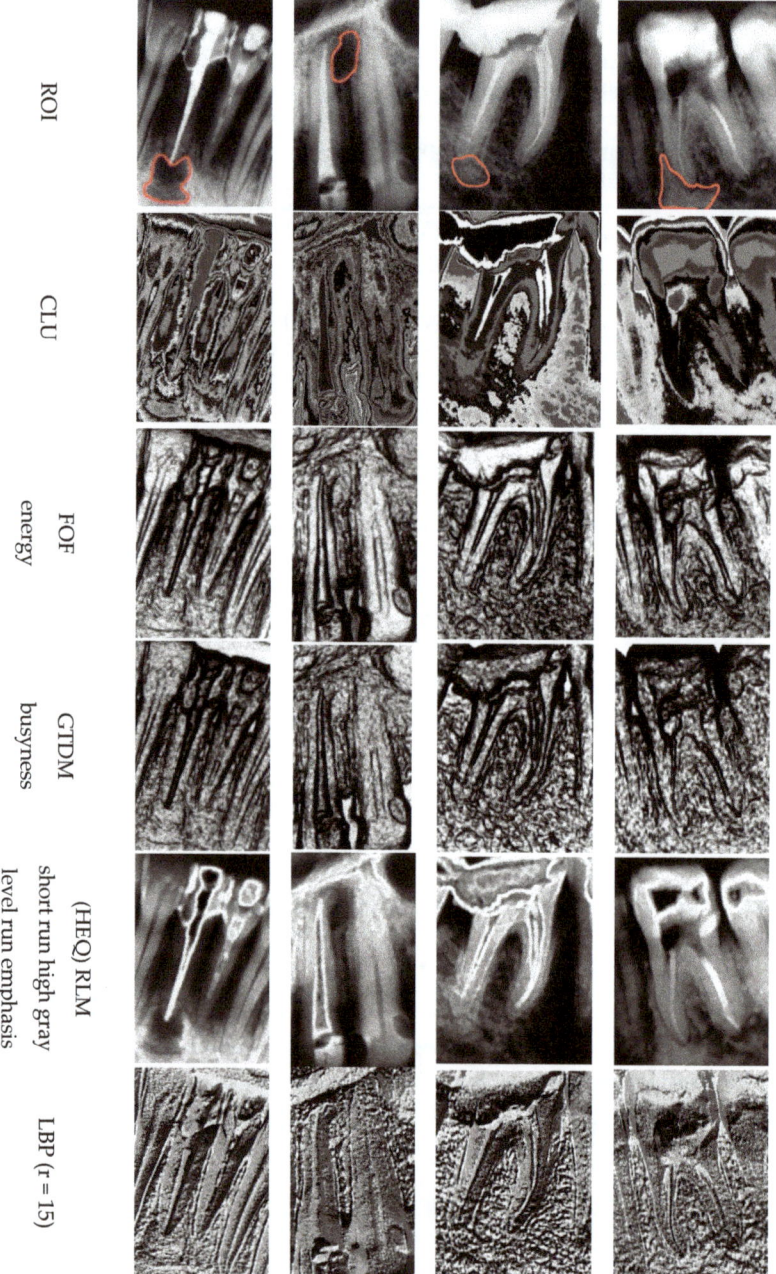

Figure 4. Example of assessment of a periapical lesion.

The best performance in terms of recognition of the features analyzed in both sclerotic and lytic lesions was achieved for the (HEQ) RLM texture feature map, although the FOF texture feature map showed acceptable specificity for recognition of these parameters. CLU was also a well-performing texture feature map for detection of lytic changes, with high sensitivity but lower specificity in comparison with the FOF texture feature map (see the comparison of parameters in Figure 6).

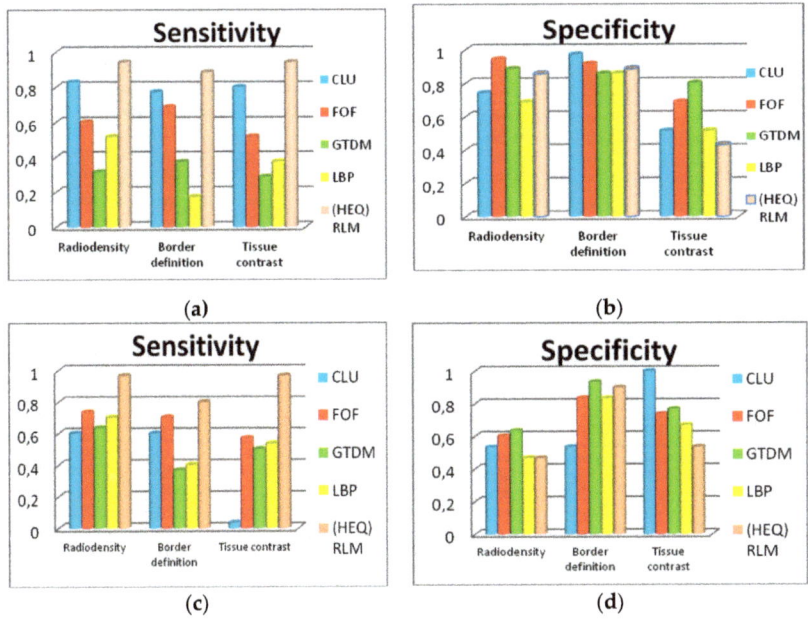

Figure 5. Sensitivity and specificity of different texture feature maps for detection of lytic (**a,b**) and sclerotic (**c,d**) lesions. CLU, clustering with k-means approach; FOF, first-order features; GTDM, gray-tone difference matrices; HEQ, histogram equalization; LBP, local binary patterns; RLM, run-length matrices.

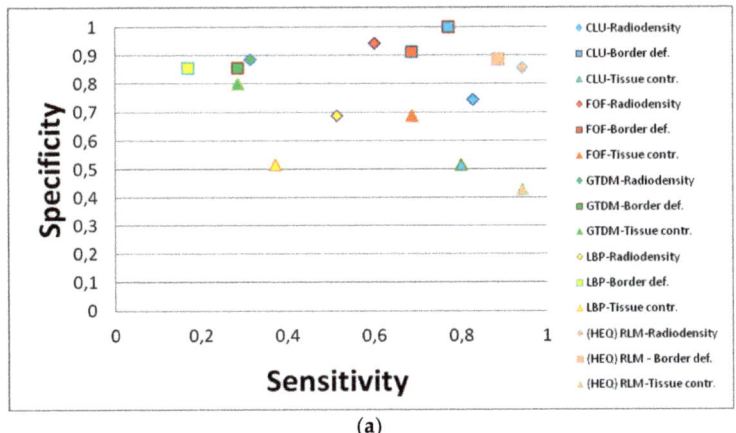

(**a**)

Figure 6. *Cont.*

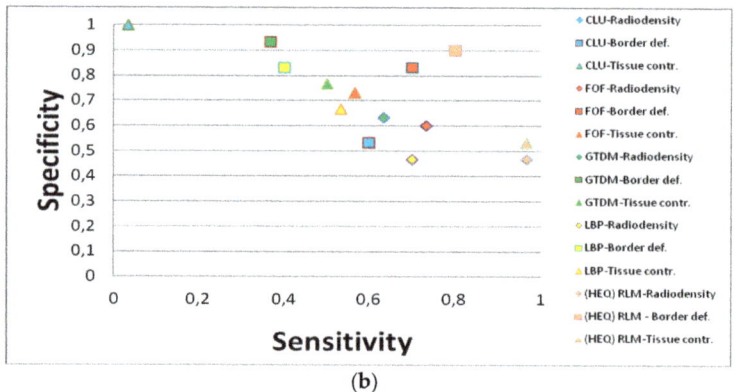

(b)

Figure 6. Relationship between specificity and sensitivity for (**a**) lytic lesions and (**b**) sclerotic lesions. Generally, the higher both values are, the better the parameter. CLU, clustering with k-means approach; FOF, first-order features; GTDM, gray-tone difference matrices; HEQ, histogram equalization; LBP, local binary patterns; RLM, run-length matrices.

4. Discussion

During image analysis, the observer first notices whole objects, anything that has delineated edges, and anything that is in contrast with the surrounding area. However, when the image quality is very low, the content is blurred, the image has no contrast, or an object's texture may appear very similar to that of the background. In such cases, only analysis of local differences that may be hard to discern on standard radiography may unmask some of the vital information that can be derived from the image. Although it may be impossible to detect differences by assessment of plain radiographic images, transforming visual data derived from an image with algorithms designed to map structural differences may reveal hidden content. The process used in our study is presented in Figure 2 in a sequence of images depicting how a substantial amount of additional information about a lesion was gained in comparison to the initial DIR image.

The method of using texture feature maps obtained by texture analysis of the DIR images used in the current study is consistent with that used in some previous studies [19,20]. Those methods are applicable to wide range of problems e.g., for identification of macerals [28], defect detection [29]. Moreover, there exists a MaZda system [30], which implements some of the described texture feature maps and returns similar results when compared to our implementation. Systems for differentiating cysts, ameloblastomas, and keratocysts on DIR images are described in those reports. The general approach consists of image preprocessing (e.g., opening, contrast stretching), obtaining similarity measures, and texture analysis.

In the current study, a much broader set of texture features based on different approaches to image analysis was utilized. FOF feature maps yielded significant improvements in delineation of lytic changes in comparison with DIR. GTDM enhanced visualization of the internal structure within a lytic area and important details of adjacent trabeculation with preserved lesion contours. LBP yielded a surface scene with a clearly differentiated surface pattern at the site of the lytic area. CLU increased tissue contrast in areas of lytic changes. Importantly, (HEQ) RLM increased differentiation of the border and contrast in lytic lesions but did not perform so well for sclerotic changes. Performance of different texture feature maps expressed in terms of sensitivity, specificity, F1 score, and accuracy were summarized in the Table 1 (for the lytic lesions recognition) and the Table 2 (for sclerotic lesions recognition).

Table 1. Sensitivity, specificity, F1 score, and accuracy in the differentiation of the different diagnostic parameters of the lytic lesions after use of texture feature maps.

		Sensitivity	Specificity	F1	Accuracy
CLU	radiodensity	0.8286	0.7429	0.7945	0.7857
	border def	0.7714	0.9714	0.8571	0.8714
	tissue contr	0.8000	0.5143	0.7000	0.6571
FOF	radiodensity	0.6000	0.9429	0.7241	0.7714
	border def	0.6857	0.9143	0.7742	0.8000
	tissue contr	0.5143	0.6857	0.5625	0.6000
GTDM	radiodensity	0.3143	0.8857	0.4400	0.6000
	border def	0.3714	0.8571	0.4906	0.6143
	tissue contr	0.2857	0.8000	0.3846	0.5429
LBP	radiodensity	0.5143	0.6857	0.5625	0.6000
	border def	0.1714	0.8571	0.2609	0.5143
	tissue contr	0.3714	0.5143	0.4000	0.4429
HISTEQ-RLM	radiodensity	0.9429	0.8571	0.9041	0.9000
	border def	0.8857	0.8857	0.8857	0.8857
	tissue contr	0.9429	0.4286	0.7500	0.6857

Table 2. Sensitivity, specificity, F1 score, and accuracy in the differentiation of the different diagnostic parameters of the sclerotic lesions after use of texture feature maps.

		Sensitivity	Specificity	F1	Accuracy
CLU	radiodensity	0.6000	0.5333	0.5806	0.5667
	border def	0.6000	0.5333	0.5806	0.5667
	tissue contr	0.0333	1.0000	0.0645	0.5167
FOF	radiodensity	0.7333	0.6000	0.6875	0.6667
	border def	0.7000	0.8333	0.7500	0.7667
	tissue contr	0.5667	0.7333	0.6182	0.6500
GTDM	radiodensity	0.6333	0.6333	0.6333	0.6333
	border def	0.3667	0.9333	0.5116	0.6500
	tissue contr	0.5000	0.7667	0.5769	0.6333
LBP	radiodensity	0.7000	0.4667	0.6269	0.5833
	border def	0.4000	0.8333	0.5106	0.6167
	tissue contr	0.5333	0.6667	0.5714	0.6000
HISTEQ-RLM	radiodensity	0.9667	0.4667	0.7733	0.7167
	border def	0.8000	0.9000	0.8421	0.8500
	tissue contr	0.9667	0.5333	0.7945	0.7500

Few studies of texture analysis have evaluated periapical changes. Possible differentiation of lytic lesions for granulomas and periapical cysts on the basis of radiograms with use of radiometric analysis by histogram calculation and histogram equalization were proposed by Shrout and White, respectively [31,32]. In another study, after some classic image processing methods (top hat, erosion, and opening) were performed, the skeleton was extracted, and textural features were calculated for the region of interest. A pair of pre-treatment and post-treatment values was then tested in an evaluation of the healing process [33]. The scope of such studies has typically been the detection of areas of alveolar bone on periapical dental radiographs [12–15]. These considerations were focused on segmentation, detection of lesions, and measuring the degree of alveolar bone loss rather than a textural analysis in cases requiring differentiation of lesions. A similar approach has been reported for the analysis of panoramic images in order to enhance the recognition of caries [34]. Another study assessed the treatment effectiveness of guided bone regeneration in cases of post-resectal and post-cystal bone loss on DIR images obtained using RVG 6100 digital radiography equipment (Kodak) [16]. Fractal dimension measurements (power spectral density, triangular prism surface area, blanket, intensity difference scaling, and variogram methods) were performed, and the images became smoother during the healing process after bone loss [35].

Despite the progress to date in the quality of plain radiographic images, DIR still has a number of limitations. In addition to "anatomical noise" and small differences in bone density, there are other phenomena that impede image quality and hinder the anatomical recognition of potentially pathological structures [36,37]. Current technical developments in informatics hardware have made it possible to perform very complicated calculations in a relatively short time at low cost.

The approach used in the current study provides significantly more radiological information than standard DIR images. This comprehensive study is the first where usability of such a broad set of approaches to image texture analysis was presented. Based on our findings, from a radiological perspective, we consider these techniques a step forward in the recognition and precise localization of periapical cystic lesions. A limitation of this retrospective study is the lack of histological verification of the lesions analyzed; however, comparison of texture feature map analysis with histological results will be the issue of an upcoming study of our team. The strengths of this study include evaluation of the most popular textures known to date (mathematical transformations of image analyses applicable to DICOM-based high-resolution DIR images). Future studies should investigate the development of new image processing algorithms based on the current study, and correlations with histopathological specimens in order to evaluate their ability to predict different histologically depicted lesions on the basis of image texture.

5. Conclusions

The RLM texture feature map significantly improves recognition of lytic and sclerotic lesions, albeit with lower specificity for sclerotic lesions, in comparison to DIR images. CLU, in comparison to the DIR images, markedly increases visualization of lytic lesions with high sensitivity and specificity but is less able to detect the radiological features associated with sclerotic changes. FOF texture feature maps significantly improve detection of the radiological features of both sclerotic and lytic lesions, compared to DIR, with good sensitivity and specificity.

Author Contributions: Conceptualization, R.O., K.N. and A.P.; Data curation, R.O. and B.O.; Formal analysis, K.N.; Investigation, R.O., K.N. and A.P.; Methodology, K.N.; Project administration, A.P.; Resources, B.O.; Software, K.N.; Supervision, R.O.; Validation, R.O., B.O. and A.U.; Writing—original draft, R.O., K.N. and A.P.; Writing—review & editing, R.O., K.N., A.U. and A.P.

Funding: This publication was funded by AGH University of Science and Technology, Faculty of Electrical Engineering, Automatics, Computer Science and Biomedical Engineering.

Acknowledgments: The authors would like to thank B. Wawrzykowska, head of the Denta-Med Kraków Clinic.

Conflicts of Interest: The authors declare no conflict of interest.

References

1. Natkin, E.; Oswald, R.J.; Carnes, L.I. The relationship of lesion size to diagnosis, incidence, and treatment of periapical cysts and granulomas. *Oral Surg. Oral Med. Oral Pathol.* **1984**, *57*, 82–94. [PubMed]
2. Shrout, M.K.; Hall, J.M.; Hildebolt, C.E. Differentiation of periapical granulomas and radicular cysts by digital radiometric analysis. *Oral Surg. Oral Med. Oral Pathol.* **1993**, *76*, 356–361. [PubMed]
3. Lofthag-Hansen, S.; Huumonen, S.; Gröndahl, K.; Gröndahl, H.G. Limited cone-beam CT and intraoral radiography for the diagnosis of periapical pathology. *Oral Surg. Oral Med. Oral Pathol. Oral Radiol. Endod.* **2007**, *103*, 114–119. [PubMed]
4. Ramachandran Nair, P.N.; Pajarola, G.; Schroeder, H.E. Types and incidence of human periapical lesions obtained with extracted teeth. *Oral Surg. Oral Med. Oral Pathol. Oral Radiol. Endod.* **1996**, *81*, 93–102. [PubMed]
5. Patel, S.; Dawood, A.; Mannocci, F.; Wilson, R.; Pitt Ford, T. Detection of periapical bone defects in human jaws using cone beam computed tomography and intraoral radiography. *Int. Endod. J.* **2009**, *42*, 507–515.

6. Kolacinski, M.; Kozakiewicz, M.; Materka, A. Textural entropy as a potential feature for quantitative assessment of jaw bone healing process. *Arch. Med. Sci.* **2015**, *11*, 78–84. [PubMed]
7. Campello, A.F.; Gonçalves, L.S.; Guedes, F.R.; Marques, F.V. Cone-beam computed tomography versus digital periapical radiography in the detection of artificially created periapical lesions: A pilot study of the diagnostic accuracy of endodontists using both techniques. *Imaging Sci. Dent.* **2017**, *47*, 25–31.
8. De Paula-Silva, F.W.; Wu, M.K.; Leonardo, M.R.; Da Silva, L.A.; Wesselink, P.R. Accuracy of periapical radiography and cone-beam computed tomography scans in diagnosing apical periodontitis using histopathological findings as a gold standard. *J. Endod.* **2009**, *35*, 1009–1012.
9. Kavitha, M.S.; An, S.Y.; An, C.H.; Huh, K.H.; Yi, W.J.; Heo, M.S.; Lee, S.S.; Choi, S.C. Texture analysis of mandibular cortical bone on digital dental panoramic radiographs for the diagnosis of osteoporosis in Korean women. *Oral Surg. Oral Med. Oral Pathol. Oral Radiol.* **2015**, *119*, 346–356.
10. Harrar, K.; Jennane, R. Quantification of trabecular bone porosity on X-ray images. In Proceedings of the 4th International Conference on Industrial and Intelligent Information (ICIII 2015), Roma, Italy, 18–19 May 2015.
11. Roberts, M.G.; Graham, J.; Devlin, H. Image texture in dental panoramic radiographs as a potential biomarker of osteoporosis. *IEEE Trans. Biomed. Eng.* **2013**, *60*, 2384–2392.
12. Lin, P.L.; Huang, P.Y.; Huang, P.W.; Hsu, H.C.; Chen, P. Alveolar bone-loss area localization in periapical radiographs by texture analysis based on fBm model and GLC matrix. In Proceedings of the IEEE 2014 International Symposium on Bioelectronics and Bioinformatics, Chung Li, Taiwan, 11–14 April 2014.
13. Lin, P.L.; Huang, P.Y.; Huang, P.W.; Hsu, H.C.; Chen, C.C. Teeth segmentation of dental periapical radiographs based on local singularity analysis. *Comput. Methods Programs Biomed.* **2014**, *113*, 433–445.
14. Lin, P.L.; Huang, P.Y.; Huang, P.W. Automatic methods for alveolar bone loss degree measurement in periodontitis periapical radiographs. *Comput. Methods Programs Biomed.* **2017**, *148*, 1–11.
15. Huang, P.W.; Huang, P.Y.; Lin, P.L.; Hsu, H.C. Alveolar bone-loss area detection in periodontitis radiographs using hybrid of intensity and texture analyzed based on FBM model. In Proceedings of the International Conference on Machine Learning and Cybernetics, Lanzhou, China, 13–16 July 2014; Volume 2, pp. 487–492.
16. Borowska, M.; Szarmach, J.; Oczeretko, E. Fractal texture analysis of the healing process after bone loss. *Comput. Med. Imaging Graph.* **2015**, *46*, 191–196.
17. Borowska, M.; Bębas, E.; Szarmach, J.; Oczeretko, E. Multifractal characterization of healing process after bone loss. *Biomed. Signal. Process. Control* **2019**, *52*, 179–186.
18. Koca, H.; Ergun, S.; Guneri, P.; Boyacıoglu, H. Evaluation of trabecular bone healing by fractal analysis and digital subtraction radiography on digitized panoramic radiographs: A preliminary study. *Oral Radiol.* **2010**, *26*, 1–8.
19. Vijayakumari, B.; Ulaganathan, G.; Banumathi, A.; Banu, A.F.S.; Kayalvizhi, M. Dental cyst diagnosis using texture analysis. In Proceedings of the 2012 International Conference on Machine Vision and Image Processing, Taipei, Taiwan, 4–15 December 2012; pp. 117–120.
20. Banu, A.F.S.; Kayalvizhi, M.; Arumugam, B.; Gurunathan, U. Texture based classification of dental cysts. In Proceedings of the 2014 International Conference on Control, Instrumentation, Communication and Computational Technologies, Kanyakumari, India, 10–11 July 2014; pp. 1248–1253.
21. Haralick, R.M.; Shanmugam, K.; Dinstein, I. Textural features for image classification. *IEEE Trans. Syst. Man Cybern. Syst.* **1973**, *SMC-3*, 610–621.
22. Amadasun, M.; King, R. Textural features corresponding to textural properties. *IEEE Trans. Syst. Man Cybern. Syst.* **1998**, *19*, 1264–1274.
23. Galloway, M.M. Texture analysis using grey level run lengths. *Comput. Graph. Image Process.* **1975**, *4*, 172–179.
24. Ojala, T.; Pietikäinen, M.; Mäenpää. Grey scale and rotation invariant texture classification with local binary patterns. In *Computer Vision-ECCV 2000*; Lecture Notes in Computer Science; Springer: Berlin/Heidelberg, Germany, 2003; Volume 1842, pp. 404–420.
25. Ojala, T.; Pietikäinen, M.; Mäenpää. Multiresolution grey-scale and rotation invariant texture classification with local binary patterns. *IEEE Trans. Pattern Anal. Mach. Intell.* **2002**, *24*, 971–987.
26. Obuchowicz, R.; Nurzynska, K.; Obuchowicz, B.; Urbanik, A.; Piórkowski, A. Caries detection enhancement using texture feature maps of intraoral radiographs. *Oral Radiol.* **2019**. [CrossRef]
27. Strzelecki, M.; Kociołek, M.; Materka, A. On the influence of image features on texture classification. In *Information Technology in Biomedicine (ITIB 2018)*; Springer: Cham, Switzerland, 2018; Volume 762, pp. 15–26.

28. Skiba, M.; Mlynarczuk, M. Identification of macerals of the inertinite group using neural classifiers, based on selected textural features. *Arch. Min. Sci.* **2018**, *63*, 827–837.
29. Nurzynska, K.; Czardybon, M. Defect detection in textiles with co-occurrence matrix as a texture model description. In *International Workshop on Combinatorial Image Analysis*; Springer: Cham, Switzerland, 2018; pp. 216–226.
30. Szczypinski, P.M.; Strzelecki, M.; Materka, A.; Klepaczko, A. MaZda-A software package for image texture analysis. *Comput. Methods Programs Biomed.* **2009**, *94*, 66–76.
31. Shrout, M.K.; Hildebolt, C.E.; Vannier, M.W. Effects of region of interest (ROI) outline variations on gray-scale frequency distributions for alveolar bone. *Oral Surg. Oral Med. Oral Pathol.* **1993**, *75*, 638–644.
32. White, S.C.; Sapp, J.P.; Seto, B.G.; Mankovich, N.J. Absence of radiometric differentiation between periapical cysts and granulomas. *Oral Surg. Oral Med. Oral Pathol.* **1994**, *78*, 650–654.
33. Kim, D.; Jeong, H.; Kim, M.; Kim, C.; Lee, B.D. Multiscale image analysis for the quantitative evaluation of periapical lesion healings. In Proceedings of the 2010 3rd International Conference on Biomedical Engineering and Informatics, Yantai, China, 16–18 October 2010; Volume 1, pp. 424–427.
34. Veena, D.K.; Jatti, A.; Joshi, R.; Deepu, K.S. Characterization of dental pathologies using digital panoramic X-ray images based on texture analysis. In Proceedings of the 2017 39th Annual International Conference of the IEEE Engineering in Medicine and Biology Society (EMBC), Seogwipo, Korea, 11–15 July 2017; pp. 592–595.
35. Leite, A.F.; De Souza Figueiredo, P.T.; Caracas, H.; Sindeaux, R.; Guimaraes, A.T.B.; Lazarte, L.; De Melo, N.S. Systematic review with hierarchical clustering analysis for the fractal dimension in assessment of skeletal bone mineral density using dental radiographs. *Oral Radiol.* **2015**, *31*, 1–13.
36. Gröndahl, H.G.; Huumonen, S. Radiographic manifestations of periapical inflammatory lesions. *Endod. Topics* **2004**, *8*, 55–67.
37. Becconsall-Ryan, K.; Tong, D.; Love, R.M. Radiolucent inflammatory jaw lesions: A twenty-year analysis. *Int. Endod. J.* **2010**, *43*, 859–865.

© 2019 by the authors. Licensee MDPI, Basel, Switzerland. This article is an open access article distributed under the terms and conditions of the Creative Commons Attribution (CC BY) license (http://creativecommons.org/licenses/by/4.0/).

Article

Machine Vision System for Counting Small Metal Parts in Electro-Deposition Industry

Rocco Furferi *[ID], Lapo Governi[ID], Luca Puggelli, Michaela Servi and Yary Volpe[ID]

Department of Industrial Engineering, University of Florence, 50134 Firenze, Italy; lapo.governi@unifi.it (L.G.); luca.puggelli@unifi.it (L.P.); michaela.servi@unifi.it (M.S.); yary.volpe@unifi.it (Y.V.)
* Correspondence: rocco.furferi@unifi.it; Tel.: +39-055-2758741

Received: 20 May 2019; Accepted: 13 June 2019; Published: 13 June 2019

Featured Application: The present work has application in the field of galvanic coating for the fashion industry by proposing a method and a machine able to count the number of items attached to a galvanic frame.

Abstract: In the fashion field, the use of electroplated small metal parts such as studs, clips and buckles is widespread. The plate is often made of precious metal, such as gold or platinum. Due to the high cost of these materials, it is strategically relevant and of primary importance for manufacturers to avoid any waste by depositing only the strictly necessary amount of material. To this aim, companies need to be aware of the overall number of items to be electroplated so that it is possible to properly set the parameters driving the galvanic process. Accordingly, the present paper describes a simple, yet effective machine vision-based method able to automatically count small metal parts arranged on a galvanic frame. The devised method, which relies on the definition of a rear projection-based acquisition system and on the development of image processing-based routines, is able to properly count the number of items on the galvanic frame. The system is implemented on a counting machine, which is meant to be adopted in the galvanic industrial practice to properly define a suitable set or working parameters (such as the current, voltage, and deposition time) for the electroplating machine and, thereby, assure the desired plate thickness from one side and avoid material waste on the other.

Keywords: Machine vision; image analysis; item counting device; electro-deposition industry

1. Introduction

As widely recognized, electroplating (more precisely electrodeposition) is a chemical process that uses electric current to transfer metal from a cation to an electrode (i.e., the object to be treated), to form a coherent thin metal coating [1]. The amount of mass deposit derives from Faraday's laws on electrolysis [2] and directly depends on current intensity and time:

$$m = \frac{M \cdot I \cdot t}{Z \cdot F} \qquad (1)$$

where:

m = mass deposited on electrode [g];
M = molar mass of the material to be deposited [g/mol];
I = current intensity (A);
t = time (s);
Z = valence of material's ions;
F = Faraday's constant (96485.33 C mol^{-1}).

The process used by manufacturers working in the electrodeposition of fashion accessories consists of arranging the (usually) small parts to be plated on a frame by using hooks or, more often, metal wires, as shown in Figure 1.

Figure 1. Typical galvanic frame.

Therefore, electrodeposition simultaneously occurs on a number of parts. Since the material to be deposited on electrode (multiple items to be plated) is required to form a uniform thin layer, the overall mass is given by:

$$m = n \cdot s \cdot T \cdot \rho \qquad (2)$$

where: where:

n = number of items to be electroplated;
s = surface of a single item;
T = coating thickness;
ρ = material mass density.

Hence, in order to obtain the desired coating thickness, both the number and surface of the items to be plated need to be known. While the item's surface is retrievable by means of possibly available items, Computer Aided Design (CAD) models, or by using 3D scanning, the number of items arranged on the galvanic frame is not straightforwardly available in order to compute the overall surface to be electroplated.

To date, the parts attached to the frame are manually counted, however, the reliability of the process is limited by ensuing weakness and inattentiveness; in other words, it is inevitably prone to errors due to the operators' tiredness and lack of attention, etc.

In scientific literature, several papers specifically address the topic of designing counting systems with reference to a variety of industrial fields [3,4]. In addition, many counting machines have been available on the market for years [5,6]. Unfortunately, regardless of the technology adopted (e.g., weight measurement, free-fall, optical scan lines), almost all the machines available on the market require items to be physically separated one from each other (i.e., not disposed on package or frames), or to be arranged upon a moving tray. Therefore, such solutions are not suitable or adaptable to count items that are already arranged on a galvanic frame. Fortunately, machine vision (MV) systems have the potential to solve this issue by implementing a combination of optical devices and proper image processing algorithms, finalized to determine the overall number of objects captured in a scene. Not by chance, a relevant number of MV systems have been proposed in the scientific literature to address the object counting issue [7]. However, the main issue for devising a system for counting metal parts attached to a galvanic frame is related to the high reflectivity of the items themselves, which presents an incredible challenge for any kind of optical acquisition system. For this reason, to the best of the authors' knowledge, no automatic counting system has been devised so far for the galvanic industry. Accordingly, the present paper proposes a machine vision-based method to automatically count small metal parts arranged on a galvanic frame. The devised method, relying on the definition of a proper acquisition system and on the development of image processing-based routines, is implemented on a counting machine to be adopted in the galvanic industrial practice. The machine architecture is designed to discard the undesired reflections due to the metal surface so to properly detach all attached items from the background. This allows a set of simple, yet effective image processing algorithms to correctly determine the number of items to be coated in the galvanic bath. Finally, the knowledge of the number of items will allow companies to define a suitable set or working parameters (such as the current, voltage and deposition time) for the electroplating machine and, thereby, assure the desired plate thickness from one side and avoid material waste on the other.

2. Materials and Methods

As shown in Figure 1, the galvanic frame is formed by 4 tubular beams welded to compose a rectangle. In the general configuration, on the shorter sides, several hooks are joined. The workers use inert metal wires to knot together a variable number of items. Successively, each wire is linked to a couple of corresponding hooks (i.e., the nth on the upper side with the nth on the lower) so that the wire results in an arrangement on the frame along an approximately vertical direction. Once all the couples of hooks are filled, the galvanic frame is sent to the electroplating bath. Considering that items can be very small in size (down to 10 mm on the shorter side), the variability in length (thus in mass) of the wire itself precludes the adoption of any weight-based approach for the counting system. Moreover, two consecutive items can be attached at a relative distance down to 20 mm.

Consequently, attention has been focused on computer vision-based approaches. The main idea is to properly acquire a 2D digital image of the frame, on which to detect each item by means of computer vision (CV) tools [7].

2.1. Literature Methods

According to scientific literature, several different CV approaches can be adopted. Considering the task, three among them seem to deserve further investigation: Deterministic template-matching, neural network-based algorithms or brightness-based segmentation. The applicability, effectiveness, and robustness of each of them strictly depend on the typology of the image to be analyzed. It has to be noted that other approaches, such as color-based ones, are not applicable since the wire color can be very close to that of the items.

In more detail, deterministic template-matching algorithms are intended to find, into an image, instances of a given template. For example, OCR (optical character recognition) procedures—which recognize text within pictures (e.g., a PDF file)—are usually built based on template matching algorithms. This approach would be optimal to solve our problem if only the items were arranged on a rigid grid, so that each item was oriented in the same way with respect to the camera. In fact, some companies use galvanic frames, where items are placed into a fixed position on the frame, as shown in Figure 2.

Figure 2. Galvanic frame with items placed in a fixed position.

However for such cases, the operators usually fill all the available slots with items, therefore, the number of items on the galvanic frame is known a priori. However, as already mentioned, in the general configuration described previously, the items are knotted on wires and, consequently, their orientation in space is far from being equal. This issue inevitably limits the applicability of deterministic template-based algorithms and makes their adoption inconvenient for the specific case analyzed in this paper.

With respect to this limitation, an evolution of the template-based algorithm, as defined above, can be found in the neural network (NN)-based approaches [8–11]. Some of them, in fact, are able to detect a specific object independently from its orientation and position in the scene. Among them, YOLO (you only look once) [12] is a state-of-the-art real-time object detection system, targeted for real-time processing. Object detection is a computer technology related to computer vision and image processing that deals with detecting instances of semantic objects of a certain class (humans, cars, etc.) in digital images and videos.

Differently from prior approaches, which apply the model to an image at multiple locations and scales and then high scoring regions of the image are considered detections, YOLO applies the network to the full image. Specifically, the image is divided into an S x S grid and the algorithm returns bounding boxes and predicted probabilities for each of these regions. The method used to compute these probabilities is logistic regression [13]. This way, other than performing a very fast detection, predictions are informed by the global context in the image.

Off-the-shelf YOLO nets with pre-trained weights, but is not able to give predictions on our subject of interest, as they have not been trained in detecting these particular objects (see Figure 3).

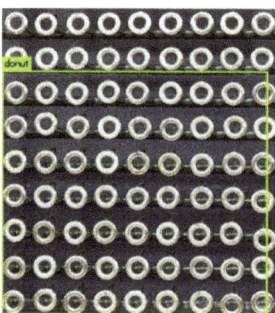

Figure 3. Prediction result obtained with an off-the-shelf pre-trained YOLO (you only look once) net.

On the other hand, to achieve a proper result using these networks, it is not sufficient to provide a limited number of training images (e.g., ten to twenty images). In the light of these considerations, and given that the shape of the items can change frequently (each month, all lots can be completely new), it is not practical for users to train the algorithm each time.

For this reason, another approach has been explored: The classical brightness-based segmentation [14–16]. Assuming that the item brightness (or range of brightness) is different from that of the background, it is possible to isolate the background itself. The resulting image contains only pixels belonging to the items and connecting wires. Unfortunately, wire and item colors can be so close that color segmentation cannot be used to separate the one from the other. Fortunately, since wires are thinner than items, their pixels can be removed by means of CV tools such as pixel erosion/dilation. The number of separate clusters of pixels describing the items can then be easily retrieved by means of labeling tools.

In detail, a threshold value (or at least a range) must be used in order to isolate on the image only pixels relative to the items to be counted. Supposing that the threshold operation works flawlessly, a binary image can be obtained, where white pixels represent items to be counted and black pixels are the background and the wires. Afterward, many well-known algorithms can be used to count the number of isolated regions in binary images.

2.2. Image Acquisition Requirements

The brightness-based segmentation approach seems the most promising for the specific application but needs to be tailored to the peculiarities entailed by the small dimensions and high reflectivity of the items to be counted. Consequently, the definition of a proper input image has primary importance for the success of the method.

Depending on the finishing and on the material of which the items are made, their aspect is rarely opaque but rather it is highly reflexive. Obviously, even color may change, varying from copper-like to silver and gold, thus resulting in different brightness. All these characteristics make it very difficult (or even impossible) to obtain satisfactory threshold values or ranges, on which set the segmentation.

To make it more complicated, the silhouette of the same items knotted to the galvanic frame varies significantly, due to their almost-random orientation. Moreover, the placement is far from being equally spaced.

Therefore, in order to make the segmentation algorithms effective, it is of critical importance to obtain a suitable image, where it is possible to separate the items from the background. To this purpose, three different lighting settings have been considered in order to evaluate their efficacy in favoring image segmentation operations:

1. Frontal lighting with a black uniform background;
2. Lighted white background;
3. Light rear-projection.

2.3. Light Settings

In industrial practice, a single galvanic frame is filled with a number of identical items. In order to make the performed tests representative and speed-up the testing process, we chose to use typical galvanic frames filled with a variety of items of different shapes and dimensions arranged on vertical metal wires, instead of using multiple frames (each one with a single item typology). Image acquisition was carried out by using a Fujifilm T1 SLR camera (APS-C sensor format, Fujifilm Holdings Corporation, Tokio, Japan) and an 18 mm focal length lens. The acquired images had a resolution of 15.8 megapixels (4826 × 3264 pixels).

2.3.1. Frontal Light with Black Uniform Background

The first tested layout setting is meant to physically isolate items from the rest of the scene by putting an opaque black canvas behind the frame. The frame, containing four different item typologies, is positioned approximately perpendicular to the camera optical axis at a distance of 500 mm, so that the frame occupies completely the field of view. The set (see Figure 4) is illuminated by a frontal lighting source (800 lm focusable LED torch, Essilor International S.A., Paris, France). This layout allows obtaining an almost-uniform black background on which the item shape appears enhanced.

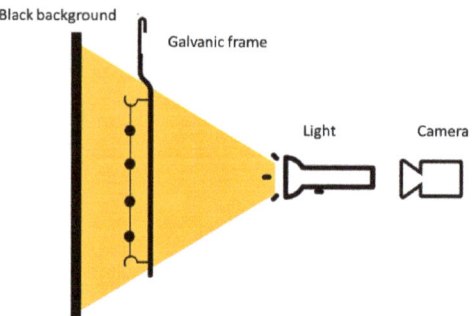

Figure 4. Frontal light with black uniform background setup.

In the resulting digital image, background pixels are characterized by low brightness. On the contrary, the items pixel brightness has, as expected, high values due to the frontal and strong illumination (see for instance Figure 5a, referred to four different kind of items). However, this layout is not optimal for a number of reasons. First, the background subtraction (i.e., to subtract from the image to be analyzed a reference image of the background canvas acquired prior to positioning the galvanic frame) is not applicable, since shadows/reflections projected on the canvas by the items and the wires make the background itself different from the reference.

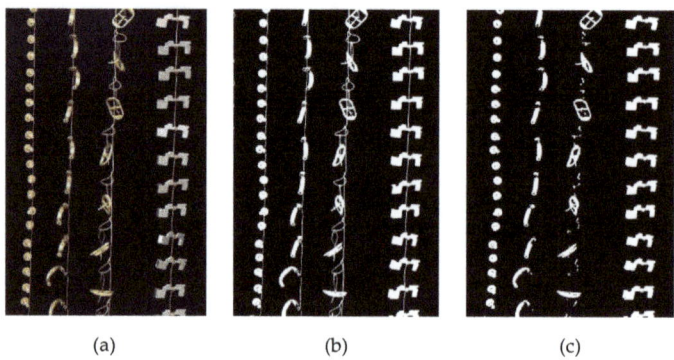

Figure 5. (**a**) Acquired image using the first setup: (**b**) Image after thresholding and (**c**) image after morphological opening.

In addition, brightness-based segmentation leads to two additional main issues, as explained below:

- Items and background pixels may be incorrectly detected/assigned;
- The wire and item brightness are similar and thus difficult to separate.

Starting from Figure 5a, it was not possible to isolate the items from the wires by thresholding, as shown in Figure 5b, since the resulting binary image also contained also the wires. The only filtering operation that could allow wires deletion is image erosion. Unfortunately, since the item dimensions and wire thicknesses are similar, the operation (even if combined with successive dilation filtering, thus performing a morphological opening) led to sub-fragmentation of single items into multiple pixel clusters (see Figure 5c), thus invalidating the successive counting operation.

Observing in detail the image of two items (named item A and B, respectively) right after thresholding (Figure 6a,c), the first issue became evident. It was noted that, for some items, the darkest pixels were mistakenly assigned to background. Consequently, some of them already fragmented into multiple parts (see Figure 6b,d). In other cases, the effects were less evident but equally dangerous, due to the successive (required) erosion operation. As shown in Figure 6c, the items may result so thinned that successive operations unavoidably cause fragmentation.

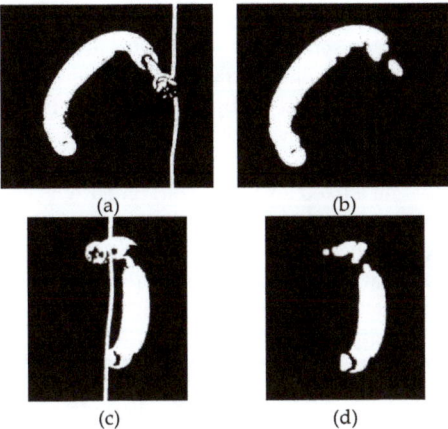

Figure 6. (a) Item A, fragmented after thresholding; (b) item A, after morphological opening; (c) item B, thinned after thresholding; (d) item B, fragmented after morphological opening.

In more detail, sub-fragmentation can occur in two possible scenarios: In the presence of bridges (Figure 6a) or in the case of inner holes (Figure 6c). In the first case, the thickness of the bridge may be similar to the thickness of the wires to be removed. Consequently, a common occurrence was that the morphological opening on the binary image (i.e., the erosion followed by dilation) removed both the wires and bridges, thus causing undesired fragmentation of the cluster (Figure 6d). Similarly, in the case of the inner holes—given by the actual shape of the item or caused by thresholding—morphological opening may cause fragmentation.

This issue can be possibly avoided by using a morphological image closure (i.e., the dilation followed by erosion) followed by an additional erosion. Figure 7a demonstrates the result of such an operation applied to Figure 6c.

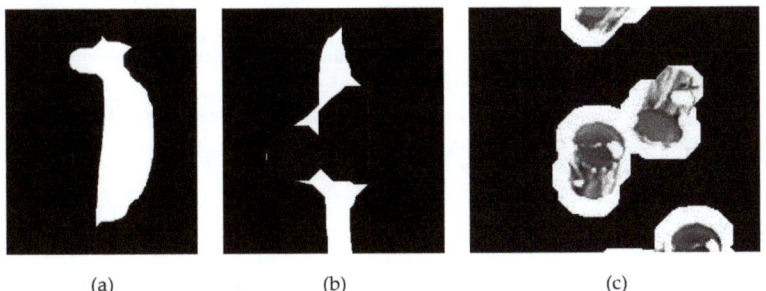

Figure 7. (a) Item B after morphological image closing plus image erosion; (b) an example of wires loop after morphological closing followed by image erosion; (c) clusters merged after dilation. The image is intentionally left in grayscale to show actual original items.

However, this solution may also lead to some unwanted side effects that make this alternative unsuitable. In fact, in Figure 7b, it can be noted that, in some cases, the wires formed closed loops in the image. Such loops may result in being completely closed by a morphological closing filtering. If sufficiently large, they can be easily mistaken for items in the counting phase. In addition, if a couple of items are sufficiently close, the filter may cause the fusion of the relative clusters into one (see Figure 7c).

Between the two alternatives proposed above, the better performance proved to be the first one (i.e., the morphological opening-based solution). Starting from the morphologically opened image (see Figure 5c), the connected regions representing the actual items needed to be discriminated from the ones representing small wire portions and/or item fragments. To this aim, an elective method could be to perform area-based discrimination, carried out by imposing an appropriate area threshold.

Since the item dimensions are widely variable and unknown a priori, a fixed area threshold value cannot be based on the item dimension itself. On the contrary, the wire dimension is constant. Accordingly, it is possible to define a fixed area threshold under which clusters are considered too small to be an item, thus must be ignored. Considering that the wire thickness is approximately 1 mm—corresponding to 7 pixels in the image (based on the shooting setup described in the previous section), a limit dimension was set at 4 mm^2—corresponding to 200 pixels. In the example shown in Figure 8a, this method allowed appropriate discarding of small clusters.

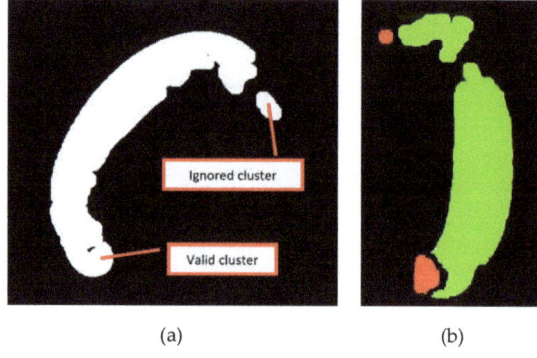

(a)   (b)

Figure 8. (**a**) Discrimination among valid and ignored clusters; (**b**) misclassification of small clusters: Red-colored clusters are discarded since their area is lower than the selected threshold (i.e., 200 pixels); green-colored clusters are counted, thus leading to counting error since both belongs to a single item.

However, in several other situations, such as the one depicted in Figure 8b, this criterion led to misclassification. This was mainly due to the heavy image cluster fragmentation induced by the acquisition setup and the subsequent image filtering. Other than the simple criterion described above, other more complex techniques have been tested in order to cluster pixel regions, namely k-means clustering and Support Vector Machine (SVM) [17,18]. The results, not detailed in the present paper, show that this misclassification still occurs.

2.3.2. Lighted Background

Moving from the issues faced with the first setting, the second layout makes use of backlighting. The galvanic frame, containing a set of identical items, is arranged between the camera and an approximately uniformly illuminated white background (see Figure 9).

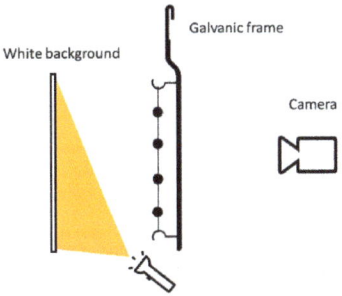

Figure 9. Lighted background setup.

Under the proper camera settings, the lights saturate the brightness for the background, while the pixels belonging to the items generally appear darker (Figure 10a).

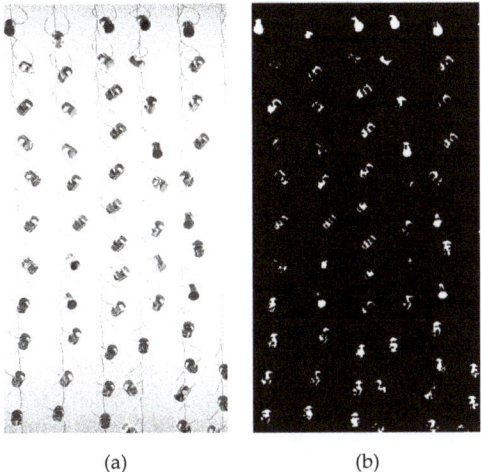

(a)   (b)

Figure 10. (**a**) Original image acquired with backlighting; (**b**) image of Figure 10a after thresholding showing that some items portions are mistakenly assigned to the background.

Awkwardly, many item regions appeared bright due to specular reflections/inter-reflections among the items themselves. Similarly to the configuration described in the previous section, over-fragmentation issues arose. In fact, despite the setup, the entire background was better detected and isolated. Some item portions, which appeared light due to the inter-reflections mentioned above, were mistakenly assigned to the background (see Figure 10b). Even using morphological operators similar to the ones described in Section 2.3.1, the fragmentation issue persisted, making it practically unfeasible to correctly classify the pixel clusters.

2.3.3. Rear Projection

To overcome all the discussed drawbacks related to direct backlighting, a third solution was developed and tested. In detail, a 0.5 mm thickness white canvas for rear-projection (100% polyvinyl chloride - PVC) was placed at a 20 mm distance from the galvanic frame, containing seven item typologies while the light source (in this case, an overhead projector with a 3300 lumen light source) and camera were arranged as depicted in Figure 11. This architecture allowed acquiring, from the scene, the projected item shadows rather than the items themselves. This enabled discarding of any kind of reflection. The light source came from an LCD overhead projector with 1000 lumens.

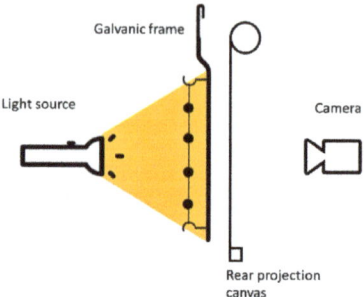

Figure 11. Rear projection setup.

Experimental tests showed that, due to the item thickness and shape, the minimum distance between the projector and the frame needed to be set at 1.8 m, in order to avoid shadow blurring. As depicted in Figure 12a, the items cast a very sharp and uniform dark shadow on the canvas. At the same time, the wires appeared thinner than (for instance) the ones shown in Figure 5b.

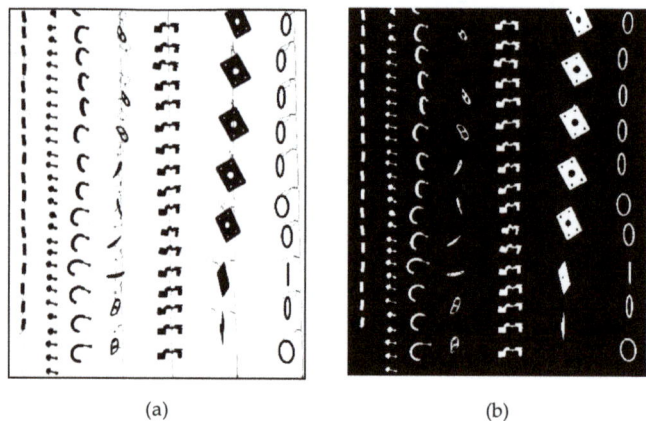

Figure 12. (a) Original image acquired with the rear projection setup; (b) binary image after morphological opening process.

Starting from the acquired image (see Figure 12a) a binary image was obtained by thresholding, using the Otsu method [18]. Subsequently, the resulting image was filtered using a 3 × 3 morphological image opening. As shown in Figure 12b, this approach led to minimally fragmented pixel clusters. Therefore, by using an area threshold equal to 200 pixels, it was possible to correctly count the item number.

In summary, the rear projection setup proved to be the most suitable among the ones tested in order to correctly isolate items to be counted. In fact, the key point of the procedure resided in the very sharp native image, in which the shadows were extremely defined. Consequently, the required filtering operations were far less aggressive than needed in previous cases.

For this reason, this method has been selected to design the counting machine, as described in the next Session.

3. Rear Projection-Based Counting Machine Prototype

Though the preliminary tests performed using the seven different item typologies, shown in Figure 12a, were deemed representative, a prototypal rear projection-based counting machine was designed in order to perform extensive testing in an industrial environment. As shown in Figure 13, the system comprises:

- An image acquisition device (industrial monochrome camera IDS UI 3200-SE-M with a 6 mm lens with 12-megapixel resolution (4104 × 3006 pixels);
- An LCD overhead light projector, to assure uniform lighting;
- A couple of orientable mirrors (used to extend the light path up to the 1.8 m, mentioned in Section 3). Such mirrors are used to reduce the overall dimensions of the counting machine, which must not exceed 1.5 × 1.0 × 1.0 m, in order to not to be excessively cumbersome for an industrial environment;
- An enclosure system, to assure the environmental light does not affect the scene.

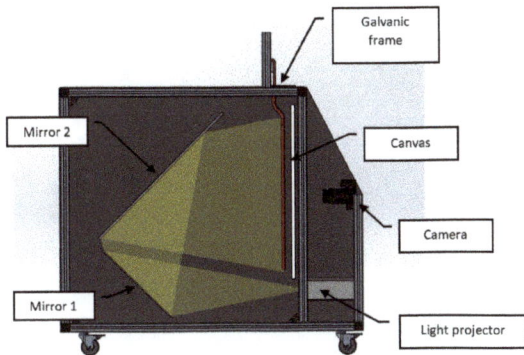

Figure 13. Counting system main components.

The projector is placed backward, on the frontal part of the machine. Light is reflected by the first mirror upwards towards the second one; this last mirror reflects it forward to hit the galvanic frame. Its shadow is projected on the rear-projection canvas, which is arranged parallel to the frame. In the prototypal implementation of the system, the setup described above did not show any unevenness in the screen illumination using a perfectly white projected image. However, in case this should happen, it is possible to compensate by projecting an image appositely designed in order to feature slightly darker or lighter regions in correspondence to more or less illuminated areas of the screen, respectively. This is a major advantage entailed by the use of the overhead LCD projector rather than a conventional light source (i.e., a lamp).

In Figure 14a, a rendering of the designed counting machine architecture shows the arrangement of the above-mentioned components. The final design of the machine is shown in Figure 14b.

A Surface Go tablet (Microsoft Corporation, Washington, U.S.)—on which ran the designed application (developed in Matlab®, MathWorks, Inc., Natick, Massachusetts, U.S., 2019)—was then used to command the industrial camera. By means of a dedicated Graphical User Interface (GUI), the operator could check the position of the frame and can start the acquisition when such a position was considered correct (see Figure 15).

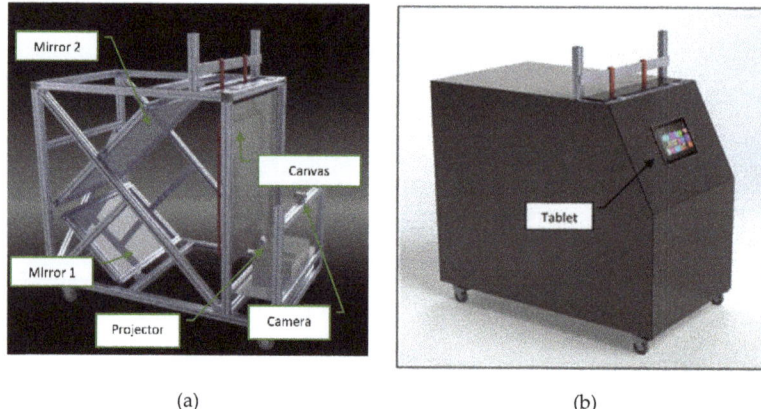

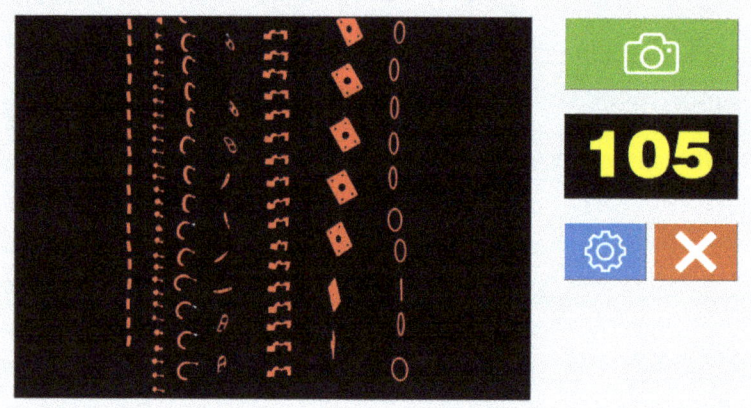

Figure 14. (a) Counting machine architecture; (b) final design of the counting machine.

Figure 15. Dedicated GUI implemented for controlling the counting machine performance. On the left, it is possible to read the number of items. Moreover, it is possible to save the screened image (green camera icon) and to close the application (red X in figure). The user can also access a setting panel (blue gear icon) in case they want to set a different threshold value for the algorithm.

The procedure was then able, in approximately 0.3s, to provide the number of the detected items. Simultaneously, it showed a control picture, on which clusters that had been considered were colored red, while those that were ignored were white. In this way, the operator could rapidly check the effectiveness of the procedure and make corrections if needed.

4. Discussion and Conclusions

In this paper, a method and a machine for counting the number of small metal parts randomly arranged on a galvanic frame was proposed. A priori knowledge of the area of each item that will be treated by the galvanic bath made it possible to estimate, with satisfying accuracy, the overall area to be treated and, consequently, optimize the settings of the treatment itself. Especially in the high fashion field, in which precious materials are often used to realize plates, this enables minimization of material waste, thus leading to a significant cost saving. Considering all the limitations that the application imposes (e.g., pieces already mounted on the frame and high reflectivity), many of the approaches usually adopted for counting machines (e.g., free fall and weight analysis) cannot be followed. The procedure is hence based on machine vision and makes use of rear-projection on a canvas to obtain a sufficiently sharp and easy to elaborate image with simple morphological operators. A counting machine, which implements the devised system was designed.

This prototypal counting machine was pre-tested with 20 different galvanic frames, hosting 20 different kinds of objects, with maximum dimensions spanning from 10 to 80 mm. In Table 1, the results obtained for 5 of the 20 tests are listed.

Table 1. The results obtained with the three developed systems for counting the items attached to a galvanic frame.

Test #	Actual # of Items	Minimum Item Dimension (mm)	Frontal Light* (# of items)	Lighted Background (# of items)	Rear Projection (# of items)
1	60	100	62	61	60
2	120	40	124	124	120
3	40	200	40	40	40
4	100	30	104	105	100
5	20	300	20	20	20

* Without image closure

Referring to the entire set of 20 tests, the frontal light architecture showed proper counting of the number of the items in nine cases (45%), while the lighted background-based architecture was successful in eight cases (40%). For both the systems, over-fragmentation led to an excessive number of items counted. This was particularly true when the number of items increased and the minimum dimensions decreased down to 50 mm. Therefore, their use is not recommended for this kind of application, since an overestimation of the number of attached items may cause a coating thickness lower than the desired one. By using the frontal light-based architecture with the addition of the image morphological closure algorithm, the percentage of correctly counted items increased to 60%. However in two cases out of 12 correctly counted frames, the number of items erroneously counted twice, due to clusters fragmentation, were compensated by the erroneous counting of two adjacent items merged together by the image closure. Quite the reverse, for all the test cases, the number of counted objects was exactly equal to the number of actual objects mounted on the frames (100%). As already mentioned, since the surface of each item is a technical specification for the galvanic companies, by multiplying it by the number of items, it is possible to know the overall surface to be coated.

Despite these encouraging results, the system will undergo an extensive test campaign in an Italian company working in the galvanic coating industry to increase the number of test cases up to 1000 different frames.

Accordingly, future work will extensively test the devised procedure, both in terms of the performance (i.e., the counted number of items vs. the actual number, verified by visually inspecting the frames) and of the usability.

Author Contributions: Conceptualization, L.P, L.G. and R.F; methodology, L.G. and R.F.; software, M.S. and L.P.; validation, Y.V., L.P. and M.S.; data curation, M.S. and Y.V.; writing—original draft preparation, L.P.; writing—review and editing, L.G., R.F and Y.V.; supervision, L.G. and R.F.; project administration, L.G.

Funding: This work has been carried out thanks to the decisive regional contribution from the Regional Implementation Programme co-financed by the FAS (now FSC) and the contribution from the FAR funds made available by the MIUR.

Conflicts of Interest: The authors declare no conflict of interest.

References

1. Bard, A.J.; Faulkner, L.R. *Electrochemical Methods: Fundamentals and Applications*; Wiley: Hoboken, NJ, USA, 1980; p. 718.
2. Barker, D.; Walsh, F.C. Applications of Faraday's Laws of Electrolysis in Metal Finishing. *Trans. IMF* **1991**, *69*, 158–162. [CrossRef]
3. Phromlikhit, C.; Cheevasuvit, F.; Yimman, S. Tablet counting machine base on image processing. In Proceedings of the 5th 2012 Biomedical Engineering International Conference, Ubon Ratchathani, Thailand, 5–7 December 2012.
4. Nudol, C. Automatic jewel counting using template matching. In Proceedings of the IEEE International Symposium on Communications and Information Technology 2004 (ISCIT 2004), Sapporo, Japan, 26–29 October 2004; Volume 2.
5. Sokkarie, A.; Osborne, J. Object counting and sizing. In Proceedings of the SoutheastCon'94, Miami, FL, USA, USA, 10–13 April 1994.
6. Courshee, R.J. Testing a counting machine. *Br. J. Appl. Phys.* **1954**, *5*. [CrossRef]
7. Barbedo, J.G.A. A Review on Methods for Automatic Counting of Objects in Digital Images. *IEEE Latin Am. Trans.* **2012**, *10*, 5.
8. Chauhan, V.; Joshi, K.D.; Surgenor, B. Machine Vision for Coin Recognition with ANNs: Effect of Training and Testing Parameters. In Proceedings of the Engineering and Applications of Neural Networks: 18th International Conference, Athens Greece, 25–27 August 2017.
9. Bremananth, R.; Balaji, B.; Sankari, M.; Chitra, A. A New Approach to Coin Recognition using Neural Pattern Analysis. In Proceedings of the 2005 Annual IEEE India Conference-Indicon, Chennai, India, 11–13 December 2005.
10. Furferi, R.; Governi, L. Prediction of the spectrophotometric response of a carded fiber composed by different kinds of coloured raw materials: An artificial neural network-based approach. *Color Res. Appl.* **2011**, *36*, 179–191. [CrossRef]
11. Sharma, T.; Rajurkar, S.D.; Molangur, N.; Verma, N.K.; Salour, A. Multi-faced Object Recognition in an Image for Inventory Counting. *Adv. Intell. Syst. Comput.* **2019**, *799*, 333–346.
12. Redmon, J.; Divvala, S.; Girshick, R.; Farhadi, A. You Only Look Once: Unified, Real-Time Object Detection. In Proceedings of the 2016 IEEE Conference on Computer Vision and Pattern Recognition (CVPR), Las Vegas, NV, USA, 27–30 June 2016.
13. Peduzzi, P.; Concato, J.; Kemper, E.; Holford, T.R.; Feinstein, A.R. A simulation study of the number of events per variable in logistic regression analysis. *J. Clin. Epidemiol.* **1996**, *49*, 1373–1379. [CrossRef]
14. van den Boomgaard, R.; van Balen, R. Methods for fast morphological image transforms using bitmapped binary images. *CVGIP: Graph. Models Image Process.* **1992**, *54*, 252–258. [CrossRef]
15. Gonzalez, R.C.; Richard, R.E.; Woods, E.; Eddins, S.L. *Digital Image Processing Using MATLAB*; Dorsing Kindersley: London, UK, 2004; p. 620.
16. Haralick, R.M.; Shapiro, L.G. *Computer and Robot Vision*; Addison-Wesley Pub Co.: Boston, MA, USA, 1992.
17. Wang, J.; Wu, X.; Zhang, C. Support vector machines based on K-means clustering for real-time business intelligence systems. *Int. J. Bus. Intell. Data Min.* **2005**, *1*, 54–64. [CrossRef]
18. Furferi, R.; Governi, L.; Volpe, Y. Modelling and simulation of an innovative fabric coating process using artificial neural networks. *Text. Res. J.* **2012**, *82*, 1282–1294. [CrossRef]

© 2019 by the authors. Licensee MDPI, Basel, Switzerland. This article is an open access article distributed under the terms and conditions of the Creative Commons Attribution (CC BY) license (http://creativecommons.org/licenses/by/4.0/).

applied sciences

MDPI

Article

An Improved MB-LBP Defect Recognition Approach for the Surface of Steel Plates

Yang Liu, Ke Xu * and Jinwu Xu

Collaborative Innovation Center of Steel Technology, University of Science and Technology, Beijing 100083, China; liuyang_ustb_1988@163.com (Y.L.); jwxu@ustb.edu.cn (J.X.)
* Correspondence: xuke@ustb.edu.cn; Tel.: +86-10-6233-2159

Received: 27 August 2019; Accepted: 30 September 2019; Published: 10 October 2019

Abstract: The detection of surface defects is very important for the quality improvement of steel plates. In actual production, as the steel plate production line runs faster, the steel surface defect detection algorithm is required to meet the requirements of real-time detection (less than 100 ms/image), and the detection accuracy is improved (at least 90%). In this paper, an improved multi-block local binary pattern (LBP) algorithm is proposed. This algorithm not only has the simplicity and efficiency of the LBP algorithm, but also finds a suitable scale to describe the defect features by changing the block sizes, thus ensuring high recognition accuracy. The experiment proves that the method satisfies the requirements of online real-time detection in terms of speed (63 ms/image), and surpasses the widely-used scale invariant feature transform (SIFT), speeded up robust features (SURF), gray-level co-occurrence matrix (GLCM), and LBP algorithms in recognition accuracy (94.30%), which prove that the MB-LBP has practical application value in an online real-time detection system.

Keywords: MB-LBP; surface defect detection; feature extraction; defect recognition

1. Introduction

Steel plates are widely used in engineering fields such as ships, bridges, machinery, construction, and automobile manufacturing. In the production process of a plate, due to the rolling process, various types of defects are easily formed on the surface of steel plates such as cracks, scratches, indentations, pits and scales [1]. These defects have a great impact on the appearance and performance of the product, so it is extremely important to detect the surface defects on the plate.

The surface defect recognition algorithm for steel plates is the core part of the entire surface defect detection system. Yun et al. [2] proposed a new defect detection algorithm, which is based on Gabor filters. The Gabor filters are optimized using a new optimization algorithm known as the univariate dynamic encoding algorithm for searches. The algorithm finds the minimum value of the cost function related to the energy separation criteria between the defect and the defect-free regions. Xu et al. [3] proposed a classifier based on Ads Boosting algorithms for the classification of defects with textural features that employed non-sampling wavelet decomposition to the scale co-occurrence matrix of the low-pass component and the grayscale co-occurrence matrix of the high-pass component. Pan et al. [4] proposed an engineering-driven rule-based detection (ERD) method. The ERD consists of three detection stages using the pixel features of bleeds, which are transferred from the physical features generated via engineering knowledge. Yu et al. [5] proposed a surface trait extraction method based on complex Contourlet decomposition, which has the characteristics of shift invariance, excellent directional selectivity, and a higher retrieval rate. Industrial test results show that the accurate rate of classifying surface features is about 90%, and it can be used in image feature extraction and slab defect detection. Miyamoto et al. [6] proposed a one-shot measurement using the plane-wave with the time-of-flight (TOF) based transmission method and validated the method using wave propagation simulations. The defect in billets can be detected regardless of the defect position even in the vicinity of the surfaces of a billet. Yan et al. [7] proposed a mathematical morphology detection method based on multi-scale element, which can not only filter the noise effectively, but can also delete the false characteristics of the cracks, scales, and slag. Lin et al. [8] proposed a robust detection method based on the vision attention mechanism and deep learning of feature map in order to relieve the problem of the false and missed detection of casting defects in x-ray detection. The experimental results showed that the false rate and missed rate for the detection of casting defects were less than 4%, and the accuracy of the defect detection was more than 96%. Di et al. [9] proposed a new semi-supervised learning method based on convolutional auto encoder (CAE) and semi-supervised generative adversarial networks (SGAN) to classify the surface defects of steels. Jiang et al. [10] suggested a method for detecting the appearance defect of castings based on a deep residual network. This method divides the casting into multiple regions, preprocesses the image of each region, and then inputs the processed image into the convolutional neural network to extract the features, before finally determining whether the sample has defects.

All of the above methods effectively solve the problem of recognition accuracy. However, in actual production, the steel production line runs very fast, so the surface defect detection algorithm needs to meet the real-time requirements while ensuring the accuracy.

In this paper, we propose the multi-block local binary pattern (MB-LBP) algorithm to extract the features of the surface defects of steel plates. The experimental results show that the MB-LBP algorithm meets the requirements of the online detection of steel plate defects in terms of both speed and accuracy.

The rest of this paper is organized as follows. Section 2 introduces the surface defects of steel plates. Section 3 introduces the principle of MB-LBP. Section 4 introduces the experiments and analysis of surface defect detection based on MB-LBP. Finally, the conclusions are discussed in Section 5.

2. Surface Defects of Steel Plates

Based on our observations and research, the surface defects of slabs can be divided into five types: cracks, scratches, indentations, pits, and scales. The following are several common defect samples collected from the surface defect online detection system.

2.1. Cracks

Cracks are the most serious defect on the surface of steel plates, as shown in Figure 1. Cracks may cause tremendous damage in the following rolling procedure [11].

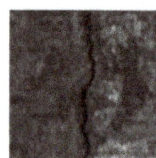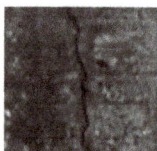

Figure 1. Cracks.

2.2. Scratches

Scratches are generally due to friction between the mechanical equipment or the relative motion of the plates on the roller table [12]. Scratches mostly appear as bright stripes in the image, as shown in Figure 2. Since scratches are mostly caused by mechanical equipment, they appear periodically at most times. That is, scratches appear continually in adjacent images, and their positions and features are similar, so scratches are easier to classify.

  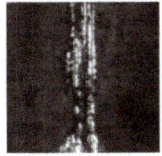  

Figure 2. Scratches.

2.3. Indentations

The occurrence of indentations is due to inclusions in the plates during continuous casting, and pits appear on the surface of the billet, or as depressions on the surface of the roller table. The typical images of indentation are shown in Figure 3. Casting temperature, improper control of casting speed, entrapment of protective slag, etc., can cause defects on the surface and inside of the strand, especially surface flaws, pits, and buckling defects. During the heating and rolling process, it is possible to further form an indentation. The indentation seriously affects the quality of the casting blank as the indentation size varies, the direction is uncertain, and the manual inspection is difficult.

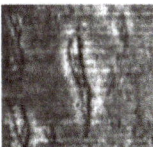  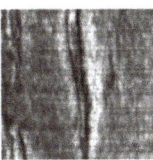 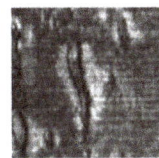 

Figure 3. Indentations.

2.4. Pits

Pits are pit-shaped defects with a certain depth due to the periodical vibration of the mold during the production of the plates. Typical images of a pit are shown in Figure 4. The pits easily cause indentations during the subsequent hot rolling process, which is one of the important factors affecting the surface quality of the plates. In addition, cracks may occur at the bottom of deep pits. Pits are common on the surfaces of the plates. Shallow pits have little effect on the surface quality of the plates, but deeper pits require more attention.

 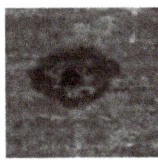  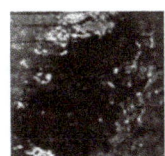 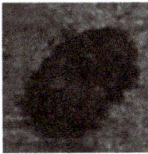

Figure 4. Pits.

2.5. Scales

The surface temperature of the continuous plates during production is usually around 1000 °C, so the surface is easily oxidized to form oxide scales. The typical images of scales are shown in Figure 5. Scales do not seriously affect the quality of the plate, but they have a great negative effect on the recognition of the surface defects of the plates.

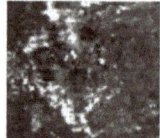

Figure 5. Scales.

3. Principle of Multi-Block Local Binary Pattern

3.1. Principle of Local Binary Pattern

Due to the high ambient temperature and complex image background, the detection of surface defects for steel plates becomes a puzzling problem. The change of illumination will cause a linear change in the gray level of the image. The local binary pattern (LBP) is a method to describe the texture of the image that can eliminate linear illumination by comparing the gray values of pixels. LBP was introduced by Ojala [13,14] and has been widely applied in many fields [15] such as metal surface quality detection [16], paper quality detection [17], image texture analysis [18], and target detection [19].

The LBP descriptor was initially proposed as an effective grayscale-invariant texture descriptor based on image gray levels. The basic LBP descriptor is defined in a 3 × 3 pixel area. The gray value of the center pixel is s g_c. The gray values of its 8 neighborhood pixels are g_0... g_7, respectively. The texture description T of the center pixel can be expressed as Equation (1):

$$T \sim (g_0 - g_c, \cdots, g_7 - g_c) \qquad (1)$$

Then, compare the gray values of the eight neighborhood pixels with g_c. If the value is greater than g_c, the binarization value of the pixel is 1, otherwise the binarization value of the pixel is 0. Then, the texture description T after binarization can be expressed as Equation (2).

$$T \sim (s(g_0 - g_c), \cdots, s(g_7 - g_c)) \qquad (2)$$

s(x) is calculated by Equation (3):

$$s(x) = \begin{cases} 1, x > 0 \\ 0, x \ll 0 \end{cases} \qquad (3)$$

After binarization, randomly choose one of eight neighborhood pixels as a starting point. It is worth noting that the choice of starting point is random, but the choice of starting point for the entire image should be consistent, for convenience, this paper consistently selected the top left point as the starting point. Then, encode all binarization values clockwise to a binary number. Therefore, the LBP value of pixel (x_c, y_c) is calculated in Equation (4).

$$LBP(x_c, y_c) = \sum_{i=0}^{7} s(g_i - g_c) 2^i \qquad (4)$$

Figure 6 shows the complete process of the local binary pattern (LBP).

The LBP value for each pixel could be calculated by the above calculation process. In this way, an LBP image can be obtained [20].

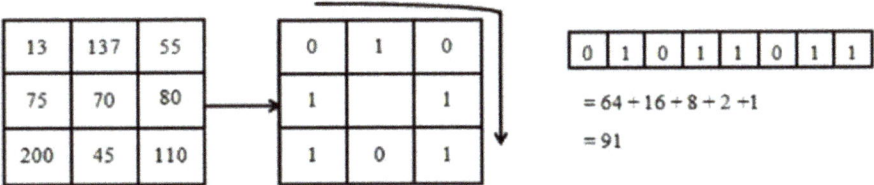

Figure 6. Complete process of the local binary pattern (LBP).

3.2. Principle of MB-LBP

A series of improved LBP algorithms have been proposed such as the LBP uniform pattern and LBP rotation-invariant pattern. All of these LBP algorithms are always focused on a single point. However, different defects have different sizes, and the appropriate scale to describe their texture features is different (larger defects should be described by more points), so different block sizes (with different number of points) should be selected. Therefore, this paper proposed the multi-block LBP (MB-LBP) to describe features of various defects with different sizes [21].

The specific steps of the MB-LBP are as follows:

(1) Division:

The source image is divided into small blocks, and each block contains n × n pixels (take 2 × 2 as an example).

(2) Binarization:

Calculate the mean gray value of all blocks, then compare the mean gray value of each block with the mean gray value of its neighborhood block. If the mean gray value of the neighborhood block is 50% larger than the center block, then set it as 1, otherwise set it as 0, as shown in Equation (5):

$$s(g_i, g_c) = \begin{cases} 1, & \frac{|g_i - g_c|}{g_c} > 50\% \\ 0, & \frac{|g_i - g_c|}{g_c} \leq 50\% \end{cases} \quad (5)$$

where g_i is the mean gray value of each neighborhood block and g_c is the mean gray value of the center block.

Another improvement of the MB-LBP to LBP is that, when calculating $s(g_i, g_c)$, it no longer simply compares the mean value of the neighborhood block with the center block, but compares the percentage of the difference of the mean value with the center block. Therefore, the robustness of the MB-LBP can be enhanced.

The binarization process is shown in Figure 7.

(3) Computation of MB-LBP pattern value:

First, randomly choose one of the eight neighborhood blocks as a starting point. It is worth noting that the choice of starting point is random, but the choice of starting point for the entire image should be consistent, for convenience, this paper consistently chose the top left point as the starting point.

Then, encode all binarization values clockwise to a binary number as calculated by Equation (6)

$$\text{MB} - \text{LBP}_P^R(x_c, y_c) = \sum_{p=0}^{p-1} s(g_i, g_c) 2^p \quad (6)$$

(4) Get MB-LBP pattern value of entire image:

Calculate the MB-LBP pattern value of each block traversing the entire image, from left to right, top to bottom. For the edge block, its neighborhood block is missing, and it has little effect on the texture feature of defects located inside the image, so THE MB-LBP pattern value of the edge block is ignored.

In this way, an entire MB-LBP image can be obtained, as shown in Figure 8.

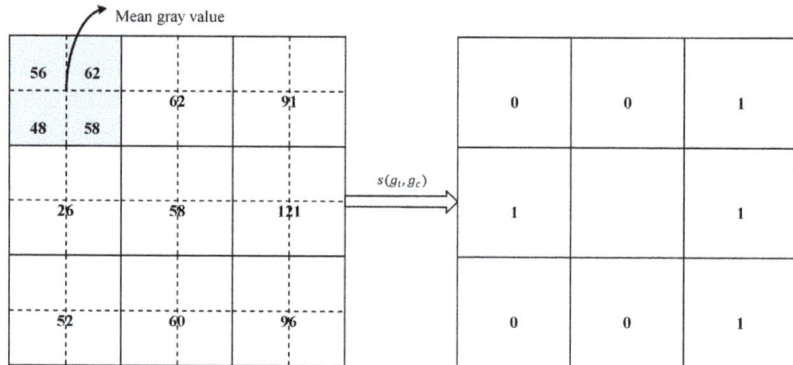

Figure 7. Binarization process of the MB-LBP.

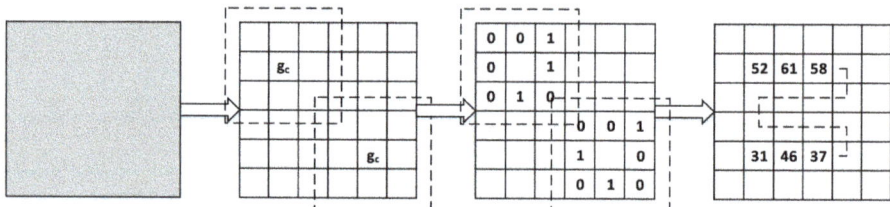

Figure 8. Obtain the MB-LBP pattern value of the entire image.

4. Experiment and Analysis of Surface Defects Detection Based on MB-LBP

4.1. Experimental Design

4.1.1. Architecture of Surface Defect Detection System

A complete online surface defect detection system for steel plates should include image capturing, image preprocessing, defect feature extraction, and defect classification [22], which is shown in Figure 9. In pre-processing, the image segmentation algorithm is used to mark suspected defect areas. It can not only make preparations for identifying the sizes and locations of defects, but can also greatly reduce the amount of calculation for defect feature extraction and defect classification and identification. After the defect areas are determined, the defect features are extracted by the feature extraction algorithm. Then, the obtained feature vectors will be input into the classifier for classification and identification. Finally, a series of information such as class, size, location, and severity of the defect can be obtained.

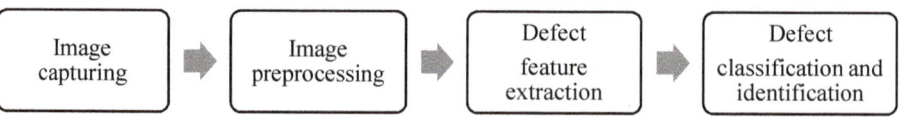

Figure 9. The flow chart of the surface defect detection system for steel plates.

4.1.2. Algorithms of Image Segmentation in Image Preprocessing

The steps of image segmentation in image preprocessing is as follows:

(1) Integral image segmentation.

The concept of integral images was proposed by Viola and Jones [23]. With the integral image, the grayscale sum of pixels in a rectangular area can be quickly calculated, which represents the features of this area. The value of any point in the integral image is the grayscale sum of all pixels from the upper left corner to this point, that is:

$$p(i,j) = \sum_{i'<=i \& j'<=j} pix(i', j') \qquad (7)$$

where $pix(i,j)$ is the grayscale of the pixel located at (i,j).

As shown in Figure 10, first, the original image is equally divided into blocks. The length and width of the block can be changed as a power of two (eight pixels in this paper). With the integral image, the mean grayscale of each block can be quickly calculated, which represents the features of this block.

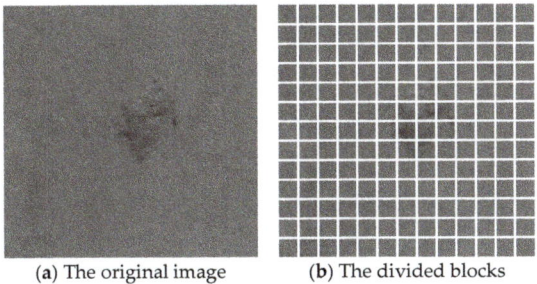

(a) The original image   (b) The divided blocks

Figure 10. Divide original image into blocks.

(2) Differential calculation

Compare each block with four neighbor blocks and obtain all difference values. Then, select the difference value with the largest absolute value as the feature value of the current block. All feature values form a block difference matrix. The result is shown in Figure 11. The difference value in the block difference matrix reflects the change of the grayscale in the image. Blocks with a small difference value and gentle grayscale changes represent a no defect area; while blocks with a large difference value and sharp grayscale changes mean a suspected defect area.

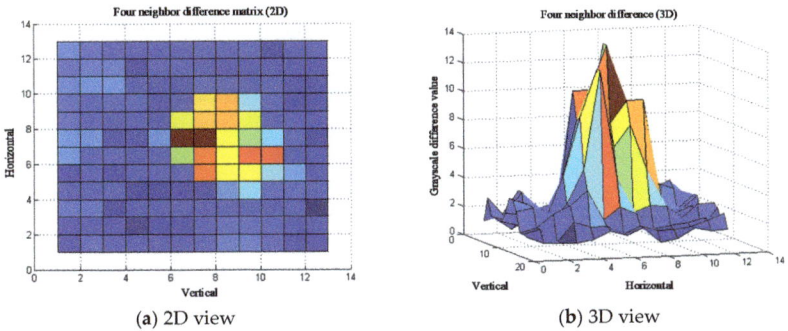

(a) 2D view   (b) 3D view

Figure 11. Results of the four neighbor block difference.

(3) Threshold setting

Now, the question is how to select a reasonable threshold to determine whether the difference value is large or small. Set two thresholds, high and low, to divide the grayscale difference. As shown in Figure 12, the upper plane represents the high threshold and the lower plane represents the low threshold. The blocks with a difference value greater than the high threshold is identified as "master", blocks with a difference value less than the low threshold is identified as "abort", and blocks with a difference value between high and low threshold is identified as "candidate". The "master" blocks are considered to be a defect area; the "candidate" block is a candidate defect area, and only when it has a "master" eight-neighbor block, will it be promoted to a "master" block.

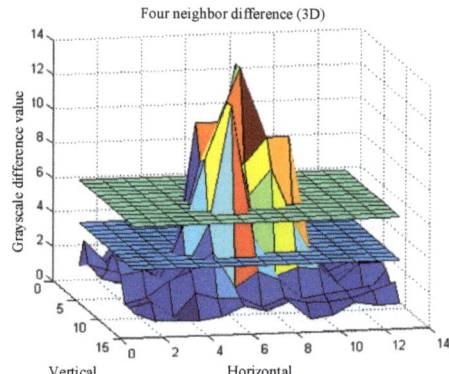

Figure 12. Division of difference with high and low threshold.

(4) Regional growth

Usually, a complete defect area is composed of multiple blocks. Therefore, it is necessary to merge adjacent "master" blocks into larger areas. That is, when there is a "master" block in eight-neighbor blocks of a "master" block, merge them as a new "master" block. Repeat the operations above until the new "master" block has no "master" block in its neighborhoods.

4.1.3. Experiment of Defect Feature Extraction and Defect Classification in this Paper

In this paper, MB-LBP was employed in feature extraction and support vector machine (SVM) was employed in the defect classification and identification.

The complete experiment process is shown in Figure 13.

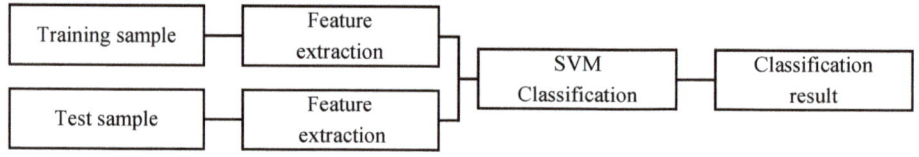

Figure 13. Feature extraction based on support vector machine (SVM).

The process of the MB-LBP + SVM experiment is detailed as follows:

(1) Image division

The image is divided into small blocks with n × n (n = 2, 4, 8, and so on) pixels, and then the mean gray value of the blocks are calculated as the gray value of the block.

(2) MB-LBP pattern value calculation

For each block, compare its gray value to each of its eight neighbors clockwise. When the gray value of center block is 50% greater than the gray value of neighbor block, write "0". Otherwise, write "1". This gives an 8-digit binary number, which is usually converted to decimals for convenience.

(3) Feature vector generation

Now, we have binary numbers of all blocks in the entire image. Then, compute the histogram of the frequency of each "binary number" occurring in the entire image.

An 8-digit binary number has $2^8 = 256$ different numbers at most. Therefore, this histogram can be seen as a 256-dimensional feature vector.

Such a large number of patterns will result in a long calculation time. Furthermore, a large pattern number will lead to a sparse feature vector that is difficult to compute. To solve these two problems, referring to the LBP uniform pattern [24], we proposed the MB-LBP uniform pattern. This idea is motivated by the fact that some binary patterns occur more commonly in texture images than in others [25]. These common binary patterns are called a uniform pattern if the binary pattern contains at most two 0–1 or 1–0 transitions. For example, 00010000 (two transitions) is a uniform pattern, 01010100 (six transitions) is not. Other uncommon binary patterns are called a non-uniformed pattern if it contains more than two transitions. The maximum MB-LBP uniform patterns is 59, much smaller than 256 of the original MB-LBP.

Figure 14 shows the comparison of the histogram between the original MB-LBP and MB-LBP uniform pattern.

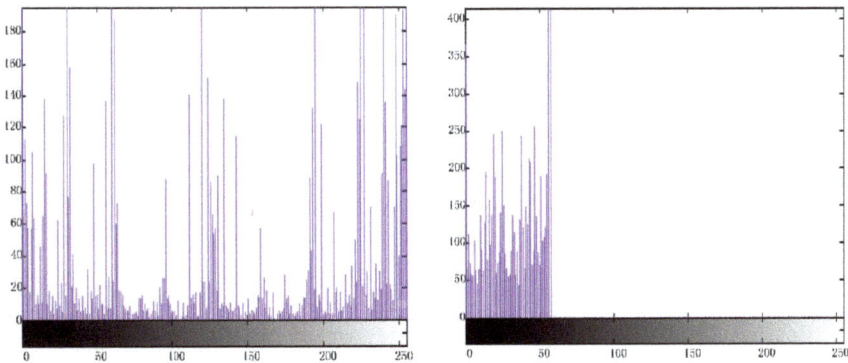

Figure 14. The comparison of the histogram between the original LBP and LBP uniform pattern.

(4) The generated feature vectors are sent to the SVM for classification and identification

The original SVM is a binary classifier. In this paper, the one-vs-one multi-classification SVM method [26] was adopted. Every two classes construct a SVM, and there are a total of n(n − 1)/2 SVM. The parameter n is the number of classes. In this paper, n = 5 and there were 5(5−1)/2 = 10 SVM in total.

The specific construction process is as follows:

(a) Select two image classes, i and j in the training set as the positive and negative sample sets;
(b) Extract the MB-LBP features of all images in class i and j, then represent all images by the histogram vector;
(c) Train the classifier between class i and j;
(d) Repeat steps (a)~(c) until the training between any two classes is completed. Obtain the optimal classification surface of all classifiers.

When a test image I is classified, vote to determine which class the image I belongs to according to the minimum sum of the distances from image I to the optimal classification surface.

4.1.4. Samples Sets of Experiment

The surface defect samples of the plates used in this paper were collected from a surface defect detection system in a steel company. The type composition of the sample sets is shown in Table 1.

Table 1. Type composition of surface defect sample sets.

	Crack	Scratch	Indentation	Pit	Scale	total
Training	318	117	100	30	511	1076
Testing	296	175	174	139	913	1697

4.1.5. Parameter Selection of MB-LBP

There is a certain fluctuation in the recognition accuracy, as shown in Figure 15. When the block size increased from 2 × 2 to 4 × 4, the accuracy increased, and when it increased to 8 × 8, it decreased. This is because the block enlargement was equivalent to image compression, and too large a block size lost too much defect feature information, so the accuracy decreased. Through analysis, this experiment selected 4 × 4 block size for calculation.

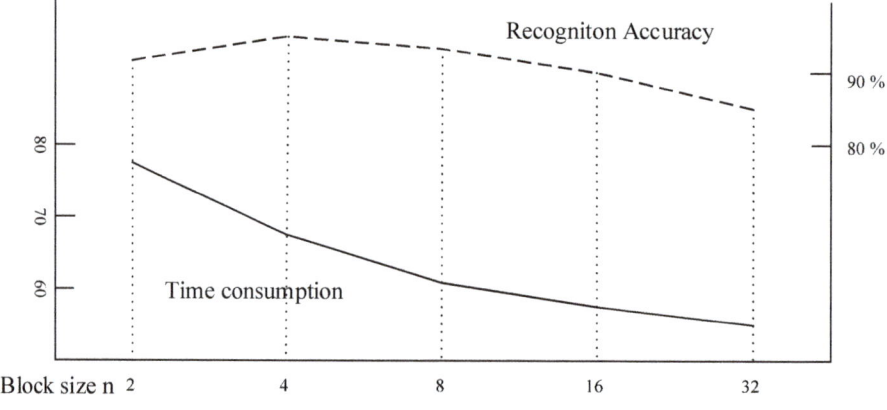

Figure 15. Variation of time consumption and recognition accuracy with a change of block size n.

4.1.6. Hardware Configuration of Experimental Platform

All algorithms were tested in the same platform. The hardware configuration was as follows: CPU Intel i5-490, main frequency 3.3 GHz, memory 8 GB.

4.2. Comparison of Experiment Result between MB-LBP and Other Algorithms

To validate the performance of the MB-LBP, a comparison with several other algorithms was made from the perspective of classification accuracy and time efficiency. The classification accuracy is the percentage of identified defects to the total defects. The time efficiency is the required time to compute the same size image on the same platform.

In this paper, the scale invariant feature transform (SIFT), the speeded up robust feature (SURF), the gray-level co-occurrence matrix (GLCM), and convolutional auto encoder-semi-supervised generative adversarial networks (CAE-SGAN) were selected for the comparison. The SIFT [27] algorithm is widely used in the fields of image matching, image recognition, and defect detection. It constructs a Gaussian scale space to describe the local features for the purpose of scale invariability, rotation invariability, and illumination invariability. This algorithm has a certain stability on overcoming the influence of noise and affine. The SURF [28] algorithm is improved based on the SIFT algorithm, which has not computed the Gaussian scale space. On the basis of retaining the main information of the feature points, SURF employs the box filtering technique to simulate the Gaussian scale space to improve the feature extraction speed. As a classic texture feature descriptor, GLCM [29] reflects the comprehensive information of a grayscale image with respect to direction, adjacent interval, and transform amplitude, and is the basis for analyzing the local pattern structure and arrangement rules of the images. The dimension of the feature vector extracted by GLCM is 256 × 256, hence the computation amount is too large. Aiming at this problem, the KLPP [30] was adopted to reduce the dimensions of the feature vector. CAE-SGAN [9] is a deep learning method based on convolutional auto encoder (CAE) and semi-supervised generative adversarial networks (SGAN). It first trains a stacked CAE through massive unlabeled data. After CAE is trained, the encoder network of CAE is reserved as the feature extractor and is fed into a softmax layer to form a new classifier. SGAN is introduced for semi-supervised learning to further improve the generalization ability of the new method. Same as the other deep learning methods, this method uses the original defect images as the network input.

The surface defect detection system for steel plates is an online real-time detection system, time efficiency is its key indicator. According to experience, for an online detection system, in order to ensure real-time requirements, the maximum time efficiency is 100 ms/image. On this basis, the recognition accuracy of the algorithm is as high as possible. For practicality, the accuracy should be at least 90%.

Table 2 shows the comparison of the recognition accuracy and time efficiency between the MB-LBP and other algorithms. The time efficiency of SIFT, SURF, and GLCM speed was separately 583 ms, 142 ms, and 248 ms, which is too slow to meet the requirements of online real-time detection. The time efficiency of LBP was 41 ms, which not only meets the real-time requirements, but is also the fastest algorithm. However, its recognition accuracy was only 89.05%, lower than the minimum practical requirements. The time efficiency of MB-LBP was 63 ms. Although it was slightly slower than the LBP, it still meets the real-time requirement. Additionally, it had the highest recognition accuracy, reaching 94.40%. As for CAE-SGAN, as a deep learning method, it had the highest recognition accuracy of 96.21%. However, it uses the original image as input, and the dimension of input vector was as high as 224 × 224. Additionally, its network structure is too complicated. Therefore, its time efficiency was only 434 ms/image, much higher than the minimum requirement (less than 100 ms/image) of online real-time detection in this paper.

Table 2. Comparison of the recognition accuracy and time efficiency between the MB-LBP and other algorithms.

	Feature Dimension	Recognition Accuracy	Time Efficiency (ms/image)
SIFT	128	91.43%	583
SURF	128	86.89%	142
GLCM	20	84.27%	248
LBP	256	89.05%	41
MB-LBP	59	94.40%	63
CAE-SGAN	224 × 224 (original image)	96.21%	434

Table 3 shows the recognition accuracy of each class of defects with the MB-LBP. The recognition accuracy of all defects met the requirement of online detection (at least 90%). The recognition accuracy of most classes was above 94%, except for scales. This is because scales have complex morphological features and are difficult to classify. Recognition of scales is also one of the future improvement directions.

Table 3. Recognition accuracy of each class of defects with the MB-LBP.

Defects	Recognition Accuracy with MB-LBP
Crack	95.61%
Scratch	94.29%
Indentation	94.25%
Pit	95.68%
Scale	93.87%
Total	94.40%

Figure 16 shows some examples of false matching. Cracks vary in size and shape, in particular, the detection of small cracks is difficult. Figure 16(a1,a2) show false-matching cracks, as they were too small and not obvious enough. Scratches were relatively easy to detect, but were greatly affected by light. Figure 16(b1,b2) show false-matching scratches. Figure 16(b1) was misclassified as a crack as it shows a thin dark stripe with weak light; and Figure 16(b2) was misclassified as a scale, as it shows an irregular area shape with strong light. Indentations vary in size and direction. Figure 16(c1,c2) show false-matching indentations. Two indentations with different sizes and directions were classified as different types of defects. Pits have irregularly shaped edges and complex backgrounds, like scales. Figure 16(d1,d2) show false-matching pits that were misclassified as scales. Scales have complex shapes and great grayscale changes, as such, they are the main misclassification defects. Scales have the lowest recognition accuracy, many other types of defects are misclassified as scales, or vice versa. Figure 16(e1,e2) show false-matching scales. Figure 16(e1,e2) were separately misclassified as a pit and crack due to their similar shapes.

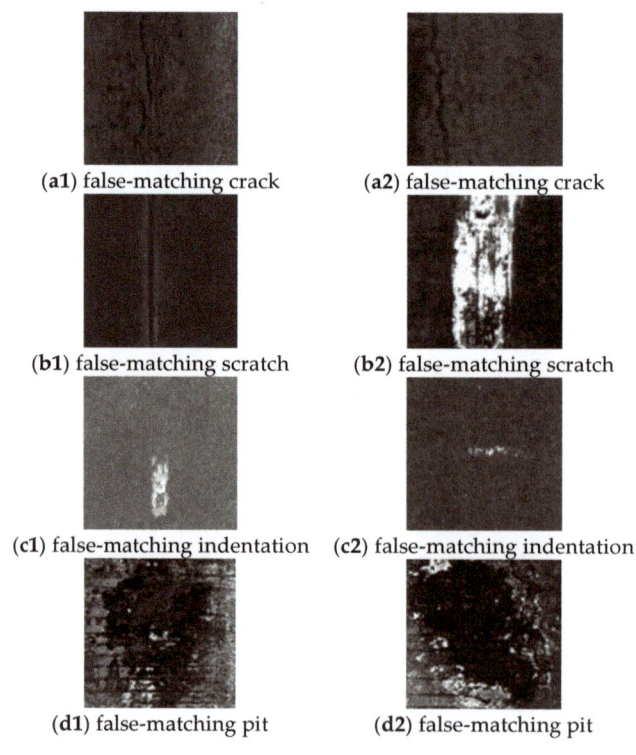

(**a1**) false-matching crack (**a2**) false-matching crack

(**b1**) false-matching scratch (**b2**) false-matching scratch

(**c1**) false-matching indentation (**c2**) false-matching indentation

(**d1**) false-matching pit (**d2**) false-matching pit

Figure 16. *Cont.*

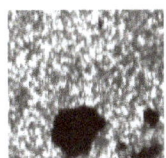 

(e1) false-matching scale　　　(e2) false-matching scale

Figure 16. Examples of misclassified defects of cracks, scratches, indentations, pits, and scales.

5. Conclusions

Online detection of the surface defects of steel plates requires an algorithm that simultaneously satisfies fast recognition speed and high recognition accuracy.

In this paper, we proposed a surface defect detection algorithm for steel plates, which adopted the MB-LBP algorithm to extract the defect features. The MB-LBP algorithm divides the image into small blocks. After the binarization and MB-LBP value calculation, grayscale histograms were generated as the feature vectors of the image.

To verify the performance of the MB-LBP algorithm, a comparison with several other algorithms was made. The experimental results show that the recognition accuracy of the MB-LBP algorithm was better than the other algorithms, and the time efficiency was fast enough to meet the real-time requirements of the online surface defect detection of steel plates.

Author Contributions: Conceptualization, Y.L.; Data curation, Y.L.; Funding acquisition, K.X.; Investigation, Y.L.; Methodology, Y.L.; Project administration, J.X.; Resources, K.X.; Supervision, K.X.; Validation, J.X.; Writing—original draft, Y.L.; Writing—review & editing, K.X.

Funding: This research was funded by the National Natural Science Foundation of China, grant numbers 51674031 and 51874022.

Conflicts of Interest: The authors declare no conflicts of interest.

References

1. Suresh, B.R.; Fundakowski, R.A.; Levitt, T.S.; Overland, J.E. A real-time automated visual inspection system for hot steel slabs. *IEEE Trans. Pattern Anal. Mach. Intell.* **1983**, *PAMI-5*, 563–572. [CrossRef]
2. Yun, J.P.; Choi, S.H.; Kim, J.W.; Kim, S.W. Automatic detection of cracks in raw steel block using Gabor filter optimized by univariate dynamic encoding algorithm for searches (uDEAS). *NDT E Int.* **2009**, *42*, 389–397. [CrossRef]
3. Xu, K.; Yang, C.L.; Zhou, P.; Yang, C.; Li, X.G. On-line detection technique of surface cracks for continuous casting billets based on linear lasers. *J. Univ. Sci. Technol. Beijing* **2009**, *31*, 1620–1624.
4. Pan, E.; Ye, L.; Shi, J.; Chang, T.S. On-line bleeds detection in continuous casting processes using engineering-driven rule-based algorithm. *J. Manuf. Sci. Eng.* **2009**, *131*, 061008. [CrossRef]
5. Yu, J.; Wang, Z.; Li, P. Continuous casting slab surface feature classification method based on complex Contourlet feature vectors. *J. Comput. Appl.* **2014**, *34*, 3660–3664.
6. Miyamoto, R.; Mizutani, K.; Wakatsuki, N.; Ebihara, T. Defect Detection in Billet Using Plane-Wave and Time-of-Flight Deviation with Transmission Method. In Proceedings of the 2018 IEEE International Ultrasonics Symposium (IUS), Kobe, Japan, 22–25 October 2018; pp. 1–4.
7. Yan, J.; Li, Z.; Wang, Z. Research on surface defects of continuous casting billets based on mathematical morphology. *Opt. Tech.* **2018**, *44*, 8.
8. Lin, J.; Yao, Y.; Ma, L.; Wang, Y. Detection of a casting defect tracked by deep convolution neural network. *Int. J. Adv. Manuf. Technol.* **2018**, *97*, 573–581. [CrossRef]
9. He, D.; Xu, K.; Zhou, P.; Zhou, D. Surface defect classification of steels with a new semi-supervised learning method. *Opt. Lasers Eng.* **2019**, *117*, 40–48.
10. Jiang, X.; Wang, X.; Chen, D. Research on Defect Detection of Castings Based on Deep Residual Network. In Proceedings of the 2018 11th International Congress on Image and Signal Processing, BioMedical Engineering and Informatics (CISP-BMEI), Beijing, China, 13–15 October 2018; pp. 1–6.

11. Vascotto, M. High Speed Surface Defect Identification on Steel Strip. *Metall. Plant Technol. Int.* **2005**, *4*, 70–73.
12. Maki, H.; Tsunozaki, Y.; Matsufuji, Y. Magnetic On-Line Defect Inspection System for Strip Steel. *Iron Steel Eng.* **1989**, *66*, 26–33.
13. Ojala, T.; Pietikäinen, M.; Harwood, D. A Comparative Study of Texture Measures with Classification Based on Feature Distributions. *Pattern Recognit.* **1996**, *3*, 51–59. [CrossRef]
14. Ojala, T.; Pietikäinen, M.; Maenpaa, T. Multiresolution grayscale and rotation invariant texture analysis with local binary patterns. *IEEE Trans. Pattern Anal. Mach. Intell.* **2002**, *7*, 971–987. [CrossRef]
15. Xu, Q.; Yang, J.; Ding, S. Texture segmentation using LBP embedded region competition. *ELCVIA Electron. Lett. Comput. Vis. Image Anal.* **2005**, *5*, 41–47. [CrossRef]
16. Pietikaeinen, M.; Ojala, T.; Nisula, J.; Heikkinen, J. Experiments with two industrial problems using texture classification based on feature distributions. *Proc. SPIE* **1994**, *2354*, 197–205.
17. Marzabal, A.; Torrens, C.; Grau, A. Texture-based characterization of defects in automobile engine valves. In Proceedings of the Ninth Symposium on Pattern Recognition and Image Processing, Ed. Univ. Jaume I Castellón, Spain, 2001; pp. 267–272.
18. Bishop, C.M. *Neural Networks for Pattern Recognition*; Oxford University Press: Oxford, UK, 1995.
19. Wang, H.; Bell, D.; Murtagh, F. Axiomatic approach to feature subset selection based on relevance. *IEEE Trans. Pattern Anal. Mach. Intell.* **1999**, *21*, 271–277. [CrossRef]
20. Pietikainen, M.; Hadid, A.; Zhao, G.Y.; Ahonen, T. *Computer Vision Using Local Binary Patterns*; Springer: Berlin/Heidelberg, Germany, 2011; pp. 193–202.
21. Zhang, L.; Chu, R.; Xiang, S.; Liao, S.; Li, S.Z. Face detection based on multi-block LBP representation. In Proceedings of the Advances in Biometrics, International Conference, ICB 2007, Seoul, Korea, 27–29 August 2007; pp. 11–18.
22. Song, K.; Yan, Y. A noise robust method based on completed local binary patterns for hot-rolled steel strip surface defects. *Appl. Surf. Sci.* **2013**, *285*, 858–864. [CrossRef]
23. Viola, P.; Jones, M. Rapid Object Detection using a Boosted Cascade of Simple Features. In Proceedings of the IEEE Computer Society Conference on Computer Vision & Pattern Recognition, Kauai, HI, USA, 8–14 December 2001; pp. 511–518.
24. Nanni, L.; Lumini, A.; Brahnam, S. Survey on Lbp Based Texture Descriptors for Image Classification. *Expert Syst. Appl.* **2012**, *39*, 3634–3641. [CrossRef]
25. Shan, C. Learning Local Binary Patterns for Gender Classification on Real-world Face Images. *Pattern Recognit. Lett.* **2012**, *33*, 431–437. [CrossRef]
26. Huang, X.; Wei, S. An improved K-means clustering algorithm. In Proceedings of the World Automation Congress, Xi'an, China, 27–29 May 2016.
27. Li, Y.; Liu, W.; Li, X.; Huang, Q.; Li, X. GA-SIFT: A new scale invariant feature transform for multispectral image using geometric algebra. *Inf. Sci.* **2014**, *281*, 559–572. [CrossRef]
28. Bay, H.; Ess, A.; Tuytelaars, T.; Tuytelaars, T.; Van Gool, L. Speeded-up robust features (SURF). *Comput. Vis. Image Underst.* **2008**, *110*, 346–359. [CrossRef]
29. Mohanaiah, P.; Sathyanarayana, P.; GuruKumar, L. Image texture feature extraction using GLCM approach. *Int. J. Sci. Res. Publ.* **2013**, *3*, 1.
30. He, X.; Niyogi, P. Locality preserving projections. In Proceedings of the Advances in neural information processing systems, Whistler, BC, Canada, 9–11 December 2003; pp. 153–160.

© 2019 by the authors. Licensee MDPI, Basel, Switzerland. This article is an open access article distributed under the terms and conditions of the Creative Commons Attribution (CC BY) license (http://creativecommons.org/licenses/by/4.0/).

Article

Measurement of Period Length and Skew Angle Patterns of Textile Cutting Pieces Based on Faster R-CNN

Lei Geng [1,2], **Qinglei Meng** [1,2], **Zhitao Xiao** [1,2,*] **and Yanbei Liu** [1,2]

1. School of Electronics and Information Engineering, Tianjin Polytechnic University, NO.399 Binshui West Street Xiqing District, Tianjin 300387, China
2. Tianjin Key Laboratory of Optoelectronic Detection Technology and Systems, NO.399 Binshui West Street Xiqing District, Tianjin 300387, China
* Correspondence: xiaozhitao@tjpu.edu.cn

Received: 18 June 2019; Accepted: 25 July 2019; Published: 26 July 2019

Abstract: The skew angle and period length of the multi-period pattern are two critical parameters for evaluating the quality of textile cutting pieces. In this paper, a new measurement method of the skew angle and period length is proposed based on Faster region convolutional neural network (R-CNN). First, a dataset containing approximately 5000 unique pattern images was established and annotated in the format of PASCAL VOC 2007. Second, the Faster R-CNN model was used to detect the pattern to determine the approximate location of the pattern (the position of the whole pattern). Third, precise position of the pattern (geometric center points of pattern) are processed based on the approximate position results using the automatic threshold segmentation method. Finally, the four-neighbor method was used to fill the missing center points to obtain a complete center point map, and the skew angle and period length can be measured by the detected center points. The experimental results show that the mean average position (mAP) of the pattern detection reached 84%, the average error of the proposed algorithm was less than 5% compared with the error of the manual measurement.

Keywords: faster R-CNN; cutting pieces; multi-period pattern; skew angle; period length

1. Introduction

Textile cutting pieces [1], as semi-finished products, have been widely used in car seats and garments areas. Most of the finished products are stitched from these pieces, and the performance of the pieces (see Figure 1a) is a key to determining the quality of the finished product. The quality of the textile pieces depends largely on their preformed geometry structure, such as the period length and skew angle of the pattern. The skew angles θ_{weft} and θ_{warp} are defined as the angle between the line along the horizontal or vertical period direction of the pattern and the overall contour of the piece (see Figure 1b). The skew angle is a critical parameter of multi-period pattern pieces, for it can affect the overall regularity of the pattern. The period length T_{hi} and T_{vi} ($i = 1, 2 \dots$) are defined as the length of one complete pattern period distance in the horizontal and vertical directions, can be used to infer the local regularity of the pattern (see Figure 1b). These two parameters can reflect the design difference between the pattern sample and the standard template, so they can be used as a criterion for judging the quality of the pattern.

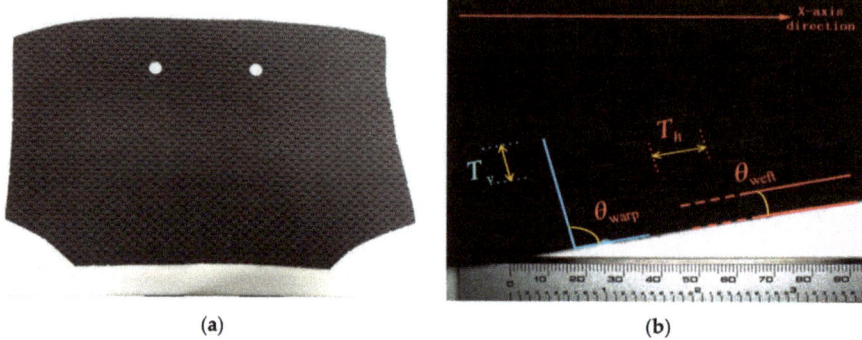

Figure 1. Textile cutting piece of car seat and pattern parameters. (**a**) The overall outline of the cutting piece with strip-shaped pattern in the global perspective. (**b**) The local part of (a) where T_h is weft period length, T_v is warp period length. θ_{weft} is weft skew angle, and θ_{warp} is warp skew angle.

The two parameters are used to check whether the cutting pieces are qualified or not in the industrial area. At present, the manual method is still the main measurement way, which is time and manpower-consuming, and due to the large amount of pattern types, only limited numbers of cutting pieces are sampled. In addition, a cutting piece with complicated patterns cannot be effectively detected by the human eyes, and this phenomenon often causes quality problems.

The periodicity of the pattern has great research significance for the pattern fabric, and is the basis for measuring the skew angle and period length of the multi-period pattern. Therefore, the period extraction of the pattern becomes the key and difficult point in measuring the parameters of a pattern. In recent years, with the rapid development of image processing technology, many approaches have been proposed for fabric periodic research. In general, these approaches can be classified into three groups: grey level co-occurrence matrix-based (GLCM) [2]; distance matching function-based (DMF) [3]; and image autocorrelation function-based [4]. The method based on GLCM is a common technique in statistical image analysis that is used to estimate image properties related to second-order statistics. Li [5] and his colleagues research the variation of the eigenvalues of four grey level co-occurrence matrices to determine the period characteristics of texture, and achieved relatively good results. Xiao, et al. [6] calculated the correlation coefficient between different regions enclosed by fabric yarns based on the grey level co-occurrence matrix method to complete the segmentation of a striped fabric. The features calculated by the co-occurrence matrix can be used for periodic detection of finite-size pattern images and the computation speed is relatively fast. However, since the quantization angle and distance are frequently used to reduce the computation time when calculating the co-occurrence matrix features, the accuracy of texture cycle extraction is significantly reduced. The method based on DMF can directly use the grey value of the texture to find the texture period and requires less computation time than the traditional co-occurrence matrix approach. Jing [7] determined the period of the printed fabrics by calculating the maximum value of the second forward difference of the two-dimensional DMF. Zhou [8] implemented an automatic measurement of the texture period of woven fabric images by combining frequency domain analysis with a distance matching function and improved the stability and computational efficiency of cycle measurement. The distance matching function is an effective method for extracting pattern period. For images of any size, the distance matching function has a faster calculation speed than the traditional co-occurrence matrix method. It is suitable for patterns of finite size. However, when the brightness and shape of the periodic pattern are inconsistent, the distance matching function cannot effectively extract the pattern period. The method based on an autocorrelation function calculates the correlation coefficient of the texture by the autocorrelation function of the image to analyze the periodicity of the texture. Wu [9] calculated the autocorrelation function of the texture edge to determine the matrix of the autocorrelation function, and then extracted the periodic and directional features of the texture. The image autocorrelation function method is easy to implement and has strong adaptability. However, the pattern period detection method based on the autocorrelation function can only reflect the periodic features of the pattern and has no other features. Moreover, it cannot efficiently acquire periodic information of a periodic pattern that has a sparsely distributed and large size in the image. Similarly, the measurement of the surface braiding angle and pitch length of the three-dimensional braided composite was realized by the corner detection-based method [10–12]. However, this algorithm based on corner detection is not suitable for the measurement of multi-period pattern parameters. When faced with complicated patterns, the corner points detected by the corner detection algorithm are disordered and the pseudo corner points are too many, and the center point of the pattern cannot be accurately found, so that the pattern period cannot be effectively extracted.

In this paper, a new measurement method based on Faster region convolutional neural network (R-CNN) [13] was proposed to measure the parameters of the multi-period pattern. At present, Faster R-CNN has been applied in many fields, such as license plate detection [14], scene text detection [15] and optical image detection [16], and achieved excellent results with its powerful performance. As Faster R-CNN has the advantages of high object detection accuracy, fast speed and strong adaptability, we used Faster R-CNN as the pattern detector to locate the pattern and extract the pattern period. The contributions of this paper are as follows:

- We have established the first multi-period pattern dataset. This dataset contains 5000 pattern images with size of 512 × 640 pixels and contains a total of six types of patterns. Moreover, each pattern in the image was annotated as an object in VOC 2007 format.
- We have proposed to use the object detection network in deep learning to locate the pattern. We selected the training model of Faster R-CNN as the pattern detector and generated the bounding boxes enclosing the patterns to achieve the approximate positioning of the pattern.
- We have proposed an automatic threshold method to extract the contour of the pattern and calculated the center points to obtain the precise positioning of the pattern. The four-neighbor-method was used to fill the missing center points to acquire the center point map that reflects the periodic characteristics of the pattern.

2. Methods

In this section, the period length and pattern skew angles of textile cutting pieces were measured based on Faster R-CNN. Firstly, original images were acquired and the pattern dataset were created. Then, the patterns were detected by a model trained by Faster R-CNN net. Secondly, the approximate location of the pattern (the position of the whole patterns) were obtained based on the detected pattern. Thirdly, the precise positions of the pattern (geometric center points of pattern) were detected based on the approximate position results using the automatic threshold segmentation method. Missing center points were filled based on the four-neighbor-method to obtain a complete center point map. Finally, the skew angle and period length were measured based on the detected center points.

2.1. Image Acquisition and Pattern Dataset Creation

In this study, the image acquisition system was composed of a dome light source, a 1.3 megapixel color industrial camera, an LCD backlight and a servo motor module (see Figure 2). A vertically installed industrial camera with a camera lens overlooked the fabric. The dome light source that illuminated the fabric surface uniformly wasplaced in front of the fabric. In order to sample multiple parts of the piece, the system contained a servo mobile module that could move industrial cameras in a flat range. Figure 3 shows the six types of pattern images F_i, F_r, F_c, F_{s1}, F_{s2} and F_{s3} (the size was 1024 × 1280 pixels) acquired by the image acquisition system, where: F_i is the pattern with irregular shape. F_r is the pattern with circular shape. F_w is the pattern with the wavy shape. F_{s1}, F_{s2}, F_{s3} are patterns with a strip shape. The acquired images were transmitted to the data processing system (see

Figure 2) to be processed by the Faster R-CNN based algorithm. The image data processing is shown in Section 2.3.

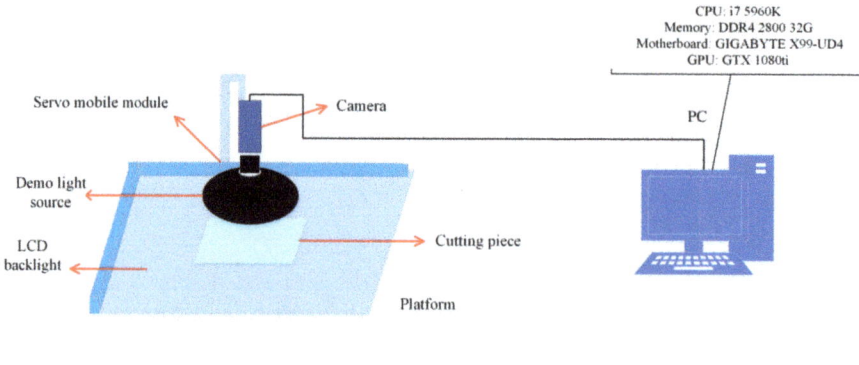

Figure 2. Image processing system. The image acquisition system on the left was used to acquire the partial texture image of the pattern, and the acquired images were three-channel RGB images and the size was 1024 × 1280 pixels. The acquired images were transmitted to the data processing system on the right for data processing.

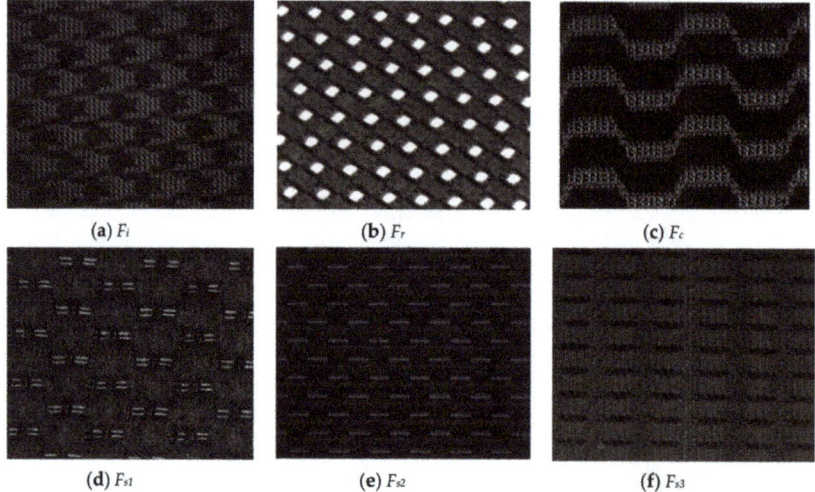

Figure 3. The sample images of different kinds of pattern piece. (**a**) irregular-shaped pattern F_i, (**b**) circular-shape pattern F_r, (**c**) wavy-shaped pattern F_c, (**d**) strip-shaped pattern F_{s1}, (**e**) strip-shaped pattern F_{s2}, (**f**) strip-shaped pattern F_{s3}.

In this paper, 5000 unique images with size of 640 × 512 × 3 pixels were contained in the pattern dataset, and each image contained approximately 20 to 50 patterns. These images were cropped from approximately 400 original images with size of 1280 × 1024 × 3 pixels. To obtain the best training effect, each image was labeled in detail. The key details for labeling each image were as follows:

- In order to avoid image over-fitting problems, each image contained 20–50 patterns, which were a critical metric for detection and recognition.
- The bounding box completely enclosed the pattern and kept the center coordinates of the pattern the same as the center coordinates of the bounding box.

- There are no overlapping regions between the bounding boxes and the dimensions of each bounding box remain the constant.

2.2. Training of Pattern Detection Model

In recent years, object detection technology has achieved rapid development, and the object detection network based on deep learning has greatly improved the ability of object detection. At present, there are two main methods: one depends on region proposal, such as R-CNN (region convolutional neural network) [17], Fast R-CNN [18], Faster R-CNN [13] and R-FCN [19]; the other does not rely on region proposal and directly estimates candidate object recommendations, such as SSD [20] and YOLO [21–23] family.

After R-CNN [17] and Fast R-CNN [18], Microsoft's Shaoqing Ren proposed Faster R-CNN [13] to optimize the running time of the detection network. The region proposal network (RPN) was proposed to generate the proposal region. RPN replaces the previous methods such as Selective Search [24] and EdgeBoxes [25] and it shares the convolution feature of the full map with the detection network so that the region proposal detection takes very little time. The Faster R-CNN structural framework consists of RPN + Fast R-CNN. The RPN network is mainly used to generate high-quality proposal region boxes, and Fast R-CNN is used to learn high-quality proposal region features and classify objects. The overall framework of Faster R-CNN is shown in Figure 4.

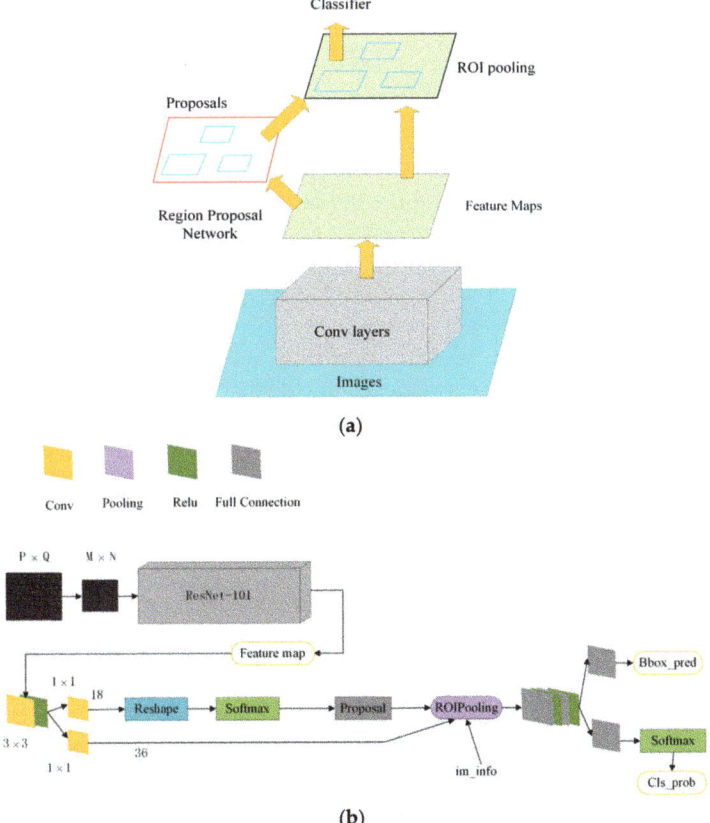

Figure 4. Faster region convolutional neural network (R-CNN) overall framework. (**a**) The streamlined flow chart of the Faster R-CNN framework [26]. (**b**) The detailed flow chart of the Faster R-CNN [13].

Faster R-CNN proposes the region proposal network and improves the efficiency of object detection. This provides feasibility for detecting multi-period patterns with Faster R-CNN. Three nets (ZF-Net [27], VGG16 [28] and Rse-Net-101 [29]) were respectively used as the pre-trained model of Faster R-CNN, where the pattern dataset contained 5382 patterns and a total of six types of patterns: F_i, F_r, F_w, F_{s1}, F_{s2} and F_{s3} (see Figure 3). The number of various pattern images was 894, 899, 902, 892, 904 and 891. Image size was 512 × 640 pixels. The pattern dataset was randomly divided into validation set, test set and train set according to the ratio of 2:2:6. Then, the divided datasets were used for training of Faster R-CNN (ZF-Net), Faster R-CNN (VGG16) and Faster R-CNN (ResNet-101), respectively. The experimental platforms included Windows 7, GPU GTX1080ti, Matlab 2014a and Visual Studio 2013, and the whole experiment was based on the deep learning framework Caffe.

The compared results are shown in Table 1, where the performance of the three pattern detection models can be seen. Precision represented the detection accuracy of the pattern. Balanced accuracy (Ba) was used to evaluate balanced accuracy of the pattern dataset. Kappa (K) was used to evaluate the accuracy of the pattern classification. Mean average precision (mAP) was the main indicator for evaluating the main detection results, because mAP was the actual metric for object detection.

Table 1. Evaluation of the Faster R-CNN with different pre-trained nets.

Pre-Trained Net	Precision						K	Ba	mAP
	F_i	F_r	F_c	F_{s1}	F_{s2}	F_{s3}			
ZF	0.876	0.801	0.832	0.887	0.857	0.824	0.798	0.842	0.78
VGG16	0.879	0.864	0.864	0.904	0.886	0.857	0.824	0.884	0.81
ResNet-101	0.885	0.881	0.901	0.972	0.958	0.898	0.836	0.910	0.84

From Table 1, it can be concluded that the precision, K, Ba and mAP of the Resnet-101 net as the pre-trained model were higher than the other two nets, and thus Resnet-101 was chosen as the pre-trained model in this paper.

Figure 5 shows the detection results of the six patterns (F_i, F_r, F_w, F_{s1}, F_{s2} and F_{s3}) on the Faster R-CNN model. The boxes of different colors represent the different pattern categories detected by the model; the upper left corner of the box represents the classification result, pattern category and category score for the object of the box region. Since the patterns F_{s1}, F_{s2}, F_{s3} and F_r had the characteristics of large pattern pitch, small volume and regular shape, we chose to completely surround the pattern with the bounding box. The patterns F_i and F_c were irregular shapes and could not be completely surrounded by the bounding box, so we regarded a part of the pattern having the periodic characteristics as the detection object. It can be concluded from Figure 5 that the Faster R-CNN model could effectively detect six types of multi-period patterns and had fewer false positive and missing alarms.

Figure 6 shows the precision–recall (P–R) curve of the Faster R-CNN pattern detection model. The precision is the vertical axis, and the recall is the horizontal axis. The area value enclosed by the curve represents the mAP. It can be concluded from the Figure 6 that the pattern detection model had high accuracy, recall rate and average precision, so this model had pretty good pattern detection ability and excellent detection accuracy.

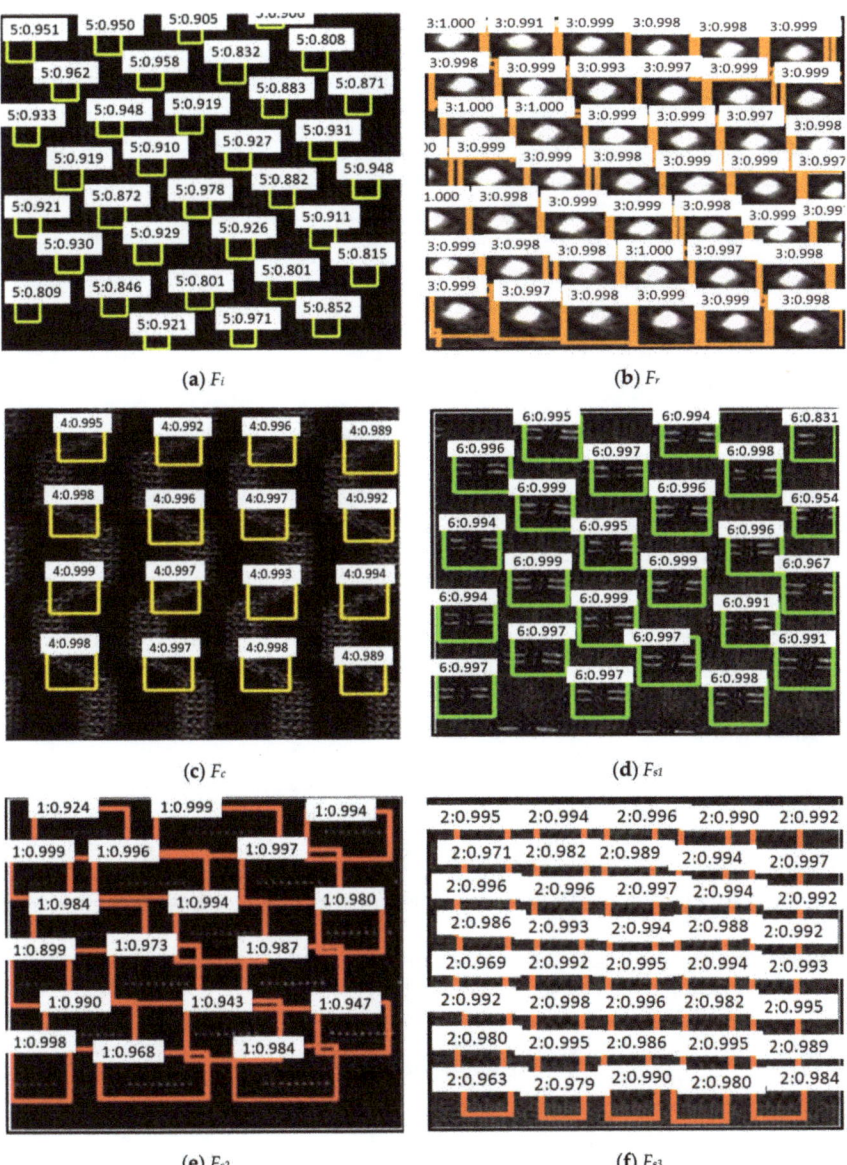

Figure 5. Detection results of the six types of patterns on the Faster R-CNN model. (**a**) Detection result of F_i. (**b**) Detection result of F_r. (**c**) Detection result of pattern F_c. (**d**) Detection result of F_{s1}. (**e**) Detection result of F_{s2}. (**f**) Detection result of F_{s3}.

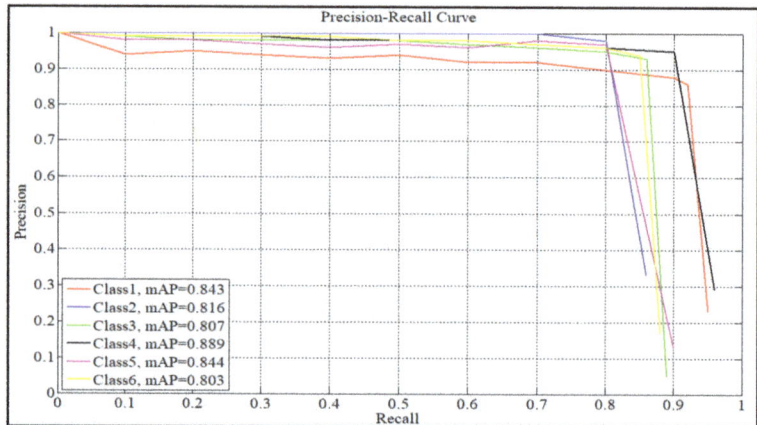

Figure 6. Precision–recall (P–R) curve of pattern detection by Faster R-CNN. The longitudinal axis indicates the detection precision, and the horizontal axis indicates the recall ratio. The area enclosed by the curve represents the mean average precision of the pattern detection.

2.3. Centre Point Extraction

The center point of the pattern is defined as the center of the region enclosed by the pattern outline. The area is defined as the number of pixels of a region. The center is calculated as the mean value of the line or column coordinates, respectively, of all pixels. The proposed method detected the approximate position of the pattern using the Faster R-CNN, and then an automatic threshold method was used to divide the pixel points of the pattern region and calculate the center coordinates. The steps are as follows:

Step 1: Image cropping method with overlapping areas is used for image cropping. The original image (Figure 7a) with size of 1208 × 1024 pixels is cropped into several sub-images (Figure 7b–e) of 640 × 512 pixels. The moving step length of the image cropping is approximately twice the length of the pattern period.

Step 2: The Faster R-CNN model is used to detect the pattern and output the classification score and categories. (Figure 7f–i).

Step 3: Merge sub-image F_{subi} according the coordinates of image cropping, obtaining image F_{new} (see Figure 8b). For example, define the coordinates of sub-image as (x_{sub}, y_{sub}), and the coordinates in image F_{new} as (x_{ori}, y_{ori}), so the coordinates (x_{ori}, y_{ori}) are computed as follows:

$$(x_{ori}, y_{ori}) = (x_{sub} + s_x \times (i-1), y_{sub} + s_y \times (j-1)) \tag{1}$$

where s_x and s_y are, respectively, the horizontal moving step length and the longitudinal moving step length. The variables i (i = 1, 2, 3 …) and j (j = 1, 2, 3 …), respectively, represent the times of horizontal and vertical cropping. Then combine the overlapping bounding boxes into one large bounding box according the maximum coordinates of overlapping bounding boxes, getting image F_{new1} (see Figure 8c).

Step 4: Correct the inaccurate bounding boxes and calculate the center points of the patterns to get the original center point map. First, the grey distribution information of the original image is counted by the grey histogram and the average grey value G_{th} is calculated. Second, G_{th} is used as a threshold to segment the patterns in the bounding boxes and calculate the area S_i (i = 1, 2 …) of segmented pattern, the average value S_{av} of S_i, the center point P_p of the pattern, and the center point P_b of the bounding box. Third, compare the size of S_i and S_{av}, and remove the bounding boxes which $S \leq S_{av}$. Fourth, adjust the positions of bounding boxes by moving the bounding boxes toward P_p to make $S > S_{av}$. Finally, segment the patterns f in the corrected bounding boxes using the threshold segmentation method and calculate them center points to obtain the original center point map (see Figure 9a).

$$f(x,y) = \begin{cases} 1 & f(x,y) > G_{th} \\ 0 & f(x,y) \leq G_{th} \end{cases}. \quad (2)$$

Step 5: Missing center points are filled based on four-neighbor-method to obtain the final center point map (see Figure 9c). First, approximately weft period length T_{hx} and approximately warp period length T_{hy} are counted from the original center point map. Second, the missing center point is between two known adjacent points A and B. If $(k + 1/2) < T_{hx} < d_m < (k + 3/2) T_{hx}$, where $k = 1, 2 \ldots$, and d_m is the distance between two adjacent corners A and B, then fill in k missing center points uniformly on the line AB. Suppose the filled point is N. Third, find two adjacent points C and D of N in the longitudinal direction. Finally, the missing point M (see Figure 9b) is the intersection between L_1 (the line formed by point A and point B) and L_2 (the line formed by point C and point D). Similarly, handle the cases with missing points in the vertical direction.

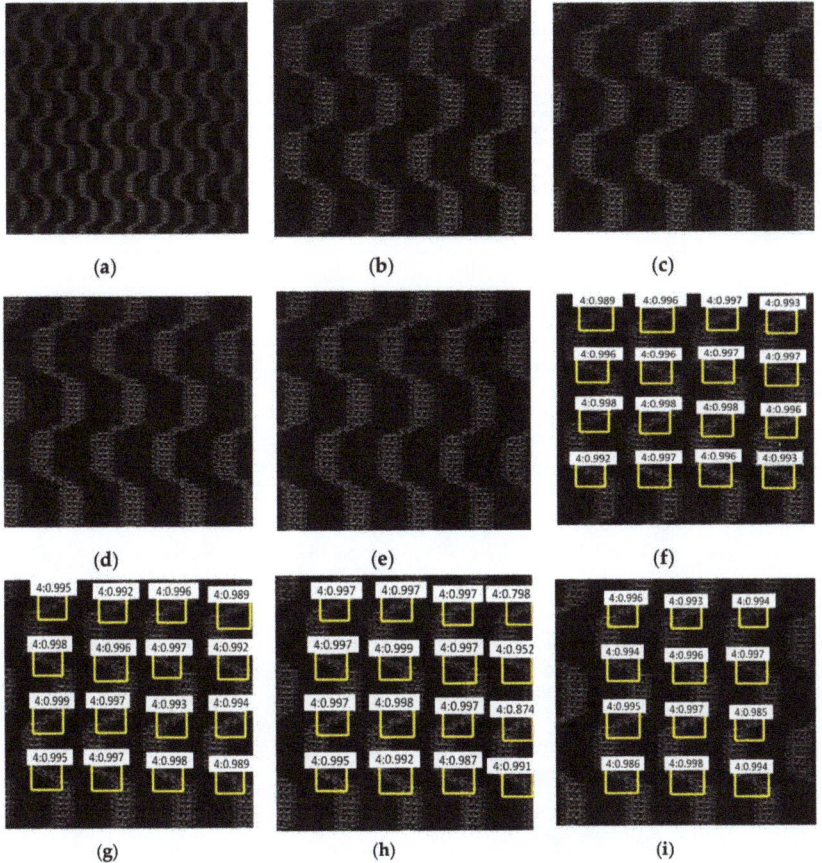

Figure 7. Intermediate process of the proposed method. (**a**) Original image. (**b**) Sub-image F_{sub1}. (**c**) Sub-image F_{sub2}. (**d**) Sub-image F_{sub3}. (**e**) Sub-image F_{sub4}. (**f**) Detection results of F_{sub1}. (**g**) Detection results of F_{sub2}. (**h**) Detection results of F_{sub3}. (**i**) Detection results of F_{sub4}.

Appl. Sci. **2019**, *9*, 3026

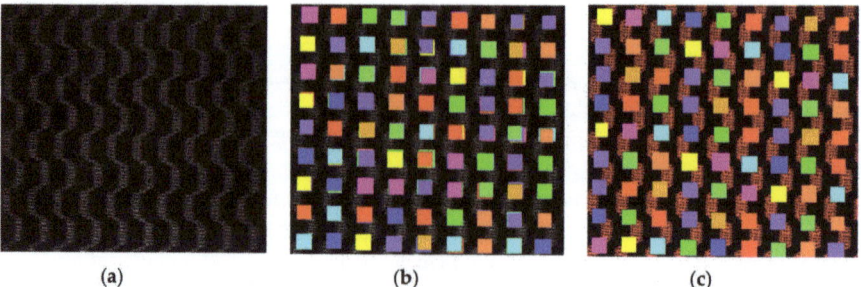

Figure 8. Bounding box mapping process. (**a**) Original image. (**b**) Bounding box mapping result F_{new}. (**c**) Bounding box merge result F_{new1}.

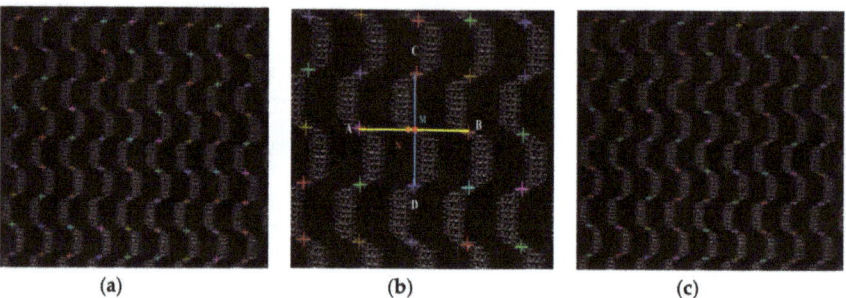

Figure 9. The procedure of the proposed method. (**a**) Original center point map. (**b**) Missing point filling process. (**c**) Final center point map.

2.4. Pattern Period Length and Skew Angle Measurement

The skew angles and period length can be measured based on the final center point map which reflects the center points distribution of the pattern.

The period lengths include the weft period length T_h and warp period length T_v. As shown in Figure 10b, the detected center points were denoted by the red points H_i, V_i, ($i = 1, 2, 3, \ldots$). One weft period length is $T_h = d_{H_i H_j}$, where $d_{H_i H_j}$ is the distance between the H_i and H_j ($j = i + 1, i = 1, 2, 3 \ldots$). Similarly, one warp period length is $T_v = d_{V_i V_j}$, where $d_{V_i V_j}$ is the distance between the V_i and V_j ($j = i + 1, i = 1, 2, 3 \ldots$).

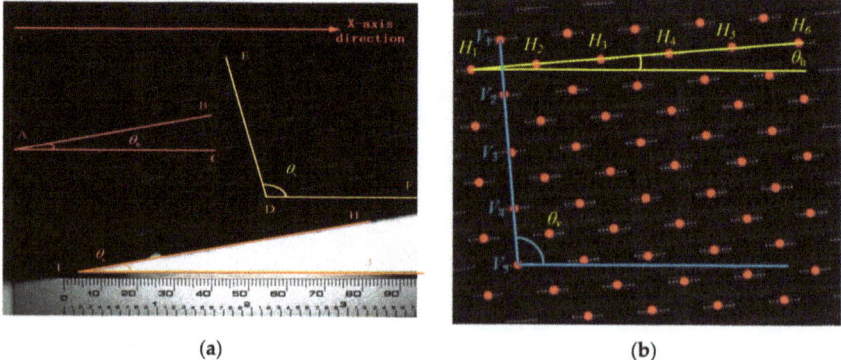

Figure 10. Measurement of the skew angles and period length. (**a**) Schematic diagram of pattern parameter measurement. (**b**) The pattern center points map of (**a**).

The skew angles also include the weft skew angle θ_{weft} and warp skew angle θ_{warp}. The θ_{weft} and θ_{warp} are calculated by θ_h, θ_v and θ_c in Figure 10a, where θ_c is the angle between the contour of the piece and the x-axis, measured by the measuring tool (see Figure 10a). Since this paper only studied the local pattern features of the piece, it is assumed that θ_c is a known angle. The way to obtain θ_h and θ_v is shown in Figure 10b. First, the least squares method is used to fit the center points H_i ($i = 1, 2, 3 \ldots$) in the weft direction as a weft period line L_h. Second, calculate the slope K_h of the line L_h. Finally, θ_h is calculated by the equation $\theta = \arctan(k)$. Similarly, the center points V_i ($i = 1, 2, 3 \ldots$) in the weft direction are used to obtain θ_v. The skew angles (θ_{weft} and θ_{warp}) can be obtained by the following equation:

$$(\theta_{wef}, \theta_{warp}) = (\theta_h - \theta_c, \theta_v - \theta_c) \qquad (3)$$

3. Results and Discussion

In this section, the proposed algorithm was used to test the six types of pattern images (F_i, F_r, F_c, F_{s1}, F_{s2} and F_{s3}) shown in Figure 11a–f. The center point maps reflecting the periodic characteristics of the pattern are shown in Figure 11g–l. Since θ_c in Figure 10a is an external angle and does not affect the overall accuracy of the angle to be measured, this paper measured and evaluated the θ_h and θ_v shown in Figure 10a.

To evaluate the proposed algorithm, we compare the proposed algorithm with manual measurement results. Manual measurement of the period length was achieved by clicking the center points of the patterns on the computer screen. As illustrated in Figure 12b, for example, we obtained the center points H_i, V_i, ($i = 1, 2, \ldots$) and then calculated the weft period length as $T_h = d_{HiHj}$, where d_{HiHj} is the distance between the H_i and H_j ($j = i + 1, i = 1, 2, \ldots$). And the warp period length as $T_v = d_{ViVj}$, where d_{ViVj} is the distance between the V_i and V_j ($j = i + 1, i = 1, 2, \ldots$). T_h and T_v were measured twenty times, and the final result was the average of the measurements. Similarly, manual measurement of the θ_h and θ_v also was accomplished by clicking the center points of the patterns on the computer screen. The center points in the same period direction were fitted as a straight line and the angle between the line and the x-axis was calculated (see Figure 11a). We measured each angle twenty times and then calculated the average values to obtain the result. The standard deviation δ was used to analyze the accuracy of the manual measurements of the period length and skew angle.

(a) F_{s2} (b) F_r (c) F_{s3}

(d) F_{s1} (e) F_c (f) F_i

(g) Centre point map of F_{s2} (h) Centre point map of F_r (i) Centre point map of F_{s3}

(j) Centre point map of F_{s1} (k) Centre point map of F_c (l) Centre point map of F_i

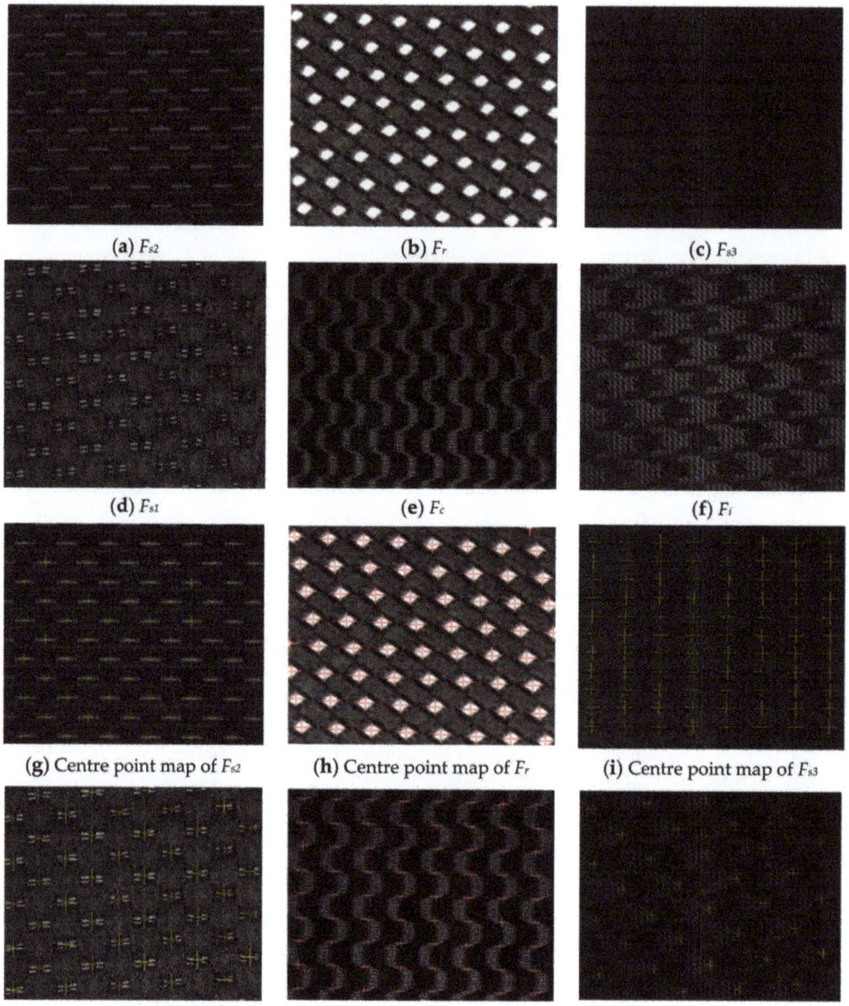

Figure 11. Six types of pattern images and their center point maps.

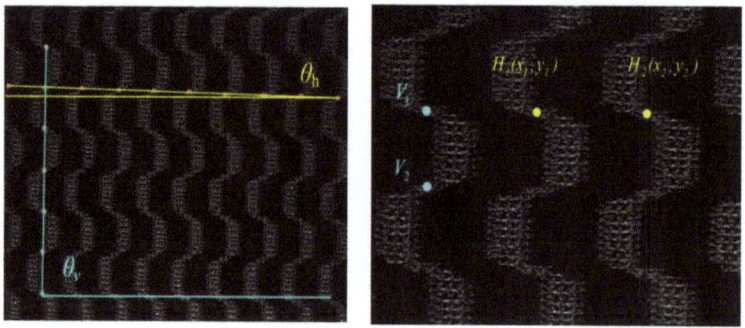

Figure 12. Manual measurement of parameters. (**a**) Skew angle, (**b**) period length.

The expression of the standard deviation is shown as follows:

$$\delta = \sqrt{\frac{1}{N}\sum_{i=1}^{N}(X_i - \bar{X})^2} \tag{4}$$

where $\bar{X}$ is the average value of X.

Table 2 shows the manual measurement standard deviation of period length of images (F_r, F_c, F_{s1}, F_{s2} and F_{s3}), where $\delta_{Th\alpha}$, $\delta_{Th\beta}$, and $\delta_{Th\gamma}$ were the minimum, maximum and average standard deviation, respectively, of the weft period length measurement. $\delta_{Tv\alpha}$, $\delta_{Tv\beta}$, and $\delta_{Tv\gamma}$ were defined similarly for the warp period length measurement. As shown in Table 2, the standard deviation obtained by manual measurement was very small. Therefore, it was reasonable to use the manual measurement results as the evaluation standard.

Table 2. Standard deviation of manual measurement of period length.

Image	Standard Deviation of T_h			Standard Deviation of T_v		
	$\delta_{Th\alpha}$	$\delta_{Th\beta}$	$\delta_{Th\gamma}$	$\delta_{Tv\alpha}$	$\delta_{Tv\beta}$	$\delta_{Tv\gamma}$
F_{s2}	0.044	0.378	0.081	0.006	0.286	0.175
F_r	0.069	0.451	0.167	None	None	None
F_{s3}	0.018	0.237	0.146	0.046	0.588	0.392
F_{s1}	0.008	0.164	0.132	0.052	0.682	0.458
F_c	0.157	0.533	0.259	0.098	0.226	0.181
F_i	0.088	0.576	0.472	0.038	0.574	0.282

Table 3 shows the period length measurement results of images (F_i, F_r, F_c, F_{s1}, F_{s2} and F_{s3} in Figure 11). The method based on image autocorrelation, the method based on distance matching function and the proposed method were compared with the manual measurement results, where T is the period length, the subscript h represents the weft direction, the subscript v represents the warp direction, m stands for manual measurement method, p stands for the proposed method, z stands for autocorrelation, d stands for distance matching function.

Table 3. Period length measurement results of various methods.

Image	T_{hm}	T_{hp}	T_{hz}	T_{hd}	T_{vm}	T_{vp}	T_{vz}	T_{vd}
F_{s2}	204.8	203.6	208.3	206.4	178.6	177.3	180.1	176.6
F_r	90.2	90.1	93.4	95.8	None	None	None	None
F_{s3}	102.2	101.8	202.8	128.5	62.8	63.7	59.7	65.4
F_{s1}	176.0	177.2	58.7	168.2	120.6	120.8	119.0	121.5
F_c	125.7	125.4	128.8	130.7	114.8	114.2	130.0	120.1
F_i	271.4	270.3	243.5	256.2	269.8	267.2	273.5	264.6

Table 4 shows the relative error value between different period length measurement methods and manual measurements of images (F_i, F_r, F_c, F_{s1}, F_{s2} and F_{s3} in Figure 11), where the e_{hpm} represents the relative error of T_{hp} with T_{hm}. e_{hzm} represents the relative error of T_{hz} with T_{hm}. e_{hdm} represents the relative error of T_{hd} with T_{hm}. Similarly, e_{vpm} is the relative error of T_{vp} with T_{vm}. e_{vzm} is the relative error of T_{vz} with T_{vm}. e_{vdm} is the relative error of T_{vd} with T_{vm}. The expression of the relative error RE is shown in the following Equation (5).

$$RE = \frac{X - T}{T} \times 100\% \tag{5}$$

where X represents the measured value and T represents the actual value.

Table 4. Relative errors of the period length measurements for the multi-period pattern piece.

Image	e_{hpm} (%)	e_{hzm} (%)	e_{hdm} (%)	e_{vpm} (%)	e_{vzm} (%)	e_{vdm} (%)
F_{s2}	0.587	1.709	0.781	0.728	0.839	1.119
F_r	0.111	3.548	6.208	None	None	None
F_{s3}	0.391	98.434	25.734	1.433	4.936	4.140
F_{s1}	0.682	66.647	4.432	0.166	1.326	0.746
F_c	0.239	2.466	3.978	0.523	13.240	4.617
F_i	0.405	10.280	5.601	0.964	1.371	1.927

From Tables 3 and 4, we could conclude that the period length measured by the proposed method had higher accuracy than the autocorrelation-based and the distance-matching function-based method. There was also a smaller relative error between the proposed method and the manual measurement result.

Similar to the evaluation of the period length, the standard deviation was used to suggest the reliability of the manual measurement of the angles. Table 5 shows the manual measurement standard deviation of θ_h and θ_v of images (F_i, F_r, F_c, F_{s1}, F_{s2} and F_{s3} in Figure 11), where the $\delta_{\theta h\alpha}$, $\delta_{\theta h\beta}$ and $\delta_{\theta h\gamma}$ are the minimum, maximum and average standard deviation of the θ_h, respectively. Similarly, $\delta_{\theta v\alpha}$, $\delta_{\theta v\beta}$ and $\delta_{\theta v\gamma}$ are the minimum, maximum and average standard deviation of the θ_v, respectively. From Table 5, we could conclude that the standard deviation of the manual measurements was small. Therefore, it was reasonable to use the manual measurement results as a benchmark to evaluate the measurement accuracy of the proposed measurement method.

Table 5. Standard deviation of manual measurement of θ_h and θ_v.

Image	Standard Deviation of θ_h			Standard Deviation of θ_v		
	$\delta_{\theta h\alpha}$	$\delta_{\theta h\beta}$	$\delta_{\theta h\gamma}$	$\delta_{\theta v\alpha}$	$\delta_{\theta v\beta}$	$\delta_{\theta v\gamma}$
F_{s2}	0.032	0.221	0.076	0.007	0.274	0.129
F_r	0.093	0.727	0.364	None	None	None
F_{s3}	0.009	0.158	0.074	0.171	0.447	0.300
F_{s1}	0.003	0.077	0.052	0.061	0.115	0.079
F_c	0.067	0.082	0.022	0.009	0.238	0.166
F_i	0.025	0.094	0.037	0.018	0.049	0.026

Table 6 shows the various methods measurement results of θ_h and θ_v of images (F_i, F_r, F_c, F_{s1}, F_{s2} and F_{s3} in Figure 11). The method based on corner detection method and the proposed method for measuring θ_h and θ_v were compared with the manual measurement results, where the subscript h represents the weft direction, the subscript v represents the warp direction, m stands for manual measurement method, p stands for the proposed method, c stands for corner detection.

Table 6. Measurement results of various methods of θ_h and θ_v.

Image	θ_{hm} (°)	θ_{vm} (°)	θ_{hp} (°)	θ_{vp} (°)	θ_{hc} (°)	θ_{vc} (°)
F_{s2}	0.182	88.502	0.188	88.424	0.397	84.186
F_r	3.578	None	3.686	None	4.858	None
F_{s3}	1.552	88.105	1.534	87.845	2.434	82.428
F_{s1}	1.784	86.238	1.779	86.173	2.379	89.327
F_c	1.874	88.386	1.824	88.693	1.467	85.486
F_i	2.836	86.332	2.868	86.546	3.542	88.248

Table 7 shows the relative error value between different angle measurement methods and manual measurements of images (F_i, F_r, F_c, F_{s1}, F_{s2} and F_{s3} in Figure 11), where the e_{hpm} represents the relative error of θ_{hp} with θ_{hm}. e_{vpm} represents the relative error of θ_{vp} with θ_{vm}. e_{hcm} represents the relative error of θ_{hc} with θ_{hm}. e_{vcm} represents the relative error of θ_{vc} with θ_{vm}.

Table 7. Relative errors of the angle measurements for the multi-period pattern piece.

Image	e_{hpm} (%)	e_{vpm} (%)	e_{hcm} (%)	e_{vcm} (%)
F_{s2}	3.297	0.088	118.131	4.877
F_r	3.018	None	35.434	None
F_{s3}	1.159	0.295	56.829	6.443
F_{s1}	0.280	0.075	98.991	3.582
F_c	2.668	0.347	21.238	3.281
F_i	1.128	0.248	24.894	2.219

The following observations were derived from Tables 6 and 7. The proposed method for measuring θ_h and θ_v achieved a smaller relative error compared to manual measurements. Compared with the corner detection-based method, the proposed method had higher accuracy and more stable performance in angle measurement.

4. Conclusions

The measurement of the skew angle and the period length is a fundamental problem in the quality inspection of multi-period pattern cutting pieces. We demonstrated a solution that Faster R-CNN efficiently detected the approximate location of the pattern and the method based on threshold achieved the precise location of pattern, which achieved the measurement of the skew angle and period length with high accuracy. We believe this work opens up exciting research opportunities to use the object detection network to extract the fabric pattern period, providing a new way to study pattern periodicity and can improve the detection accuracy of the pattern parameters.

Author Contributions: L.G. and Q.M. wrote the paper; Z.X. and Y.L. gave guidance in experiments and data analysis.

Funding: This work was sponsored by the Program for Innovative Research Team in University of Tianjin (No. TD13-5034), the Tianjin Research Program of Application Foundation and Advanced Technology under grant (No. 15JCYBJC16600) and the Textile Industry Association Applied Basic Research Program of China (J201509).

Conflicts of Interest: The authors declare no conflict of interest.

References

1. Reber, J.K.; Jill, A.R. Method and Apparatus for Cutting Pieces of Cloth for Use in Quilts or the Like. U.S. Patent No. 5,557,996, 24 September 1996.
2. Miao, Y.E. Image texture detection based on parallel gray level grade co-occurrence matrix. *Laser Infrared* **2011**, *41*, 1287–1291.
3. Sheng, J.; Guo-An, T.; Yang, T. Automatic Extraction Method for Texture Periodicity Based on Improved Normalized Distance Matching Function. *Pattern Recognit. Artif. Intell.* **2014**, *27*, 1098–1104.
4. Rocio, L.M.; Raul, E.; Sanchez, Y.; Victor, A.R. Periodicity and Texel Size Estimation of Visual Texture Using Entropy Cues. *Comput. Sist.* **2010**, *14*, 309–319.
5. Li, J.; Yang, Y.Q.; Shen, W.; Li, D.; Zhou, H. Research on fabric texture based on gray level co-occurrence matrix. *Adv. Text. Technol.* **2013**, *3*, 12–16.
6. Xiao, Z.; Nie, X.; Zhang, F.; Geng, L. Recognition for woven fabric pattern based on gradient histogram. *J. Text. Inst.* **2014**, *105*, 744–752. [CrossRef]
7. Jing, J.F.; Yang, P.P.; Li, P.F. Determination on design cycle of printed fabrics based on distance matching function. *J. Text. Res.* **2015**, *36*, 98–103.
8. Jian, Z.; Jingan, W.; Ruru, P.; Weidong, G. Periodicity Measurement for Fabric Texture by Using Frequency Domain Analysis and Distance Matching Function. *J. Donghua Univ.* **2017**, *43*, 629–633.
9. Ning, W.U.; Shengqi, G.; Shuaihua, X.U. Fabric Defect Detection Based on Periodicity and Local Directivity of Texture Edge. *Comput. Mod.* **2014**, *4*, 16–19.
10. Kahaki, S.; Nordin, M.; Ashtari, A. Contour-based corner detection and classification by using mean projection transform. *Sensors* **2014**, *14*, 4126–4143. [CrossRef] [PubMed]

11. Xiao, Z.; Pei, L.; Zhang, F.; Lei, G.; Wu, J.; Tong, J.; Xi, J.T.; Ogunbona, P. Measurement of surface parameters of three-dimensional braided composite preform based on curvature scale space corner detector. *Text. Res. J.* **2018**, *88*, 2641–2653. [CrossRef]
12. Xiao, Z.; Pei, L.; Zhang, F.; Sun, Y.; Geng, L.; Wu, J.; Tong, J.; Wen, J. Surface parameters measurement of braided preform based on local edge extreme. *J. Text. Inst.* **2019**, *110*, 535–542. [CrossRef]
13. Ren, S.; He, K.; Girshick, R.; Sun, J. Faster r-cnn: Towards real-time object detection with region proposal networks. *IEEE Trans. Pattern Anal. Mach. Intell.* **2015**, *39*, 91–99. [CrossRef] [PubMed]
14. Xu, Z.; Yang, W.; Meng, A.; Lu, N.; Huang, H.; Ying, C.; Huang, L. Towards End-to-End License Plate Detection and Recognition: A Large Dataset and Baseline. In Proceedings of the European Conference on Computer Vision (ECCV), Munich, Germany, 8–14 September 2018; pp. 255–271.
15. Tian, Z.; Huang, W.; He, T. Detecting text in natural image with connectionist text proposal network. In Proceedings of the European Conference on Computer Vision, Amsterdam, The Netherlands, 11–14 October 2016; pp. 56–72.
16. Ren, Y.; Zhu, C.; Xiao, S. Small object detection in optical remote sensing images via modified faster R-CNN. *Appl. Sci.* **2018**, *8*, 813. [CrossRef]
17. Girshick, R.; Donahue, J.; Darrell, T.; Malik, J. Rich feature hierarchies for accurate object detection and semantic segmentation. In Proceedings of the IEEE Conference on Computer Vision and Pattern Recognition, Columbus, OH, USA, 23–28 June 2014; pp. 580–587.
18. Girshick, R. Fast r-cnn. In Proceedings of the IEEE International Conference on Computer Vision, Santiago, Chile, 11–18 December 2015; pp. 1440–1448.
19. Dai, J.; Li, Y.; He, K.; Sun, J. R-fcn: Object detection via region-based fully convolutional networks. In Proceedings of the 30th International Conference on Neural Information Processing Systems, Barcelona, Spain, 5–10 December 2016; pp. 379–387.
20. Liu, W.; Anguelov, D.; Erhan, D.; Szegedy, C.; Reed, S.; Fu, C.; Berg, A.C. SSD: Single Shot MultiBox Detector. In Proceedings of the European Conference on Computer Vision, Amsterdam, The Netherlands, 11–14 October 2016; pp. 21–37.
21. Joseph, R.; Santosh, D.; Ross, G.; Ali, F. You only look once: Unified, real-time object detection. In Proceedings of the IEEE Conference on Computer Vision and Pattern Recognition, Las Vegas, NV, USA, 27–30 June 2016; pp. 779–788.
22. Yang, W.; Zhang, Z.; Wang, H.; Zhang, J. A vehicle real-time detection algorithm based on YOLOv2 framework. In Proceedings of the Real-Time Image and Video Processing 2018, Orlando, FL, USA, 15–19 April 2018.
23. Redmon, J.; Ali, F. Yolov3: An incremental improvement. *arXiv* **2018**, arXiv:1804.02767.
24. Uijlings, J.R.; Van De Sande, K.E.; Gevers, T.; Smeulders, A.W. Selective search for object recognition. *Int. J. Comput. Vis.* **2013**, *104*, 154–171. [CrossRef]
25. Lawrence Zitnick, C.; Piotr, D. Edge boxes: Locating object proposals from edges. In Proceedings of the European Conference on Computer Vision, Zurich, Switzerland, 6–12 September 2014; pp. 391–405.
26. R-CNN, Fast R-CNN, Faster R-CNN, YOLO—Comparisions. Available online: https://towardsdatascience.com/r-cnn-fast-rcnn-faster-r-cnn-yolo-object-detection-algorithms-36d53571365e (accessed on 10 February 2019).
27. Zeiler, M.D.; Rob, F. Visualizing and understanding convolutional networks. In Proceedings of the European Conference on Computer Vision, Zurich, Switzerland, 6–12 September 2014; pp. 818–833.
28. Simonyan, K.; Zisserman, A. Very deep convolutional networks for large-scale image recognition. *arXiv* **2014**, arXiv:1409.1556.
29. He, K.; Zhang, X.; Ren, S.; Sun, J. Deep residual learning for image recognition. In Proceedings of the IEEE Conference on Computer Vision and Pattern Recognition, Las Vegas, NV, USA, 26 June–1 July 2016; pp. 770–778.

© 2019 by the authors. Licensee MDPI, Basel, Switzerland. This article is an open access article distributed under the terms and conditions of the Creative Commons Attribution (CC BY) license (http://creativecommons.org/licenses/by/4.0/).

Article

Intelligent Identification of Maceral Components of Coal Based on Image Segmentation and Classification

Hongdong Wang [1], Meng Lei [1,2], Yilin Chen [3], Ming Li [1] and Liang Zou [1,2,*]

[1] School of Information and Control Engineering, China University of Mining and Technology, Xuzhou 221116, China
[2] Department of Electrical and Computer Engineering, University of British Columbia, Vancouver, BC V6T 1Z4, Canada
[3] School of Resources and Geosciences, China University of Mining and Technology, Xuzhou 221116, China
* Correspondence: liangzou@ece.ubc.ca

Received: 19 July 2019; Accepted: 6 August 2019; Published: 8 August 2019

Featured Application: Maceral Analysis; Coal Processing.

Abstract: An intelligent analytical technique which is able to accurately identify maceral components is highly desired in the fields of mining and geology. However, currently available methods based on fixed-size window neglect the shape information, and thus do not work in identifying maceral composition from one entire photomicrograph. To address these concerns, we propose a novel Maceral Identification strategy based on image Segmentation and Classification (MISC). Considering the complex and heterogeneous nature of coal, a two-level coarse-to-fine clustering method based on K-means is employed to divide microscopic images into a sequence of regions with similar attributes (i.e., binder, vitrinite, liptinite and inertinite). Furthermore, comprehensive features along with random forest are utilized to automatically classify binder and seven types of maceral components, including vitrinite, fusinite, semifusinite, cutinite, sporinite, inertodetrinite and micrinite. Evaluations on 39 microscopic images show that the proposed method achieves the state-of-the-art accuracy of 90.44% and serves as the baseline for future research on maceral analysis. In addition, to support the decisions of petrologists during maceral analysis, we developed a standalone software, which is freely available at https://github.com/GuyooGu/MISC-Master.

Keywords: maceral components; image segmentation; coal petrography; random forest; two-level clustering

---

1. Introduction

1.1. Background and Motivation

Coal is an extremely complex heterogeneous material formed from ancient wetlands over geological processes. It consists of various organic components called macerals and a lesser amount of inorganic minerals [1,2]. Different from the minerals with homogeneous internal composition and structures, the macerals derived from coalified plant tissues have distinct physical and chemical properties, and are related to the degree of coalification. In addition, the maceral composition is an important factor in evaluating the coal seam quality. Precise identification of the maceral components has a multitude of uses across various industry sectors, including hydrogenation, combustion, carbonization and gasification [2–4].

Macerals can be categorized into three basic groups through petrographic analysis, including vitrinite from coalified woody tissue, liptinite from more decay-resistant parts of plants and inertinite from hydrogen-rich plant and decomposition products. These maceral groups are subdivided into maceral subgroups and macerals [5]. Most laboratories associated with the coke-making industry Standard Test Method for Microscopical Determination of the Maceral Composition follow the standard test methods, such as International Commission for Coal Petrology (ICCP) standard [6] and American Society for Testing and Materials (ASTM) standard D2799-13 [7], and the microscopic analysis is always performed manually for the identification of maceral components. Despite being the most widely used method, it is costly and labor-intensive to identify the maceral composition due to the complicated nature and substantial diversification of the petrographical properties of coal. Even for specialists in petrography, they may arrive at different judgements in the analysis of the same microscopic image as a consequence of the subjective factors. To address these concerns, an intelligent analytical technique which is able to automatically provide objective identification of maceral components is highly desired for the growing industrial demand.

1.2. Related Work

Automatic geological identification is becoming an increasingly important technique in various fields, such as in mining and geology. Camalan et al. presented a novel strategy to estimate the liberation spectrum from optical micrographs via random forest [8]. Lei et al. proposed an autonomous classification method of rock images via unsupervised feature learning [9]. In [10], transfer learning was employed to deal with the problem of cross-region microscopic sandstone images classification. Numerous attempts have been made on the classification of microscopic rock images and achieved great success [11–13]. Considering the heterogeneous nature of coal, automatic identification of maceral components is still a challenging task [3].

In the early stage, the analytical methods to estimate the volume proportions of coal macerals were mainly based on the gray scale value of pixels [14,15] and provided interesting results. However, the liptinite and background resin have similar gray scale values, and therefore it is difficult to separate them. In addition, different maceral components in a maceral group differ only subtly in term of the reflectance, and the existing methods merely based on gray scale values are not suitable for distinguishing maceral components. Furthermore, the gray scale values of a specific component may vary over a large range with the degree of coalification. Although the gray scale descriptions of pixels remain important, the need for more quantitative features, such as shape and texture, from photomicrographs has been recognized. With the development of machine learning and image processing, it is possible to automatically make more elaborate classifications [12,16]. Over the past several years, attempts based on machine learning techniques have shown promising results in maceral analysis.

Wang et al. utilized principal component analysis (PCA) to extract primary features from texture-related and intensity-related features, and employed Support Vector Machine (SVM) to classify maceral components of the vitrinite group [17]. Skiba et al. selected 10 textural features via PCA and developed a novel strategy for automatic identification of macerals of the inertinite group. The proposed method achieved an outstanding accuracy of 93.6% based on a group of neural networks [16]. Most of the works focus on a single maceral group. So far, attempts to provide full identification of macerals have been considerably limited. Młynarczuk and Skiba evaluated the ability of three machine learning methods for identifying three maceral groups of coal and non-organic minerals [3] . They cropped the region of interest (ROI) of 41 px × 41 px and determined the label of the central pixel. Considering the morphological gradients along with the gray level features, the proposed method achieved an accuracy of 97% in classifying maceral groups. Furthermore, they analyzed six kinds of macerals of the inertinite group and an obtained accuracy is over 91%. Pearson and CSIRO have released two automated tools to identify maceral composition, including Pearson Coal Petrography (Pearson Coal Petrography—http://www.coalpetrography.com/blog1/) and CSIRO coal grain analysis (CGS) (Coal Grain Analysis—https://www.csiro.au/en/Do-business/Commercialisation/

Marketplace/Coal-Grain-Analysis), which have been successfully commercialized. However, they do not mention the detailed technologies employed in these two tools on the corresponding websites.

Despite providing inspiring performance via machine learning-based methods, there are many issues that require more scientific breakthroughs. The motivations of the proposed method derive from the following three aspects:

First and foremost, both patch-wise classification aiming to assign a label to a given region and pixel-wise classification aiming to provide a label for the central pixel of ROI neglect the shape and the size information. For instance, the micrinites are always small in size and the cutinites are very thin [18,19]. In cases where the selected regions contain two or more groups/components of macerals, it directly affects the performance. The results of the previous methods based on fixed-size window are always observed with poor generalized ability.

Second, due to the complex and heterogeneous nature of coal, the task for identifying macerals requires more parameters describing the shape, texture and morphology. Comprehensive features along with powerful machine learning techniques are required to detect the subtle differences between maceral groups/components.

Last but not least, there is no publicly available software for identifying maceral groups/components, especially targeted for geologists without strong expertise in the machine learning and image processing domains. In addition, the existing methods focus on predicting the label for a given region or pixel, whereas they do not work in the identification of maceral components from the entire photomicrograph.

To address the above-mentioned concerns, we propose a novel framework for autonomous coal macerals identification based on image segmentation and classification (MISC). A two-level coarse-to-fine clustering strategy is implemented for image segmentation, and random forest is employed to classify maceral components from the entire microscopic image. The main contributions of the proposed framework can primarily be broken down into three aspects.

1. Inspired by the distribution of maceral subgroups and the gray scale characteristics of maceral groups, we design a coarse-to-fine segmentation strategy to divide an entire photomicrograph into a number of discrete regions, providing the shape and the size information. Both coarse clustering and fine clustering are based on K-means, which is one of the most popular unsupervised image segmentation methods.
2. We extract the discriminative features from microscopic images, including geometric, grayscale and texture features, which are combined into a 172-element feature vector. A comprehensive feature combination for identification of maceral components, not limited to maceral groups, is proposed. In addition, we evaluate six kinds of machine learning classifiers, and the random forest provides the best performance with an average accuracy of 90.44%.
3. A publicly available tool, namely MISC, to identify macerals in microscopic images of coal is released. The software integrates the best segmentation and classification algorithms involved in this paper, and provides an AI-assisted autonomy algorithm for maceral components identification.

2. Experiment Dataset

The metallurgical coal samples used in the study are randomly selected from samples submitted to the laboratory of the United States Geological Survey (USGS). The selected samples were prepared through a sequence of operations, including sieving, molding and polishing. All the procedures follow protocols established by the ASTM Standard D2797 [20]. Photomicrographs are captured using a Leica DMRX microscope with a Leica DFC 480 digital camera under incident white light in oil immersion, in accordance with ASTM standard D2799-13 [7].

With the increment of coalification, the difference in term of gray levels between macerals is reduced. It will be difficult to distinguish between vitrinite and liptinite at a high degree of coalification (e.g., R0 > 1.25%). In this work, the selected coal samples are with a relatively low degree of coalification (i.e., R0 < 1%). The maceral composition of each coal sample was annotated by 5 petrographers according to ASTM Standard D2799-13, and the 39 samples out of 50 samples with consistent results were further analyzed. The resolutions of these photomicrographs are different with each other, in the range of (267–1024) × (230–768) px, with each pixel roughly corresponding to 2–4 μm. Table 1 shows all the maceral components analyzed in this study, with brief descriptions and the number of macerals (909 in total). In addition, 64 objects belonging to binder are also included in the dataset. Binder is large relative to maceral components and can hold these components together. The demonstration of each maceral is provided in Figure 1.

Table 1. Brief description of seven types of macerals used in this study.

Maceral Group	Maceral	Brief Description of Specific Maceral	The Number of Macerals
vitrinite	—	The predominate maceral in most coals of intermediate reflectance. It is always brilliantly glossy resembling vitreous. Vitrinite is derived from coalified woody tissue and occurs generally in thin bands of 2–10 mm thickness. It plays an important role in defining the properties of the whole coal.	116
liptinite	sporinite	A liptinite maceral exhibition various lenticular, oval, round forms, or small rod-like projections.	102
liptinite	cutinite	A maceral is derived from the stratum corneum of plant leaves, roots and stems. It occurs as stringers strips of varying thickness, with a smooth outer margin and serrated edges. It is not very abundant.	30
inertinite	fusinite	An inertinite maceral distinguished principally by the preservation of some features of the plant cell wall structure. It has charcoal-like structure and is commonly broken into small shards and fragments.	198
inertinite	semifusinite	It looks like fusinite in morphology. It has the largest range of reflectance. The partial size is always great than 50 μm except when occurring as a fragment within binder.	122
inertinite	inertodetrinite	Small, discrete inertinite fragments (>2 μm in size) of varying shape. Reflectance values of inertodetrinite are greater than surrounding vitrinite macerals.	141
inertinite	micrinite	Generally, micrinite is non-angular, and occurs as particles around 1 to 5 μm diameter.	200

Figure 1. Examples of maceral components and binder. Each row represents a class. They are binder, vitrinite, fusinite, semifusinite, cutinite, sporinite and inertodetrinite from top to bottom respectively. The black color areas (i.e., RGB value = 0) in each figure represent the background of the given maceral component. The micrinite is not shown here for its small size accounting for only a few pixels.

3. Methods

3.1. Image Segmentation Based on Two-Level Clustering

The main differences between maceral groups are the gray scale values. Generally, the gray scale values of liptinite, vitrinite and inertinite decrease successively, as demonstrated in Figure 2a. The gray scale distribution curves illustrate that there are noticeable differences among three maceral groups and the binder. However, it is difficult to define the boundaries between them. In addition, the boundaries corresponding to different photographs are different. The gray scale range of each maceral group varies with the degree of coalification. Figure 2b shows the distribution of gray scale values of binder and inertinite across four photomicrographs. It can be seen that the gray range of binder is relatively fixed, whereas that of inertinite group of 4 coal samples differs greatly. Therefore, it is unreliable to adopt a fixed threshold to segment microscopic photographs with different coalification degrees.

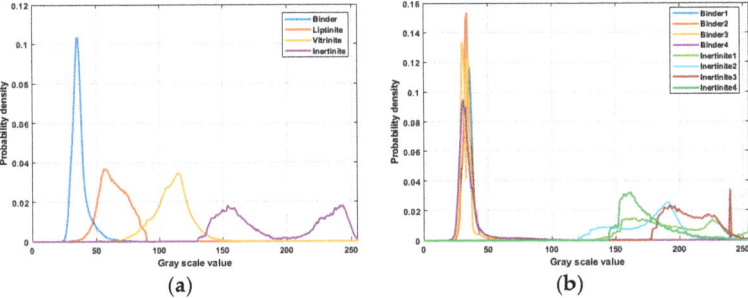

Figure 2. Comparison of gray scale distributions corresponding to different maceral groups and the binder. (**a**) Gray scale distributions of maceral groups and the binder in one coal sample; (**b**) the difference in the gray scale distributions across 4 coal samples. For simplicity, we only show the curves corresponding to binder and inertinite in (**b**).

In order to automatically detect the boundaries of each maceral, we employed image-wise segmentation which is able to divide each microscopic image into a sequence of discrete regions, each having similar attributes. It is an essential step for further maceral composition analysis. Considering the fact that the maceral components within each group mostly are not adjacent to each other and the gray scale values are the major difference between maceral groups, we adopt the gray scale values of each pixel as the features. K-means clustering, one of the most favorable clustering techniques, is utilized for its simplicity and computational efficiency [21,22].

The main steps of K-means algorithm can be summarized as follows:

(1) Choose initial cluster centroids $\mu_1, \mu_2, \mu_3, \ldots, \mu_k \in R^d$ randomly, where k represents the number of clusters and d represents the dimension of the feature space.

(2) Repeat until convergence:

For each pixel p, assign it to its nearest centroid,

$$c^p := \arg\min_j \|v^p - \mu_j\|^2 \qquad (1)$$

Update each centroid,

$$\mu_j := \frac{\sum_{i=1}^{M} 1\{c^p = j\} v^p}{\sum_{i=1}^{M} 1\{c^p = j\}} \qquad (2)$$

where c^p represents the cluster of pixel p, v^p is a vector consisting of RGB value of pixel p, and M represents the total number of pixels in each photomicrograph, respectively.

In addition, due to the complexity of coal properties, it is difficult to achieve accurate clustering results by using single-level clustering only. Single-level clustering is difficult to provide satisfactory segmentation results. Hence, a coarse-to-fine strategy is adopted. In the coarse clustering level, the regular K-means clustering algorithm is first applied to get rough clustering results, which splits a whole image into two sub-clusters. In the fine clustering level, each of the previous clusters is partitioned again into two fine sub-clusters (i.e., maceral groups and binder).

3.2. Feature Extraction

Inspired by the way that the petrologists examine photomicrographs, we extracted three types of features for maceral identification. Table 2 lists 172 features utilized in this study, such as reflectance contrasts, shape, morphology and size, which can be categorized into geometric, grayscale and texture features [23]. The detailed descriptions of these features can refer to the papers on image pattern recognition [24,25].

Table 2. Feature space utilized in this study.

Geometric Features (x1–x16)	Grayscale Features (x17–x90)	Texture Features (x91–x172)
x1: Area	x17: Mean gray value	x91–x92: Mean and standard deviation of energy
x2: Perimeter	x18: Standard deviation of gray value	x93–x94: Mean and standard deviation of entropy
x3: Rectangle degree	x19: Max gray value	x95–x96: Mean and standard deviation of inertial moment
x4: Aspect ratio	x20: Min gray value	x97–x98: Mean and standard deviation of correlative
x5: Length of long axis	x21: Gray scale median	X99–x100: Small and large gradient advantage
x6: Length of short axis	x22: Gray scale mode	x101–x102: Inhomogeneity of grayscale and gradient distribution
x7: Eccentricity	x23: Average contrast	x103: energy
x8: Solidity	x24: Smoothing Degree	x104–x105: Mean value of grayscale and gradient
x9: Extent	x25: Third-order Moment	x106–x107: Mean variance of grayscale and gradient
x10–x16: Hu' seven invariant moments	x26–x90: Grayscale probability	x108: correlative
		x109–x111: Grayscale entropy, gradient entropy, mixed entropy
		x112: inertia
		x113: Deficit moment
		x114–x172: Fifty-nine local binary pattern features

For instance, we employ the image moment as the shape descriptor. The moment invariants have been extensively exploited to characterize image patterns. Among various image moments, Hu's 7 invariant moments have been widely applied in a variety of applications for its invariant features on image translation, scaling and rotation [26]. They are defined as follows:

$$
\begin{aligned}
M_1 &= \eta_{20} + \eta_{02} \\
M_2 &= (\eta_{20} - \eta_{02})^2 + 4\eta_{11} \\
M_3 &= (\eta_{30} - 3\eta_{12})^2 + (3\eta_{21} - \eta_{03})^2 \\
M_4 &= (\eta_{30} + \eta_{12})^2 + (\eta_{21} + \eta_{03})^2 \\
M_5 &= (\eta_{03} - 3\eta_{12})(\eta_{30} + \eta_{12})[(\eta_{30} + 3\eta_{12})^2 - 3(\eta_{21} + \eta_{03})^2] + \\
&\quad (3\eta_{21} - \eta_{03})(\eta_{21} + \eta_{03})[3(\eta_{30} + \eta_{12})^2 - (\eta_{21} + \eta_{03})^2] \\
M_6 &= (\eta_{20} - \eta_{02})[(\eta_{30} + \eta_{12})^2 - (\eta_{21} + \eta_{03})^2] + 4\eta_{11}(\eta_{30} + \eta_{12})(\eta_{21} + \eta_{03}) \\
M_7 &= (3\eta_{21} - \eta_{03})(\eta_{30} + \eta_{12})[(\eta_{30} + \eta_{12})^2 - 3(\eta_{21} + \eta_{03})^2] + \\
&\quad (3\eta_{21} - \eta_{30})(\eta_{21} + \eta_{03})[3(\eta_{30} + \eta_{12})^2 - (\eta_{21} + \eta_{03})^2]
\end{aligned}
\tag{3}
$$

$$
\eta_{ab} = \frac{\mu_{ab}}{\mu_{00}^\rho}, \rho = \frac{a+b}{2} + 1 \tag{4}
$$

where μ_{ab} represents the central moment and η_{ab} stands for the normalized central moments.

The features x17–x90 are the statistical characters related to the gray scale values of the region of interest; x91–x99 are statistics for examining texture features based on the spatial relationship of pixels [27]; x99–x113 are the gray gradient features [28]; and the remaining features x114–x172 correspond to local binary patterns encoding the texture information [29].

3.3. Random Forest for Image Classification

Random forest (RF) is an ensemble machine learning method, which consists of multiple uncorrelated decision trees. It is widely used in image classification tasks due to its high accuracy, easy parameterization and robustness against overfitting [30]. Figure 3 illustrates how the random forest model works. Given the dataset with N samples $D = \{(x^1, y^1), \ldots, (x^l, y^l), \ldots, (x^N, y^N)\}$, where $x^l = [x^l(1), x^l(2), \ldots, x^l(172)]$ and $y^l \in \{1, 2, 3, 4, 5, 6, 7, 8\}$ denote the input 172 features and the output label of sample l, the general idea of random forest can be described as,

(1) Randomly select N samples with replacement from the original dataset, and obtain N subsamples for constructing each tree.
(2) Select features for constructing decision tree nodes from a random subset of all 172 features, and construct a decision tree.
(3) Repeat step (1) and (2) for B times and construct a random forest with B trees. The final prediction result is obtained by the majority vote of the trees in the forest.

As an ensemble model, random forest model fits the input data in a shorter time as each decision tree is independent, making parallel computing and modeling possible [31,32]. We also test the performance of the other five machine learning methods, including Fine Tree, Radial Basis Function kernel Support Vector Machine (RBF SVM), Weighted K-Nearest Neighbors (KNN), Linear Discriminant and Subspace KNN [33,34].

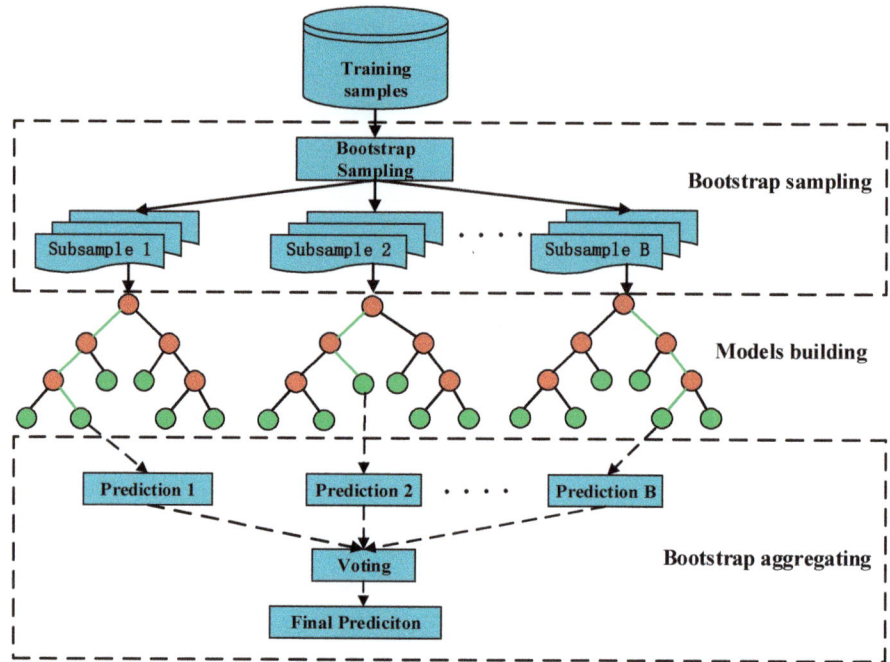

Figure 3. The scheme of random forest algorithm. The final prediction is obtained by taking a majority vote of the predictions from all the trees in the forest.

3.4. Evalutation Criteria

The results of the automated segmentation method are quantitatively evaluated by using three popular evaluation criteria including clustering accuracy, entropy and purity. We refer to the class labels as the ground truth and the results of clustering algorithms as the clusters [35,36]. We refer to class as the ground truth and cluster as the results of clustering algorithms. Clustering accuracy is the most intuitive measure to evaluate the performance of clustering, which is defined as follows

$$accuracy = \sum_{i=1}^{4} \frac{n_{ii}}{n} \quad (5)$$

where n_{ii} represents the number of common samples in cluster i and class i, $i \in \{1, 2, \ldots, 4\}$, and n is the size of the data set.

Entropy is an information theoretic measure and is defined as

$$entropy = -\sum_{i=1}^{4} \frac{n_i}{n} \sum_{j=1}^{4} \frac{n_{ij}}{n_i} \log_2 \frac{n_{ij}}{n_i} \quad (6)$$

where n_{ij} indicates the number of common samples in cluster i and class j, n_i stands for the number of samples in cluster i.

Purity is computed to measure the degree of clusters containing a single class. The purity is calculated as follows

$$purity = \sum_{i=1}^{4} \frac{n_i}{n} \max(\frac{n_{ij}}{n_i}) \quad (7)$$

4. Experimental Results and Discussion

4.1. Image Segmenation

The proposed segmentation strategy has been tested on 39 microscopic images taken by Leica DFC 480 digital camera, and the results in terms of accuracy, purity and entropy are given in Table 3. It can be observed that the proposed two-level K-means algorithm achieved significantly higher accuracy (90.82%), higher purity (90.82%) and lower entropy (0.6042) than the other clustering algorithms. In particular, the output result via two-level coarse-to-fine clustering consistently has better segmentation results as compared to the corresponding single-level clustering. For instance, regarding the K-means algorithm, the accuracy of the two-step strategy is 17.59% better than that of applying single-level K-means.

Table 3. Quantitative assessment of automatic segmentation methods.

Methods	Accuracy (%)	Purity (%)	Entropy
Fuzzy c-means	69.35	86.36	0.7291
K-medoids	68.34	89.94	0.6805
K-means	73.21	89.43	0.6748
2-level Fuzzy c-means	76.58	83.33	0.7961
2-level K-medoids	82.14	85.31	0.7306
2-level K-means	90.82	90.82	0.6042

Furthermore, in this paper, one out of those tested images was selected to visualize the performance of the proposed strategy and the other five kinds of clustering methods. The segmentation results of single-level clustering and two-level clustering are compared with the ground truth segmentations provided by five petrologists for evaluation. It can be seen from Figure 4 that the boundary of the resultant segmentation images by K-means is slightly clearer as compared to Fuzzy c-means (FCM) and K-medoids clustering. The segmented images produced by the two-level clustering are sharper and much closer to the ground truth. Overall, the proposed two-level coarse-to-fine clustering strategy based on K-means has outperformed the other clustering algorithms.

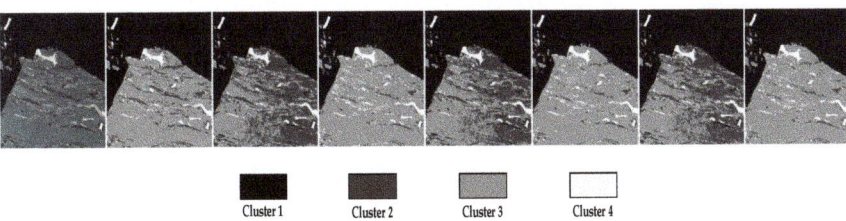

Figure 4. Comparison of automated segmentation results with ground truth. From left to right, the images represent original image, ground truth, the segmentation results of single-level Fuzzy c-means (FCM), 2-level FCM, single-level K-medoids, 2-level K-medoids, single-level K-means, and 2-level K-means.

We also compare the results of image segmentation with the results of the fixed-size window strategy. As can be seen from Figure 5, it is feasible to detect the thin sporinites (i.e., a-1) and the granular micrinites (i.e., a-2). We can obtain the shape information for each object of interest. As to the fixed-size window strategy, 41 px × 41 px was demonstrated to be the optimal size for maceral identification [3]. However, it is unrealistic to retrieve the shape information in feature extraction. In addition, micrinites distributing through the window may only account for a small amount of the area, and therefore it might be unreliable to train the classification models based on the information provided by the whole window. Similarly, sporinites are always thin and they might be misclassified based on the fixed-size window strategy. Our method is also very effective in extracting maceral composition with a large area (i.e., a-3), which is helpful to improve classification accuracy. Although the proposed strategy achieved satisfying performance in identifying the objects of interest (i.e., maceral groups), the differences in term of gray scale values between maceral components are too subtle to differentiate them. More discriminative features are required.

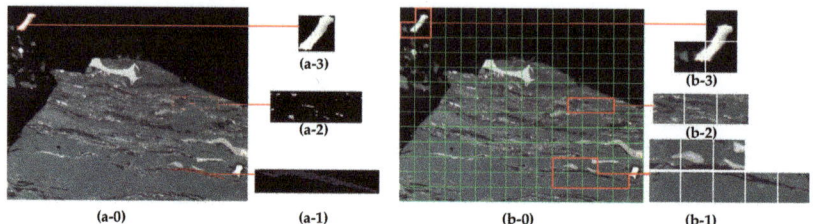

Figure 5. The results of image segmentation based on (**a**) the proposed Maceral Identification strategy based on image Segmentation and Classification (MISC) and (**b**) the fixed-size window strategy. (a-0) and (b-0) denote the original image. (a-1) and (b-1) are the enlargement of sporinites, (a-2) and (b-2) are the enlargement of micrinites, (a-3) and (b-3) are the enlargement of fusinites.

4.2. Maceral Composition Classification

Each microscopic image was divided into a series of discrete objects (i.e., macerals groups). Then we extracted geometric, grayscale, texture features from each object, and created a 172-dimensional feature vector for each object. We compared the recognition performance of random forest with five popular classification methods via a 10-fold cross validation. The experiments were repeated 10 times and the average accuracies were reported in Table 4. This summarizes the classification results of all the 973 objects obtained by image segmentation. The proposed approach yields a high accuracy of 90.44%, outperforming other classifiers. This should be regarded as a very good performance, especially when we take into account the high degree of complexity of coals. The obtained results show that the proposed strategy based on image segmentation and classification has a high potential for maceral components identification.

Table 4. Comparing the results of random forest and the other 5 machine learning methods for identifying maceral components (10-fold cross-validation, and repeat the experiment 10 times).

Classifier	Classification Accuracy (%) ± std (%)			
	Geometric Features	Grayscale Features	Texture Features	All Features
Fine Tree	72.29 ± 0.72	70.01 ± 0.83	78.79 ± 1.02	85.71 ± 1.08
Radial basis function kernel support-vector machine	72.47 ± 0.11	76.97 ± 0.59	85.54 ± 0.33	86.62 ± 0.58
Weighted K-Nearest Neighbors	55.22 ± 0.49	71.40 ± 0.56	80.44 ± 0.47	78.69 ± 0.88
Linear Discriminant	53.76 ± 0.32	70.34 ± 0.46	82.12 ± 0.40	83.36 ± 0.60
Subspace K-Nearest Neighbors	55.45 ± 0.74	71.13 ± 0.68	79.61 ± 0.56	81.26 ± 0.47
Random Forest	79.30 ± 0.47	78.86 ± 0.34	85.64 ± 0.30	90.44 ± 0.37

We further test the performance of an individual kind of features. The recognition performance using texture features is much higher than that of the other two types of features. Generally, the fusion of multiple kinds of features can achieve a significant improvement over a single kind of features, except for weighted KNN classifier.

To observe relations between the predictions of classifiers and the true labels, we also employ confusion matrices to report the results of different approaches. The confusion matrix enables us to know not only the error rates being made by a classifier but also the types of errors. More specifically, the rows of the matrix represent the predicted class, and the columns correspond to the true class (i.e., ground truth). The green diagonal cells stand for the number of correctly classified observations, while the red cells represent the misclassified observations. The precision and the recall rate corresponding to each class are also shown at the far right of each row and the bottom of each column, respectively. The overall accuracy is shown at the bottom-right corner of the matrix. As shown in Figure 6, the proposed method provides satisfying performance for most of the maceral components. The main flaws of all these six classifiers come from the misclassification of semifusinite and inertodetrinite. The following reasons could contribute to the worse performance for these two components: semifusinite and inertodetrinite belong to inertinite group, and the difference in terms of gray level is too subtle to classify them properly [37]; semifusinite is intermediate between fusinite and vitrinite, and has a similar texture and general structure to fusinite; the origin of inertodetrinite is similar to fusinite and semifusinite [38].

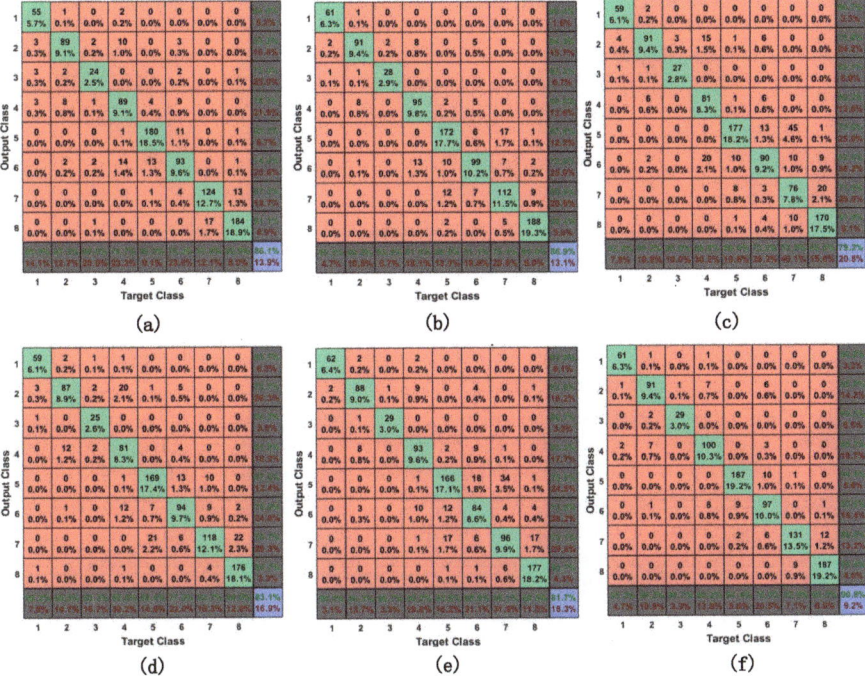

Figure 6. Confusion matrix comparison of six classifiers, including the results of (**a**) Fine Tree; (**b**) Radial basis function kernel support vector machine; (**c**) Weighted K-Nearest Neighbors (KNN); (**d**) Linear Discriminant Analysis; (**e**) Subspace KNN and (**f**) Random Forest. The labels in each subfigure include 1: binder, 2: sporinite, 3: cutinite, 4: vitrinite, 5: fusinite, 6: semifusinite, 7: inertodetrinite, and 8: micrinite.

The process of training a random forest involves the construction of multiple decision trees. In this study, we also evaluated the classification accuracy with the increase of the number of trees from 1 to 300. As shown in Figure 7, in general, more trees can provide the better performance, especially when the number of trees is smaller than 50. However, the improvement decreases as the number of trees increases from 50. Considering the tradeoff between the computational efficiency and the robustness of the developed model, in this study, we set the number of trees to be 200.

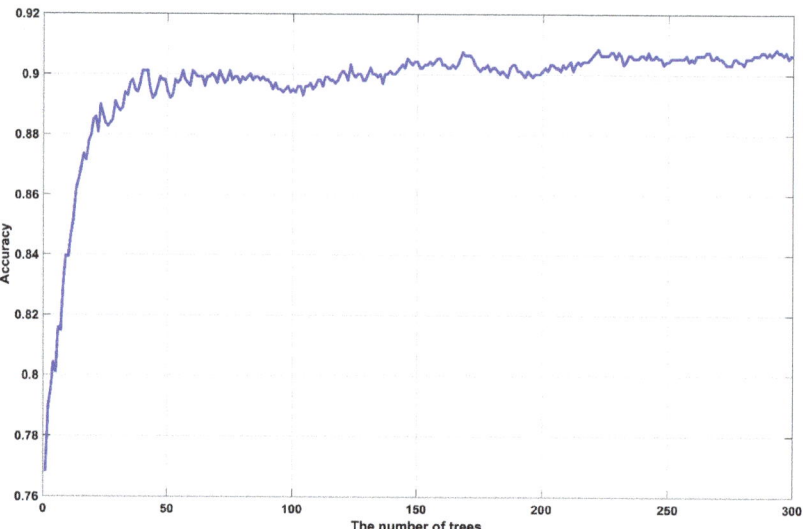

Figure 7. The identification accuracies with different number of trees in random forest.

4.3. The Platform of Automatic Coal Petrographic Analysis

The proposed maceral analysis method, MISC, based on image segmentation and classification makes it possible to identify the maceral composition automatically and intelligently. In order to facilitate the usage by petrologists, we developed a standalone software implemented in Matlab. We integrate the two-level K-means and various classification algorithms into the software for intelligent identification of maceral composition. Figure 8 is the screen snapshot of MISC software. Users can submit a microscopic image of coal with the degree of coalification R0 < 1.0%. The segmentation results are presented as four subfigures, corresponding to the binders and three maceral groups. The classification result for each object detected by image segmentation is shown with different colors for visualization. The MISC is freely available for users at the following website: https://github.com/GuyooGu/MISC-Master. It can be used to support the decisions of petrologists in classifying maceral components. To the best of our knowledge, it is the first non-commercial software for identification of maceral components. It is an efficient and effective tool for the complete analysis of maceral composition.

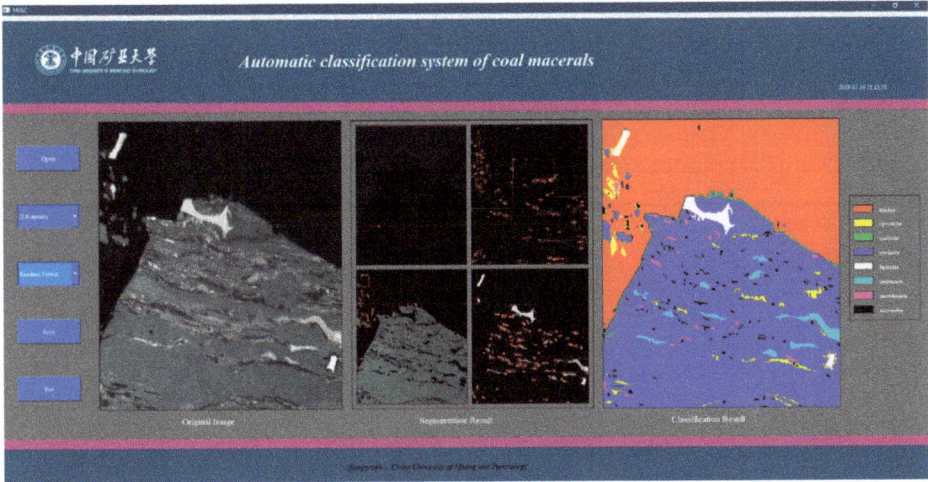

Figure 8. The user interface of MISC for automatic coal petrographic analysis.

5. Conclusions

Inspired by the way that petrologists examine photomicrographs, we proposed an automatic and effective framework for maceral classification. The proposed strategy is fundamentally different from previous attempts to classify a region and classify the central pixel of a ROI. Based on the image segmentation, rather than the fixed-size window, the regions of interest are cropped automatically. Current fixed-size window-based strategies, including both patch-wise classification and pixel-wise classification, neglect the shape and the size information and therefore do not work in identifying maceral composition from one entire photomicrograph. The utilized two-level coarse-to-fine clustering strategy achieved significantly better segmentation results as compared to the corresponding single-level clustering. In addition, considering the complicated nature of coal, it may prove difficult to identify maceral components based on a single kind of features. Our results suggest that classification approach based on multiple kinds of features, such as geometric, grayscale and texture features, will be a promising direction for identifying maceral components.

However, it should be stressed that the research described in this paper is a preliminary study which still has some limitations. First, the proposed method only works when the degree of coalification is smaller than 1.0%. With the increase of that degree, it will be difficult to distinguish vitrinite and liptinite based on gray scale values. Second, we assume that there are four maceral groups of each sample. However, this assumption may not hold in a few cases. We have tried density-based clustering methods, such as density-based spatial clustering of applications with noise (DBSCAN), which can automatically detect the optimal number of clusters. However, the performance is not better than the proposed method. We will investigate this issue in the future. In addition, 39 photomicrographs used in this study were analyzed according to definitions in ASTM Standard D 2799-13 and atlas in [39]. There are seven types of maceral components, belonging to three maceral groups (i.e., vitrinite, liptinite and inertinite). Liptinites include sporinite and cutinite. Inertinite macerals include fusinite, semifusinite, inertodetrinite, and macrinite. However, the mineral components and the other macerals, such as funginite or macrinite, are rarely found in these photomicrographs, and are not abundant in nature. Therefore, in this study, we do not consider these components, and follow a simplified classification as that in [39]. The proposed MISC strategy obtained relatively good performance, especially when we take into account a high degree of complexity of coals. Although with some limitations, to the best of our knowledge, our work is the first study aiming to provide complete analysis of maceral composition.

Author Contributions: Conceptualization, H.W. and L.Z.; funding acquisition, M.L. (Meng Lei) and L.Z.; methodology, H.W. and Y.C.; project administration, M.L. (Meng Lei); resources, M.L. (Ming Li); supervision, M.L. (Ming Li) and L.Z.; validation, Y.C. and M.L. (Ming Li); writing—original draft, H.W.; writing—review and editing, L.Z.

Funding: This research was funded by the Fundamental Research Funds for the Central Universities with grant number 2019ZDPY17.

Conflicts of Interest: The authors declare no conflict of interest.

References

1. Flores, R.M. *Chapter 5-Coal Composition and Reservoir Characterization*; Elsevier: Amsterdam, The Netherlands, 2014; pp. 235–299.
2. Chen, Y.; Yong, Q.; Wei, C.; Huang, L.; Shi, Q.; Wu, C.; Zhang, X. Porosity changes in progressively pulverized anthracite subsamples: Implications for the study of closed pore distribution in coals. *Fuel* **2018**, *225*, 612–622. [CrossRef]
3. Mlynarczuk, M.; Skiba, M. The application of artificial intelligence for the identification of the maceral groups and mineral components of coal. *Comput. Geosci.* **2017**, *103*, 133–141. [CrossRef]
4. Rallakis, D.; Michels, R.; Brouand, M.; Parize, O.; Cathelineau, M. The Role of Organic Matter on Uranium Precipitation in Zoovch Ovoo, Mongolia. *Minerals* **2019**, *9*, 310. [CrossRef]
5. Chaudhuri, S.N. Coal macerals. *Encycl. Mineral Energy Policy* **2016**, 1–5.
6. Anon New inertinite classification (ICCP System 1994). *Fuel Energy Abstr.* **2001**, *80*, 459–471.
7. ASTM. Standard Test Method for Microscopical Determination of the Maceral Composition of Coal. In *ASTM D2799-13*; ASTM International: West Conshohocken, PA, USA, 2013.
8. Camalan, M.; Çavur, M.; Hoşten, Ç. Assessment of chromite liberation spectrum on microscopic images by means of a supervised image classification. *Powder Technol.* **2017**, *322*, 214–225. [CrossRef]
9. Shu, L.; McIsaac, K.; Osinski, G.R.; Francis, R. Unsupervised feature learning for autonomous rock image classification. *Comput. Geosci.* **2017**, *106*, 10–17. [CrossRef]
10. Li, N.; Hao, H.; Gu, Q.; Wang, D.; Hu, X. A transfer learning method for automatic identification of sandstone microscopic images. *Comput. Geosci.* **2017**, *103*, 111–121. [CrossRef]
11. Aligholi, S.; Lashkaripour, G.R.; Khajavi, R.; Razmara, M. Automatic mineral identification using color tracking. *Pattern Recogn.* **2017**, *65*, 164–174. [CrossRef]
12. Młynarczuk, M.; Górszczyk, A.; Ślipek, B. The application of pattern recognition in the automatic classification of microscopic rock images. *Comput. Geosci.* **2013**, *60*, 126–133. [CrossRef]
13. Hofmann, P.; Marschallinger, R.; Unterwurzacher, M.; Zobl, F. Marble provenance designation with object based image analysis: State-of-the-art rock fabric characterization from petrographic micrographs. *Austrian J. Earth Sci.* **2013**, *106*, 40–49.
14. Goodarzi, F. The use of automated image analysis in coal petrology. *Can. J. Earth. Sci.* **1987**, *24*, 1064–1069. [CrossRef]
15. Lester, E.; Watts, D.; Cloke, M. A novel automated image analysis method for maceral analysis. *Fuel* **2002**, *81*, 2209–2217. [CrossRef]
16. Skiba, M.; MŁYNARCZUK, M. Identification of Macerals of the Inertinite Group Using Neural Classifiers, Based on Selected Textural Features. *Arch. Min. Sci.* **2018**, *63*, 827–837.
17. Wang, P.-Z.; Yin, Z.-H.; Wang, G.; Zhang, D.-L. A classification method of vitrinite for coal macerals based on the PCA and RBF-SVM. *J. China Coal Soc.* **2017**, *42*, 977–984.
18. Karayigit, A.I.; Whateley, M. Properties of a lacustrine subbituminous (k1) seam, with special reference to the contact metamorphism, Soma-Turkey. *Int. J. Coal Geol.* **1997**, *34*, 131–155. [CrossRef]
19. Chaudhuri, O.; Gu, L.; Klumpers, D.; Darnell, M.; Bencherif, S.A.; Weaver, J.C.; Huebsch, N.; Lee, H.; Lippens, E.; Duda, G.N. Hydrogels with tunable stress relaxation regulate stem cell fate and activity. *Nat. Mater.* **2016**, *15*, 326. [CrossRef]
20. ASTM. Standard Practice for Preparing Coal Samples for Microscopical Analysis by Reflected Light. In *ASTM D2797/D2797M-11a*; ASTM International: West Conshohocken, PA, USA, 2011.
21. Dhanachandra, N.; Manglem, K.; Chanu, Y.J. Image segmentation using K-means clustering algorithm and subtractive clustering algorithm. *Procedia Comput. Sci.* **2015**, *54*, 764–771. [CrossRef]

22. Peng, Y.; Liu, X.; Shen, C.; Huang, H.; Zhao, D.; Cao, H.; Guo, X. An Improved Optical Flow Algorithm Based on Mask-R-CNN and K-Means for Velocity Calculation. *Appl. Sci.* **2019**, *9*, 2808. [CrossRef]
23. Rezaei, Z.; Selamat, A.; Taki, A.; Mohd Rahim, M.; Abdul Kadir, M.; Penhaker, M.; Krejcar, O.; Kuca, K.; Herrera-Viedma, E.; Fujita, H. Thin cap fibroatheroma detection in virtual histology images using geometric and texture features. *Appl. Sci.* **2018**, *8*, 1632. [CrossRef]
24. Olson, E. Particle shape factors and their use in image analysis part 1: Theory. *J. GXP Compliance* **2011**, *15*, 85.
25. Yang, M.; Kpalma, K.; Ronsin, J. A Survey of Shape Feature Extraction Techniques. *Pattern Recogn.* **2008**, *15*, 43–90.
26. Huang, Z.; Leng, J. Analysis of Hu's moment invariants on image scaling and rotation. In Proceedings of the 2010 2nd International Conference on Computer Engineering and Technology, Chengdu, China, 16–18 April 2010; pp. V7–V476.
27. Mohanaiah, P.; Sathyanarayana, P.; GuruKumar, L. Image texture feature extraction using GLCM approach. *Int. J. Sci. Res. Publ.* **2013**, *3*, 1.
28. Gao, S.; Peng, Y.; Guo, H.; Liu, W.; Gao, T.; Xu, Y.; Tang, X. Texture analysis and classification of ultrasound liver images. *Bio-med. Mater. Eng.* **2014**, *24*, 1209–1216.
29. Li, W.; Chen, C.; Su, H.; Du, Q. Local binary patterns and extreme learning machine for hyperspectral imagery classification. *IEEE Trans. Geosci. Remote Sens.* **2015**, *53*, 3681–3693. [CrossRef]
30. Zou, L.; Huang, Q.; Li, A.; Wang, M. A genome-wide association study of Alzheimer's disease using random forests and enrichment analysis. *Sci. China Life Sci.* **2012**, *55*, 618–625. [CrossRef] [PubMed]
31. Feng, Q.; Liu, J.; Gong, J. UAV remote sensing for urban vegetation mapping using random forest and texture analysis. *Remote Sens.* **2015**, *7*, 1074–1094. [CrossRef]
32. Lin, W.; Wu, Z.; Lin, L.; Wen, A.; Li, J. An ensemble random forest algorithm for insurance big data analysis. *IEEE Access* **2017**, *5*, 16568–16575. [CrossRef]
33. Zou, L.; Wang, M.; Shen, Y.; Liao, J.; Wang, M. PKIS: computational identification of protein Kinases for experimentally discovered protein Phosphorylation sites. *BMC Bioinform.* **2013**, *14*, 247. [CrossRef] [PubMed]
34. Kotsiantis, S.B. Supervised Machine Learning: A Review of Classification Techniques. *Emerg. Artif. Intell. Appl. Comput. Eng.* **2007**, *160*, 3–24.
35. Zhao, L.; Chen, Z.; Yang, Y.; Zou, L.; Wang, Z.J. ICFS clustering with multiple representatives for large data. *IEEE Trans. Neural Netw. Learn. Syst.* **2018**, *30*, 728–738. [CrossRef] [PubMed]
36. Zhao, L.; Chen, Z.; Yang, L.T.; Deen, M.J.; Wang, Z.J. Deep Semantic Mapping for Heterogeneous Multimedia Transfer Learning Using Co-Occurrence Data. *ACM Trans. Multimed. Comput. Commun. Appl.* **2019**, *15*, 9. [CrossRef]
37. Scott, A.C.; Glasspool, I.J. Observations and experiments on the origin and formation of inertinite group macerals. *Int. J. Coal Geol.* **2007**, *70*, 53–66. [CrossRef]
38. Speight, J.G. *The Chemistry and Technology of Coal*; CRC Press: Boca Raton, FL, USA, 2012; pp. 101–128.
39. Gesserman, R.M.; Morrissey, E.A.; Hackley, P.C. Petrographic Web Atlas for Metallurgical Bituminous Coal Macerals. In Proceedings of the 2009 Portland GSA Annual Meeting, Portland, ON, USA, 18–21 October 2009.

© 2019 by the authors. Licensee MDPI, Basel, Switzerland. This article is an open access article distributed under the terms and conditions of the Creative Commons Attribution (CC BY) license (http://creativecommons.org/licenses/by/4.0/).

Article

Segmentation of River Scenes Based on Water Surface Reflection Mechanism

Jie Yu, Youxin Lin, Yanni Zhu, Wenxin Xu, Dibo Hou *, Pingjie Huang and Guangxin Zhang

State Key Laboratory of Industrial Control Technology, College of Control Science and Engineering, Zhejiang University, Hangzhou 310027, China; yu_jie@zju.edu.cn (J.Y.); yxlin@zju.edu.cn (Y.L.); yanni_z@zju.edu.cn (Y.Z.); xuwen_xin@zju.edu.cn (W.X.); huangpingjie@zju.edu.cn (P.H.); gxzhang@zju.edu.cn (G.Z.)
* Correspondence: houdb@zju.edu.cn

Received: 15 February 2020; Accepted: 30 March 2020; Published: 3 April 2020

Abstract: Segmentation of a river scene is a representative case of complex image segmentation. Different from road segmentation, river scenes often have unstructured boundaries and contain complex light and shadow on the water's surface. According to the imaging mechanism of water pixels, this paper designed a water description feature based on a multi-block local binary pattern (MB-LBP) and Hue variance in HSI color space to detect the water region in the image. The improved Local Binary Pattern (LBP) feature was used to recognize the water region and the local texture descriptor in HSI color space using Hue variance was used to detect the shadow area of the river surface. Tested on two data sets including simple and complex river scenes, the proposed method has better segmentation performance and consumes less time than those of two other widely used methods.

Keywords: river scene segmentation; local binary pattern; hue variance; surface reflection; hand-designed image descriptors

1. Introduction

Segmentation of a river scene plays an important role in many fields such as the water hazard detection of unmanned ground vehicles [1], the navigation of unmanned ships [2], river analysis or flood monitoring by remote sensing [3–6] and vision-based object monitoring on rivers. This study aims to recognize the river region in an image taken in outdoor scenes based on the water surface reflection mechanism, which is an important task in applications of intelligent video surveillance in river environments. Moreover, segmentation of the river scene is a representative case of complex image segmentation, which can serve as a reference for complex image segmentation.

For water region segmentation, researchers have explored different kinds of methods that fall into three main categories—image processing-based methods, Machine Learning-based methods (including Deep Learning, Supervised Learning, Clustering, etc.), and hardware-based methods. For image processing-based methods, Rankin et al. [7] combined the color and texture features to detect the water region according to the appearance characteristics of the river in the outdoor scene. Yao [8] used the Region Growing method firstly to separate the obvious water region based on the brightness value. Then a designed texture feature is used to perform K-Means clustering on each 9 × 9 small patch in the image, where the class with the smallest average value of texture is classified as the water region, where the detection of water region with shadow needs the aid of stereo vision. Zhao et al. [9] used the adaptive threshold Canny edge detection algorithm to detect the river boundary. The texture and structure of images are also widely used in related research in water scenes such as waterline detection [10] and maritime horizon line detection [11]. For Machine

Learning-based methods, Achar et al. [12] proposed a self-supervised algorithm to classify all the image patches in an image into water or not-water category by features of RGB, texture, and height. The results show high accuracy but this algorithm requires prior knowledge of horizon by hardware and is only applicable to images that conform to a specific structure. Moreover, with the development of Deep Learning, it has also been applied in water region segmentation. For example, Zhan et al. [13] proposed an online learning approach to recognize the water region for the USV in the unknown navigation environment using a convolutional neural network (CNN). Han et al. [14] innovatively used the Fully Connected Convolutional Network (FCN) to achieve water hazards detection on the road. Despite of high accuracy, the artificial neural network with complex structure needs to be pre-trained in many scenes before use and requires high computing power. For hardware-based methods, some studies have used various optical sensors such as laser radar [15], infrared camera [1], stereo camera [16,17], and polarized camera [16,18,19] to easily realize water hazard detection based on the optical characteristics of waters [20]. These methods are still difficult to popularize in applications due to the cost and equipment complexity.

The above methods have some defects. As Rankin [21] made the observation that the river has inhomogeneous appearance in outdoor scenes, the methods simply utilizing image features, whose underlying assumption is that river appearance is fairly uniform, remain problematic due to inhomogeneous appearance (such as shadow and changing illumination) and show bad performance. For the same reason, it is also inappropriate for Machine Learning-based methods using the global features of an image to train a classification model to segment itself. As for the hardware-based methods, they are beyond the scope of this paper.

Since image processing technology has advantages of simplicity and interpretability, this study proposes a segmentation algorithm utilizing designed image features without machine learning. To overcome the drawback that the current methods cannot well deal with inhomogeneous appearance of the river, this study designs an improved LBP feature extraction method based on the water surface reflection mechanism to detect water region in the image. A texture feature based on Hue(H) variance in HSI color space is also introduced for detection of shadow area. Compared with two other principle methods using image processing techniques, the proposed method consumes the least time, and in the complex river scenes where other methods failed, the proposed algorithm still shows satisfactory performance. Lastly in this study, the parameters in the proposed algorithm are discussed for better performance.

2. Materials and Methods

2.1. Algorithm Framework

The qualitative imaging law of the riverine water region in an image is the basis of the algorithm designed in this paper. The overall flowchart of the proposed algorithm is shown in Figure 1.

First of all, the input image needs to be pre-processed, including image down-sampling and image blurring operation. The pre-processing operations will be discussed in Section 3.1. Secondly, the improved LBP feature and local hue variance are calculated in parallel. Then the water region with and without shadow are both obtained by threshold method. The two parts are fused as the major water region. Finally, the image morphological operation is carried out on the candidate region of the water region. After obtaining the results of the morphological processing, the largest connected domain is taken as the final water region. Judging the maximum connected domain is also an important task. It is based on the common sense that the water area often occupies a main and large part of the image, which helps to eliminate the pseudo-water patches whose features are similar to those of water patches.

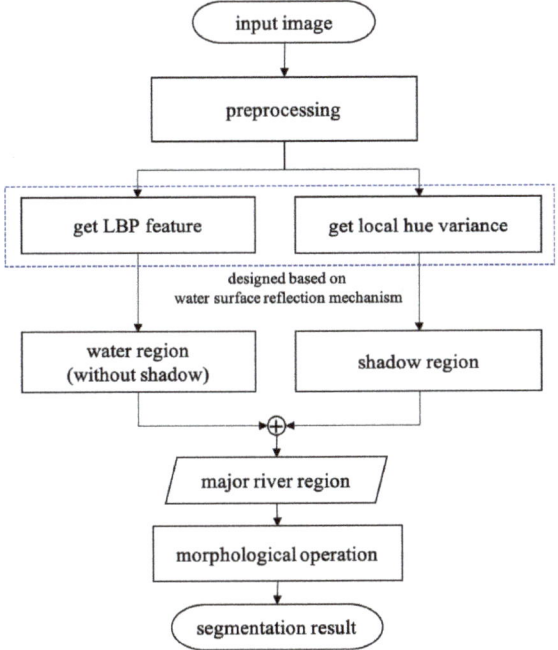

Figure 1. Flowchart of the proposed algorithm.

2.2. Light Reflection Mechanism of Water Surface

In order to study the imaging law of water in rivers, it is necessary to first understand the general reflection mechanism of objects. According to Lambert Law, the intensity of object surface through various types of reflections reaches the image sensor is [22]:

$$I(x) = \int_\omega e(\lambda)\rho_k s(x,\lambda)d\lambda, \tag{1}$$

where $e(\lambda)$ is the color of the light source, $s(x,\lambda)$ is the surface reflection value, ρ_k is the sensitive function of the camera ($k \in \{R, G, B\}$), ω represents the visible spectral range, and x denotes the corresponding space coordinates. For a particular color camera, the pixel intensity values in the image are only related to the reflected light [23]. On this basis, the relationship between the pixel value of the water and the light reflection of the water surface is further expressed as follows:

$$I = LR_{\text{total}}, \tag{2}$$

where L is the illumination factor related to the illumination condition, R_{total} is the total reflection energy. In river scenes, R_{total} is mainly composed of the following four parts [21]: the energy that reflected off the water surface R_r, that scattered by water molecules to the camera R_o, that reflected or scattered by materials suspended in the water to the camera R_s, and that reflected off the bottom of the water the camera R_p:

$$R_{\text{total}} = R_r + R_o + R_s + R_p. \tag{3}$$

Since the reflection from the water surface to the camera R_r plays a dominant role in R_{total}, that is, (2) and (3) can be further simplified as:

$$I \approx LR_r. \tag{4}$$

For light polarized perpendicular to and parallel to the plane of incidence, R_r can be respectively decomposed into $R_{r,\perp}(\theta)$ and $R_{r,\parallel}(\theta)$, where θ is the incident angle, $\theta \in (0, \frac{\pi}{2})$, as shown in (5):

$$R_r(\theta) = \frac{R_{r,\perp}(\theta) + R_{r,\parallel}(\theta)}{2}. \tag{5}$$

According to Fresnel Law:

$$R_{r,\perp}(\theta) = [\frac{n_1 \cos\theta - n_2\sqrt{1 - \left(\frac{n_1}{n_2}\sin\theta\right)^2}}{n_1 \cos\theta + n_2\sqrt{1 - \left(\frac{n_1}{n_2}\sin\theta\right)^2}}]^2, \tag{6}$$

$$R_{r,\parallel}(\theta) = [\frac{n_1\sqrt{1 - \left(\frac{n_1}{n_2}\sin\theta\right)^2} - n_2 \cos\theta}{n_1\sqrt{1 - \left(\frac{n_1}{n_2}\sin\theta\right)^2} + n_2 \cos\theta}]^2, \tag{7}$$

where n_1 is the refractive index of air, n_2 is the refractive index of water, and θ is the angle of incidence. $n_1 = 1.0$ and $n_2 = 1.33$ are taken under ideal conditions. The water region reaches the sensor through various types of reflections, as shown in Figure 2, where l is the horizontal displacement of the point to the camera lens, h is the height at which the camera is placed (in a certain scene, the image sensor used to capture images is commonly fixed). According to the simplified scenario shown in Figure 2, $\alpha \approx \theta$ can be obtained from the geometric relationship, thus $R_r(\theta)$ can be converted into a function $R_r(l)$ about the horizontal displacement, just make:

$$\theta \approx \arctan\frac{l}{h} \tag{8}$$

and then substitute (8) into (5). Since only qualitative rather than quantitative law is used in the subsequent algorithm design for the water region detection, the above equation do not need to be strictly equal. Given that the expression of the result is too complicated, and the designed algorithm only needs the qualitative law, we explored the relationship between the reflection intensity of the water and the horizontal distance to the camera by giving some different h values that indicate some conventional installation heights, as shown in Figure 3.

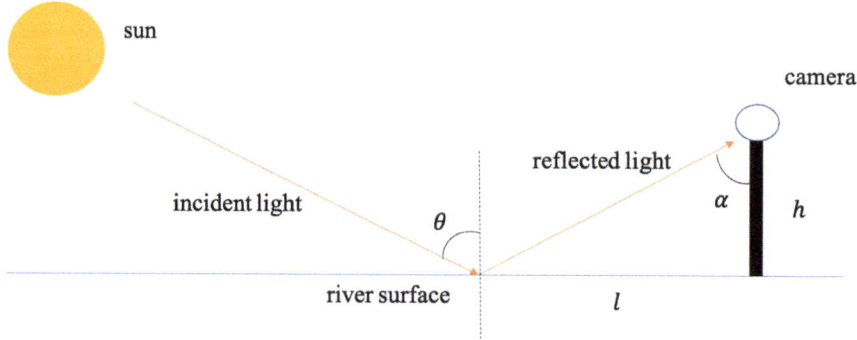

Figure 2. Light path diagram of river scene.

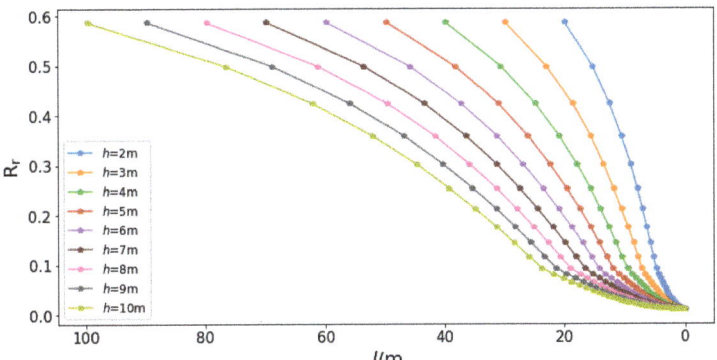

Figure 3. Relationship between water surface reflection and horizontal distance l with different heights camera.

It can be seen that the reflected energy to the image sensor from far to near is monotonically decreasing. This qualitative law of water pixels is used to design subsequent water region detection algorithm.

In addition to the above-mentioned reflection mechanism, the water surface in outdoor scenes often contains shadow caused by the occlusion of the riverside scenery. The H component in the HSI color space of the image is not sensitive to illumination, and can maintain a relatively stable state under illumination changes [24]. In order to calculate the feature of H, firstly the RGB image should be converted into an HSI one by:

$$h = \begin{cases} \beta, g \geq b \\ 2\pi - \beta, g < b \end{cases}, h \in [0, 2\pi] \tag{9}$$

$$s = 1 - 3 \cdot \min(r, g, b), s \in [0, 1] \tag{10}$$

$$i = \frac{r + g + b}{3}, i \in [0, 1], \tag{11}$$

where r, g, b, h, s, i are all normalized values, and:

$$\beta = \arccos \left\{ \frac{[(r - g) + (r - b)]/2}{[(r - g)^2 + (r - b)(g - b)]^{0.5}} \right\}. \tag{12}$$

This law is illustrated in Figure 4. By traversing I and H values of pixels on the specified column (indicated by the red line), the values are shown on the right of Figure 4. The result showed that for the water region without shadow, the distribution of I values of the pixels is closely related to the variation law shown in Figure 3. For the water region with shadow, the H values keep spatially stable.

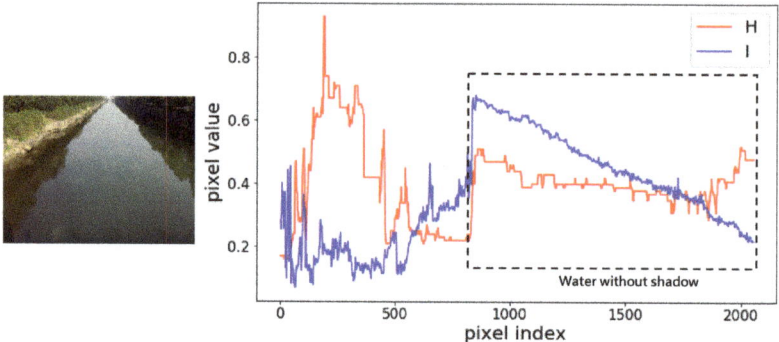

Figure 4. Pixel-wise intensity values and hue values in a column of river scene image.

2.3. Improved Local Binary Pattern Feature

The water part in an image tends to present simpler textures. Some studies utilize this characteristic to segment water bodies. The common texture features include gray-level co-occurrence matrix (GLCM) [25], Laws' Mask [26], Local Binary Pattern (LBP) [27] and so on. The study of the water surface reflection mechanism in the previous section shows that the appearance of the water region changes spatially. Consequently, the results of the textural descriptors calculated from the whole image, such as GLCM, for water region in an image would have a distinct numerical difference. LBP constructs a local feature descriptor that reflects the magnitude relationship between the center pixel and the neighborhood ones, which can effectively deal with the inhomogeneous appearance in an image, and establish a more reliable description for an image patch. Based on the water surface reflection mechanism discussed previously, an improved LBP feature is designed to describe the spatial characteristics of water appearance and then used to detect the water part of an image.

To obtain the improved LBP feature, the image is firstly divided into several patches of a specified size, and then each patch is further divided into 9 blocks. The pixel value(or the average value) in each block is denoted as I_k, $k = 1, 2, 3, ..., 9$ as shown in Figure 5.

Figure 5. Illustration of generation of image patches.

The traditional LBP feature compares the values of center block pixel I_5 with the neighborhood ones and encodes them into a binary string. However, the comparison results are in different weights for different directions. Therefore, the proposed algorithm improves traditional LBP feature, as shown in Algorithm 1. It is designed based on the qualitative law that the pixel values decrease from far to near and the pixel values at the same distance to the camera are close.

Algorithm 1 Improvd LBP feature

Input: gray-scale image patch in matrix form
Output: 8-dimension feature
1: divide the image into 9 equal-size blocks with pixel value $I_k, k = 1, 2, 3, ..., 9$
2: **if** $|I_1 - I_2| < 1\% * I_1$ **and** $|I_2 - I_3| < 1\% * I_2$ **then**
3:     $f_1 = 1$
4: **else**
5:     $f_1 = 0$
6: **if** $|I_4 - I_5| < 1\% * I_4$ **and** $|I_5 - I_6| < 1\% * I_5$ **then**
7:     $f_2 = 1$
8: **else**
9:     $f_2 = 0$
10: **if** $|I_7 - I_8| < 1\% * I_7$ **and** $|I_8 - I_9| < 1\% * I_8$ **then**
11:     $f_3 = 1$
12: **else**
13:     $f_3 = 0$
14: **if** $|I_4 - I_1| \approx |I_5 - I_2| \approx I_6 - I_3$ **then**
15:     $f_4 = 1$
16: **else**
17:     $f_4 = 0$
18: **if** $|I_7 - I_4| \approx |I_8 - I_5| \approx I_9 - I_6$ **then**
19:     $f_5 = 1$
20: **else**
21:     $f_5 = 0$
22: **if** $I_1 > I_4 > I_7$ **then**
23:     $f_6 = 1$
24: **else**
25:     $f_6 = 0$
26: **if** $I_2 > I_5 > I_8$ **then**
27:     $f_7 = 1$
28: **else**
29:     $f_7 = 0$
30: **if** $I_3 > I_6 > I_9$ **then**
31:     $f_8 = 1$
32: **else**
33:     $f_8 = 0$
34: **return** $f = [f_1, f_2, f_3, f_4, f_5, f_6, f_7, f_8]$

In the improved LBP calculation, the features f_1, f_2, and f_3 indicate that the I values of every row in the image patch are very close because the water pixels from a similar distance have almost the same reflected energy to the camera. While the pixel value differences in the vertical direction in the patch are numerically similar, as the meaning by f_4 and f_5, since the distance between adjacent pixels is small enough to neglect the gap. Moreover, the father pixel theoretically has a larger pixel value than that of a closer one, as the meaning of f_6, f_7, and f_8. Finally, to overcome the drawback that the relationship of different directions in the traditional LBP has different weights, the improved LBP sums the obtained Boolean results $f_i, i = 1, 2, 3, ..., 8$ as a score F:

$$F = \sum_{i=1}^{8} f_i. \tag{13}$$

After all, an appropriate threshold T_1 is adopted to compare with the obtained score F to decide whether the patch is part of water or not, which can be formulated as follows:

$$\begin{cases} \text{water, if } F \geq T_1 \\ \text{not water, if } F < T_1 \end{cases} \quad (14)$$

Empirically, the algorithm has satisfactory performance in most scenes when T_1 is set to 5 or 6.

2.4. Local Hue Variance in HSI Color Space

Since the shadow area may not be subject to the model of (3), after recognizing the main part of the water region, another method to recognize the water area covered by the shadow is needed to increase the recall rate of the water region's segmentation. In shadow, the lighting conditions are difficult to estimate, and the reflection law reflected by (8) is not available. However, H values keep uniformity within neighbor pixels as shown in Figure 4.

The calculation of the local hue variance is as follows: firstly, convert the original RGB input image block into an HSI image. Then the extracted H layer is divided into 9 blocks of the same size, as shown in Figure 6. Finally, calculate the mean value of H denoted as $H_k(k = 1, 2, 3 \ldots, 9)$ of each block and obtain the variance of H_k:

$$V_H = \frac{\sum_{k=1}^{9} (H_k - \overline{H})^2}{9}. \quad (15)$$

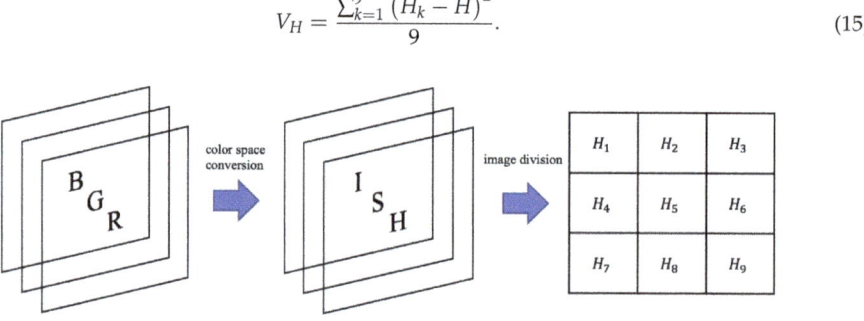

Figure 6. Color space conversion and image division for local hue variance calculation.

An appropriate threshold T_2 is then adopted to compare with the obtained V_H to identify the shadow area. The image patches that have bigger V_H than the designed threshold are labeled as part of water, which is express as:

$$\begin{cases} \text{water, if } V_H < T_2 \\ \text{not water, if } V_H \geq T_2 \end{cases} \quad (16)$$

Since H_k are normalized values in the calculation, the same T_2 can be used for different images. Empirically, T_2 can be set within $[1.5, 1.8]$ to get satisfactory performance in most scenes.

2.5. Morphological Operation

Morphological Operation [28,29] is a widely used technique for digital images. The basic idea in binary morphology is to probe an image with a simple, pre-defined shape called structuring element, drawing conclusions on how this shape fits or misses the shapes in the image. The basic operations include erosion and dilation. The erosion eliminates sporadic targets or noise, while the dilation amplifies the target area. Different size structuring elements lead to different results of Morphological Operation.

In this study, Morphological Operation is employed to eliminate potential pseudo-water patches that wrongly detected by the proposed algorithm and obtain the largest connected domain in the image as water region. Erosion is performed firstly, and then triple expansion with increasing size structuring elements is carried out to ensure the integrity of the segmenting area. This process is shown in Figure 7.

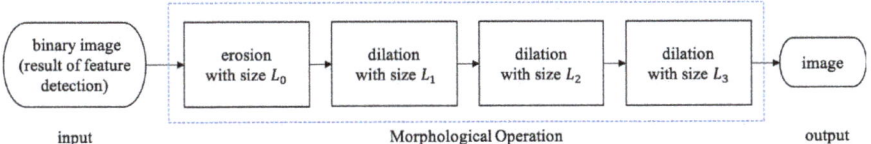

Figure 7. Morphological operation method designed in our algorithm.

The size of the structuring element can affect algorithm performance. Empirically, it was recommended to use a rectangular structuring element slightly larger than the patch size for erosion operation, since the patch shape in pre-processing was rectangular. To make the boundary of segmentation closer to the human visual system, an elliptical structural element was then used for dilation. Moreover, to eliminate the foreground outliers during the morphology process, triple times of dilation were consecutively carried out, with increasing size of the structuring elements, which is formulated as follows:

$$\begin{cases} L_0 = 10 \times 10 \\ L_1 = 15 \times 15 \\ L_2 = \frac{L_3 + L_1}{2} \\ L_3 = kL_{\text{image}} \end{cases}, \quad (17)$$

where L_{image} is the size of the input image and k is a coefficient indicating that the size of the structuring element was determined by the size of the input image, which is further discussed in Section 3.5.

3. Results and Discussion

The proposed algorithm was tested on a dataset made up of 500 images taken from different river scenes. These scenes were divided into simple scenes(110) and complex scenes(390) to test the performance of different methods under both general and special conditions. The simple scenes in this study refer to general and common outdoor river scenes that do not contain complex issues such as shadow and intense sunlight reflections, while the complex scenes are the opposite. Moreover, given that the image sensors used in different scenes are likely to be various, the images we used include different resolutions.

In the experiments, two principal river segmentation methods utilizing image features [8] and edge detection [9], respectively, were compared with our method. It should be noted that, because the proposed algorithm is designed specifically for river scenes, the general image segmentation algorithm using Deep Learning has high requirements on data sets and operational capabilities, so it is beyond the range of comparison. The running environment in this study was—Python3 in MacOS system with 2.9 GHz Intel Core i5 CPU, 16GB memory. The algorithm's parameters were set to fixed value empirically in advance. Further discussion about the parameters is in this section later.

3.1. Pre-Processing

The original input images that are too large in size need to be scaled down to reduce the time consumed by the subsequent algorithm, which was followed by denoising and blurring. Therefore, a threshold for input image size (denoted as S_o) was set in advance and circulated downsampling was likely to be performed, as shown in Figure 8.

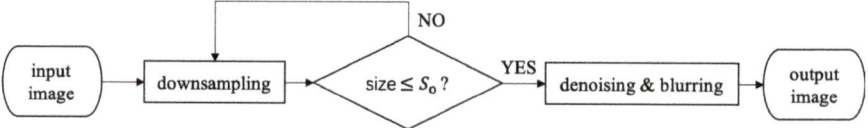

Figure 8. Pre-processing operations.

Since the spikes or glitches in the distribution signal of pixel values, which were usually caused by noise, had a great effect on local H variance, the blurring operation was significant to obtain reliable H values. Therefore, a Gaussian blur filter was introduced to reduce the influence of image noise before the image was analyzed. The results were compared in Figure 9. The picture on the right shows the distribution of H and I values of pixels lying at the column (the red line) in the image after Gaussian blurring. The H values after blurring were more suitable to use for the subsequent feature analysis.

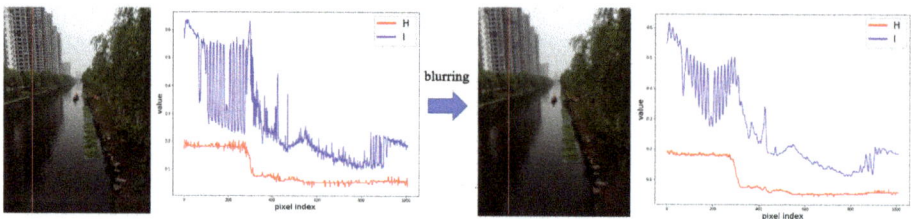

Figure 9. Pixel values before and after blurring operation.

3.2. Experiments in Simple Scenes

The performance of different methods was evaluated by Pixel Accuracy(PA), Mean Intersection over Union($MIoU$) which are two types of widely used criteria in image segmentation [30], shown as follows:

$$PA = \frac{\sum_{i=0}^{k} p_{ii}}{\sum_{i=0}^{k} \sum_{j=0}^{k} p_{ij}}, \tag{18}$$

$$MIoU = \frac{1}{k+1} \sum_{i=0}^{k} \frac{p_{ii}}{\sum_{j=0}^{k} p_{ij} + \sum_{j=0}^{k} p_{ji} - p_{ii}}, \tag{19}$$

where P_{ij} indicates the number of pixels of class i that are predicted to belong to class j, where there are $k+1$ classes in total. In this study, $k+1 = 2$. To better evaluate the overall segmentation performance, PA and $MIoU$, which may have different weights in practical applications, were merged to generate the weighted harmonic mean F_β as:

$$F_\beta = \frac{(1+\beta^2) \, PA \times MIoU}{\beta^2 \times PA + MIoU}, \tag{20}$$

where $\beta > 0$ measures the relative importance between PA and $MIoU$. When $\beta > 1$, $MIoU$ has a greater impact. In practice, $MIoU$ was slightly more important, thus $\beta = 1.5$ is adopted.

The results of some examples are shown in Figure 10 where the detected water region by "intensity + texture" method is marked with blue, while those by "edge detection" and the proposed algorithm is highlighted in red for edges. Table 1. shows the criteria values of the result.

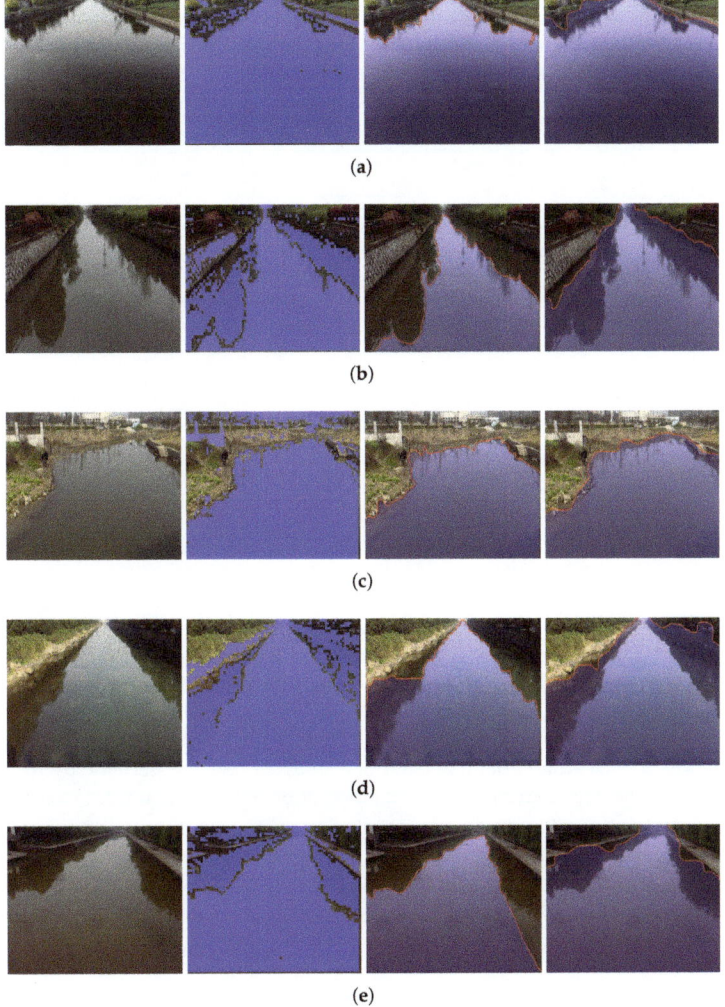

Figure 10. Segmentation results of different methods in simple scenes (**a–e**). From left to right, the first column shows the input images after pre-processing; The second and third column are the segmentation results using "intensity + texture" features and the adaptive threshold edge detection algorithm respectively; The fourth column are the results of our method.

Table 1. Performance of different methods in simple scenes.

River Scene	Intensity + Texture			Edge Detection			Our Algorithm		
	PA	MIoU	F_β	PA	MIoU	F_β	PA	MIoU	F_β
(a)	0.997	0.915	**0.939**	0.964	0.926	**0.937**	0.987	0.969	**0.967**
(b)	0.901	0.824	**0.846**	0.995	0.592	**0.676**	0.953	0.928	**0.948**
(c)	0.837	0.821	**0.826**	0.976	0.932	**0.945**	0.981	0.947	**0.956**
(d)	0.920	0.851	**0.871**	0.993	0.829	**0.873**	0.974	0.927	**0.946**
(e)	0.921	0.881	**0.893**	0.998	0.746	**0.809**	0.871	0.957	**0.969**
Average of 110 images	0.915	0.847	**0.867**	0.973	0.795	**0.842**	0.966	0.923	**0.938**

All the three algorithms achieved not bad segmentation results, which meant they were all effective for segmentation of simple river scenes. But the proposed algorithm had a more stable performance. More importantly, the proposed algorithm took the least time, as obviously shown in Figure 11.

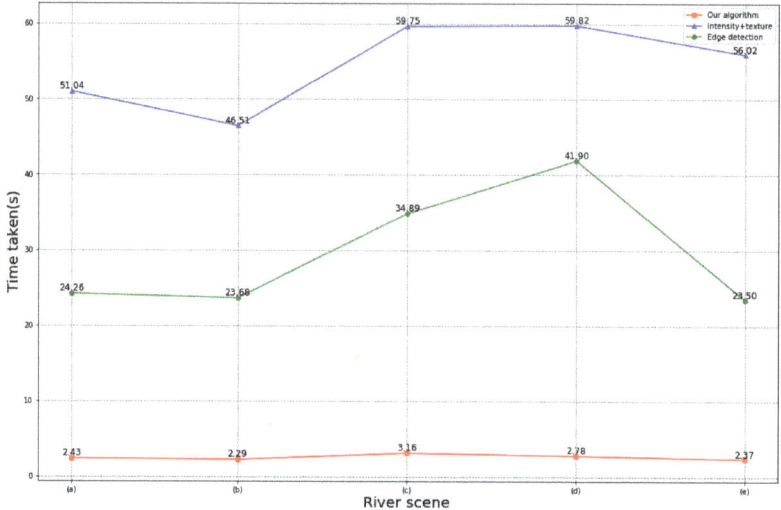

Figure 11. Speed of different methods in simple scenes.

The method utilizing "intensit + texture" features was not only to calculate the brightness and texture information of each small image patch, but also to achieve the decision by the result of the clustering algorithm. The edge detection-based method often obtained many edges at the initial time. The adaptive threshold method required a lot of calculations to pick the one that is most likely to be the edge of the river. Both of them required a large amount of computation. However, the algorithm designed in this paper was essentially a fast two-class classification process on each image patch by using a preset threshold. The improved LBP feature is based on the comparison of intensity of neighbor pixels instead of exact calculations. Therefore, the proposed algorithm consumed the least time.

3.3. Experiments in Complex Scenes

Besides the simple river scenes, there are also some complex outdoor scenes where the traditional algorithms are difficult to take effect or even fail. Tests of different methods on complex scenes were conducted, among which four typical examples are shown in Figure 12 with the corresponding criteria values shown in Table 2.

Moreover, Figure 13 shows the speed of different methods in complex river scenes.

Table 2. Performance of different methods in complex scenes.

River Scene	Intensity + Texture			Edge Detection			Our Algorithm		
	PA	MIoU	F_β	PA	MIoU	F_β	PA	MIoU	F_β
(a)	0.908	0.708	**0.759**	0.987	0.629	**0.708**	0.908	0.878	**0.887**
(b)	0.864	0.822	**0.834**	0.999	0.722	**0.789**	0.948	0.910	**0.921**
(c)	0.657	0.664	**0.662**	0.995	0.845	**0.886**	0.913	0.873	**0.885**
(d)	0.854	0.775	**0.798**	0.981	0.830	**0.871**	0.934	0.931	**0.932**
Average of 390 images	0.802	0.745	**0.762**	0.985	0.745	**0.805**	0.935	0.925	**0.929**

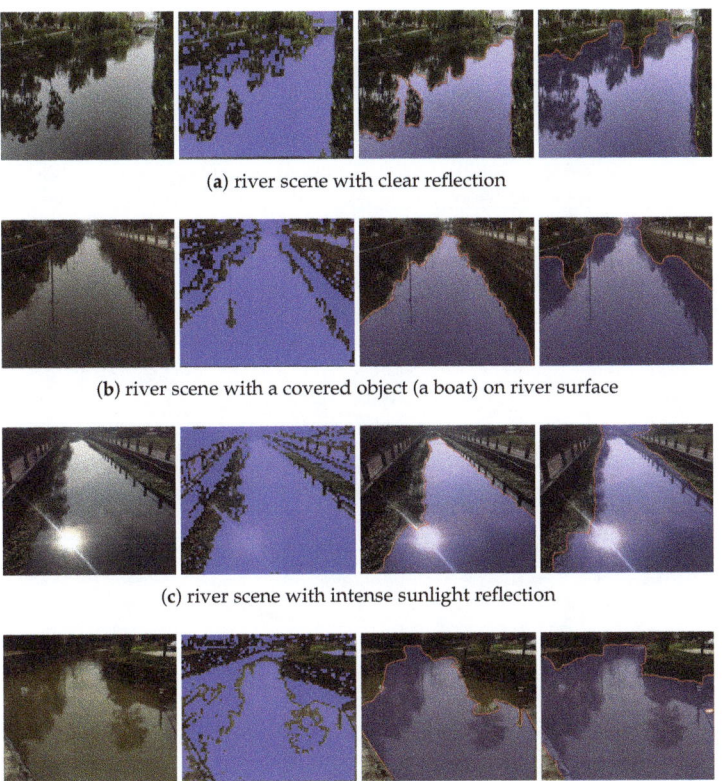

Figure 12. Segmentation results of different algorithms in complex scenes (**a–d**). From left to right: The first column are the input images after pre-processing; The second and third column are respectively the segmentation results using "intensity + texture" features and the adaptive threshold edge detection algorithm; The fourth column are the results of our method.

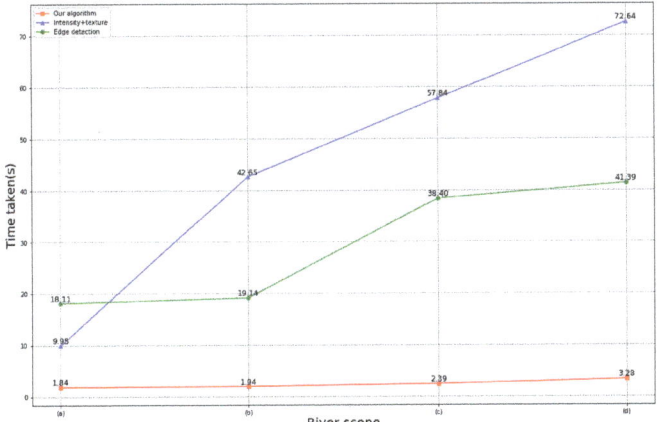

Figure 13. Speed of different methods in the complex scenes.

The proposed algorithm showed robust performance in complex scenes, and it got the highest F_β and the least time cost compared with other methods. The proposed algorithm is proved effective and has better segmentation performance of river images.

The method utilizing "intensity + texture" features was prone to false detection. As shown in Figure 12, some pixels on the riverside were also detected as water. This is because some parts of the riverside in the image had similar features to the designed one. Therefore, the method simply using global image features could be confused. As for the method based on edge detection, it was likely to miss part of the water region with shadow due to the strong edge of clear shadow. This method could not distinguish whether the detected edge was a riverbank or other edges, which resulted in mistakes. However, the improved LBP and H variance features designed in this study were local features based on the water surface reflection mechanism, which was close to characteristics of water pixels. Such features could describe not only common water part of the image, but also those with complex appearance like light and covered shadow. To illustrate this, the results of each step in our algorithm are shown in Figure 14 to show how it works.

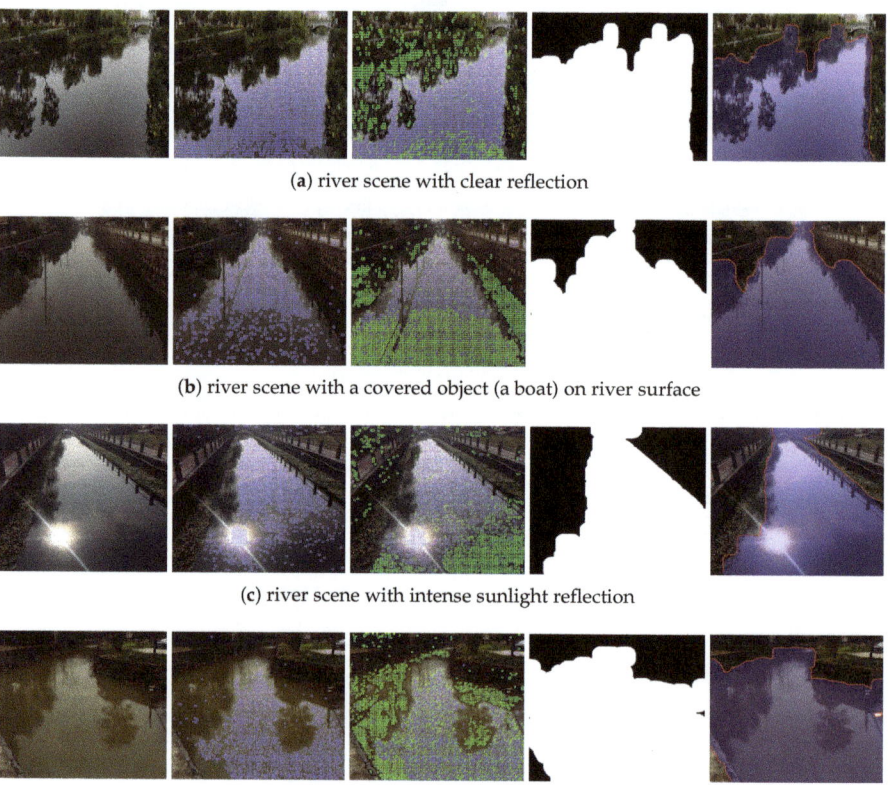

(a) river scene with clear reflection

(b) river scene with a covered object (a boat) on river surface

(c) river scene with intense sunlight reflection

(d) river scene with large area shadow

Figure 14. Performance of the proposed method in complex scenes (**a**–**d**). From left to right, the first column includes the images after pre-processing. In the second column, The blue squares in images represent the detection result of water region with the improved LBP feature. In the third column the green squares represent the detection result by H-variance feature developed from the second column images. The fourth column images are the binary mask of detection results after the designed morphological operation. The fifth column images are final segmentation results of river region indicated by a covered translucent blue area.

3.4. Discussion of Patch Size

The patch size, that is, the size of each detection window in the image, was the basic unit in the feature extraction operation in our algorithm. Theoretically, using a smaller patch size is faster in feature extraction, but the total times of feature calculation will increase, while the larger patch size made each patch contain more pixels, which might include negative samples (non-water pixels) that damaged the judgment of segmentation algorithm. Figure 15 shows the segmentation results using different patch sizes in our algorithm.

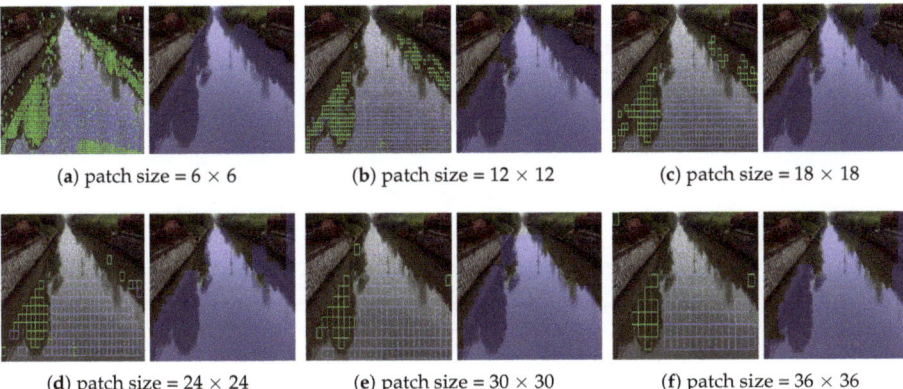

(a) patch size = 6 × 6 (b) patch size = 12 × 12 (c) patch size = 18 × 18

(d) patch size = 24 × 24 (e) patch size = 30 × 30 (f) patch size = 36 × 36

Figure 15. Segmentation results under different patch sizes from 6 × 6 to 36 × 36 pixels used in the proposed algorithm.

The F_β and speed under different patch sizes were shown in Figure 16. With the patch size increased, F_β ($\beta = 1.5$, see Equation (20)) reduced and the time cost was lower. After comprehensive consideration of the segmentation performance and time consumed, a 6 × 6 patch size is usually adopted in practice.

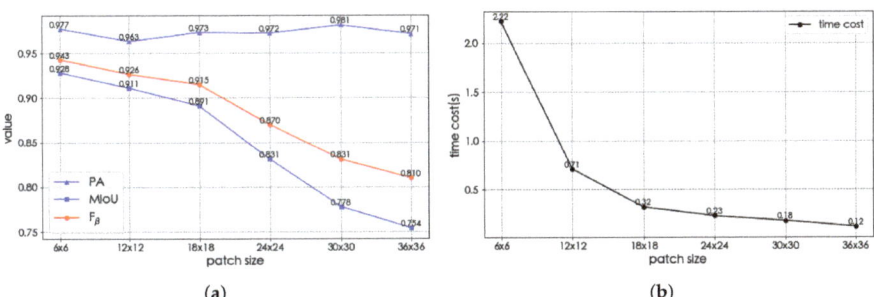

(a) (b)

Figure 16. Performance of the proposed algorithm under using different patch sizes. (a) *PA*, *MIoU* and F_β under different patch sizes. (b) Speed under different patch sizes.

3.5. Discussion of Structuring Element

The size and shape of the structuring elements affected the final segmentation result. Some tests were performed on different resolution images using different sizes of structuring elements from 1/5 to 1/30 of the input image size. Two examples with criteria measuring the segmentation performance were shown in Figure 17.

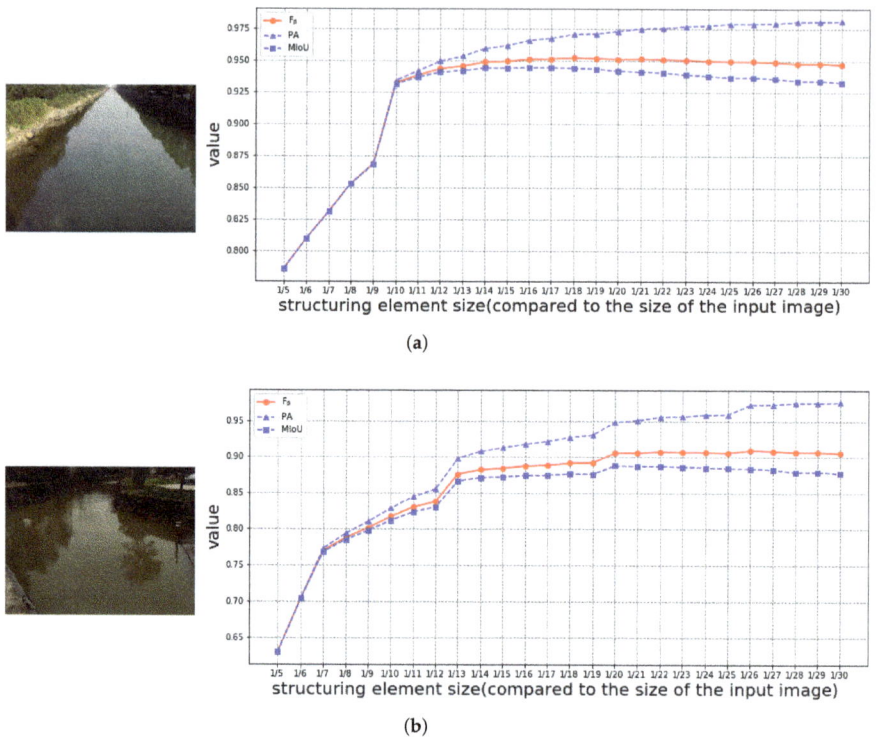

Figure 17. Performance under different sizes of structuring elements in the proposed algorithm. (**a**) An example of simple river scene; (**b**) An example of complex river scene.

As shown in the results, when the size of the structuring element grew larger than 1/15 of the input image size, the segmentation performance distinguishes little. Based on more experiments on the dataset, the size could be set to 1/15 of the size of the input image, where the algorithm was usually effective and reliable.

4. Conclusions

In this study, we focus on the image segmentation of outdoor river scenes. To solve the problem that current methods often missed detection and made false segmentations when applied to complex river scenes, this study proposed a novel segmentation method based on a reflection mechanism of the water surface. An improved LBP feature descriptor was designed for water detection and H variance was introduced to detect the shadow area of the water's surface. Morphological operation with multiple dilation was employed to eliminate pseudo-water patches wrongly detected by the proposed algorithm and to obtain the largest connected domain in the image as water region. The experiments were performed in simple and complex river scenes respectively where the proposed method was compared with two other river segmentation methods. The results showed the proposed method took the least time and had better and robust performance in both simple and complex river scenes.

At present, the proposed algorithm has only been proven to be suitable for segmenting water parts in river images. Since the algorithm is designed based on the reflection mechanism of the water surface, it remains to be further studied whether it is effective for other types of images. The design ideas of the proposed algorithm may be helpful to other segmentation algorithms.

In the future, research can be conducted on anomaly detection of water surfaces based on the proposed method. This study is also important for unmanned surface vehicles (USVs) and river mapping.

Author Contributions: Conceptualization, J.Y.; Methodology, Y.L.; Data curation, Y.Z.; Supervision, P.H., J.Y., G.Z. and D.H.; Validation, W.X.; Writing—original draft, Y.L.; Writing—review & editing, P.H., J.Y. and D.H. All authors have read and agreed to the published version of the manuscript.

Funding: This work was funded by the Fundamental Research Funds for the Central Universities (No.2019QNA5015), the National Natural Science Foundation of China (No. 61803333, 61573313), the Key Technology Research and Development Program of Zhejiang Province (No.2015C03G2010034), and the National Key R&D Program of China (No.2017YFC1403801).

Conflicts of Interest: The authors declare no conflict of interest.

References

1. Matthies, L.H.; Bellutta, P.; McHenry, M. Detecting water hazards for autonomous off-road navigation. In Proceedings of the Unmanned Ground Vehicle Technology V, Orlando, FL, USA, 21–25 April 2003; Volume 5083, pp. 231–242.
2. Yu, J.J.; Luo, W.T.; Xu, F.H.; Wei, C.Y. River boundary recognition algorithm for intelligent float-garbage ship. *Electron. Des. Eng.* **2018**, *2018*, 29.
3. Song, Y.; Wu, Y.; Dai, Y. A new active contour remote sensing river image segmentation algorithm inspired from the cross entropy. *Digit. Signal Process.* **2016**, *48*, 322–332. [CrossRef]
4. Ciecholewski, M. River channel segmentation in polarimetric SAR images: Watershed transform combined with average contrast maximisation. *Expert Syst. Appl.* **2017**, *82*, 196–215. [CrossRef]
5. Han, B.; Wu, Y. A novel active contour model based on modified symmetric cross entropy for remote sensing river image segmentation. *Pattern Recognit.* **2017**, *67*, 396–409. [CrossRef]
6. Lopez-Fuentes, L.; Rossi, C.; Skinnemoen, H. River segmentation for flood monitoring. In Proceedings of the 2017 IEEE International Conference on Big Data (Big Data), Boston, MA, USA, 11–14 December 2017; pp. 3746–3749.
7. Rankin, A.L.; Matthies, L.H.; Huertas, A. Daytime water detection by fusing multiple cues for autonomous off-road navigation. In *Transformational Science And Technology For The Current And Future Force: (With CD-ROM)*; World Scientific: Singapore, 2006; pp. 177–184.
8. Yao, T.; Xiang, Z.; Liu, J.; Xu, D. Multi-feature fusion based outdoor water hazards detection. In Proceedings of the 2007 International Conference on Mechatronics and Automation, Harbin, China, 5–8 August 2007; pp. 652–656.
9. Zhao, J.; Yu, H.; Gu, X.; Wang, S. The edge detection of river model based on self-adaptive Canny Algorithm and connected domain segmentation. In Proceedings of the 2010 8th World Congress on Intelligent Control and Automation, Jinan, China, 7–9 July 2010; pp. 1333–1336.
10. Wei, Y.; Zhang, Y. Effective waterline detection of unmanned surface vehicles based on optical images. *Sensors* **2016**, *16*, 1590. [CrossRef] [PubMed]
11. Sun, Y.; Fu, L. Coarse-fine-stitched: A robust maritime horizon line detection method for unmanned surface vehicle applications. *Sensors* **2018**, *18*, 2825. [CrossRef]
12. Achar, S.; Sankaran, B.; Nuske, S.; Scherer, S.; Singh, S. Self-supervised segmentation of river scenes. In Proceedings of the 2011 IEEE International Conference on Robotics and Automation, Shanghai, China, 9–13 May 2011; pp. 6227–6232.
13. Zhan, W.; Xiao, C.; Wen, Y.; Zhou, C.; Yuan, H.; Xiu, S.; Zhang, Y.; Zou, X.; Liu, X.; Li, Q. Autonomous Visual Perception for Unmanned Surface Vehicle Navigation in an Unknown Environment. *Sensors* **2019**, *19*, 2216. [CrossRef]
14. Han, X.; Nguyen, C.; You, S.; Lu, J. Single Image Water Hazard Detection using FCN with Reflection Attention Units. In Proceedings of the European Conference on Computer Vision (ECCV), Munich, Germany, 8–14 September 2018; pp. 105–120.
15. Hong, T.H.; Rasmussen, C.; Chang, T.; Shneier, M. Fusing ladar and color image information for mobile robot feature detection and tracking. In Proceedings of the 7th International Conference on Intelligent

Autonomous Systems, Marina del Rey, CA, USA, 25–27 March 2002; Gini, M., Shen, W.-M., Torras, C., Yuasa, H., Eds.; IOS Press: Amsterdam, The Netherlands, 2002; pp. 124–133.
16. Nguyen, C.V.; Milford, M.; Mahony, R. 3D tracking of water hazards with polarized stereo cameras. In Proceedings of the 2017 IEEE International Conference on Robotics and Automation (ICRA), Singapore, 29 May–3 June 2017; pp. 5251–5257.
17. Kim, J.; Baek, J.; Choi, H.; Kim, E. Wet area and puddle detection for Advanced Driver Assistance Systems (ADAS) using a stereo camera. *Int. J. Control Autom. Syst.* **2016**, *14*, 263–271. [CrossRef]
18. Pandian, A. Robot Navigation Using Stereo Vision and Polarization Imaging. Master's Thesis, Institut Universitaire de Technologie IUT Le Creusot, Universite de Bourgogne, Le Creusot, France, 2008.
19. Yang, K.; Wang, K.; Cheng, R.; Hu, W.; Huang, X.; Bai, J. Detecting traversable area and water hazards for the visually impaired with a pRGB-D sensor. *Sensors* **2017**, *17*, 1890. [CrossRef] [PubMed]
20. Iqbal, M.; Morel, M.; Meriaudeau, F. A survey on outdoor water hazard detection. In *Skripsi Program Studi Siste Informasi*; University of Southampton: Southampton, UK, 2009.
21. Rankin, A.; Matthies, L. Daytime water detection based on color variation. In Proceedings of the 2010 IEEE/RSJ International Conference on Intelligent Robots and Systems, Taipei, Taiwan, 18–22 October 2010; pp. 215–221.
22. KOEN, E. Evaluation of color descriptors for object and scene recognition. In Proceedings of the IEEE Conference on Computer Vision and Pattern Recognition, Anchorage, AK, USA, 23–28 June 2008.
23. Xu, M.; Ellis, T. Illumination-Invariant Motion Detection Using Colour Mixture Models. In *BMVC*; Citeseer: Princeton, NJ, USA, 2001; pp. 1–10.
24. Gonzalez, R.C.; Wintz, P. *Digital Image Processing*; Number 13; Addison-Wesley Pub. Co.: Boston, MA, USA, 1977; p. 451.
25. Haralick, R.M.; Shanmugam, K. Textural features for image classification. *IEEE Trans. Syst. Man Cybern.* **1973**, *3*, 610–621. [CrossRef]
26. Laws, K.I. Rapid texture identification. In Proceedings of the Image Processing for Missile Guidance, San Diego, CA, USA, 29 July–1 August 1980; Volume 238, pp. 376–381.
27. Guo, Z.; Zhang, L.; Zhang, D. A completed modeling of local binary pattern operator for texture classification. *IEEE Trans. Image Process.* **2010**, *19*, 1657–1663. [PubMed]
28. Serra, J. *Image Analysis and Mathematical Morphology*; Academic Press, Inc.: Cambridge, MA, USA, 1983.
29. Haralick, R.M.; Sternberg, S.R.; Zhuang, X. Image analysis using mathematical morphology. *IEEE Trans. Pattern Anal. Mach. Intell.* **1987**, *9*, 532–550. [CrossRef] [PubMed]
30. Garcia-Garcia, A.; Orts-Escolano, S.; Oprea, S.; Villena-Martinez, V.; Martinez-Gonzalez, P.; Garcia-Rodriguez, J. A survey on deep learning techniques for image and video semantic segmentation. *Appl. Soft Comput.* **2018**, *70*, 41–65. [CrossRef]

© 2020 by the authors. Licensee MDPI, Basel, Switzerland. This article is an open access article distributed under the terms and conditions of the Creative Commons Attribution (CC BY) license (http://creativecommons.org/licenses/by/4.0/).

Article

Spectrogram Classification Using Dissimilarity Space

Loris Nanni [1,*], Andrea Rigo [1], Alessandra Lumini [2] and Sheryl Brahnam [3]

1 DEI, Via Gradenigo 6, 35131 Padova, Italy; andrea.rigo.5@studenti.unipd.it or rigoandrea96@gmail.com
2 DISI, University of Bologna, Via dell'Università 50, 47521 Cesena, Italy; alessandra.lumini@unibo.it
3 Department of Information Technology and Cybersecurity, Missouri State University, 901 S. National Street, Springfield, MO 65804, USA; sbrahnam@missouristate.edu
* Correspondence: loris.nanni@unipd.it

Received: 19 May 2020; Accepted: 9 June 2020; Published: 17 June 2020

Abstract: In this work, we combine a Siamese neural network and different clustering techniques to generate a dissimilarity space that is then used to train an SVM for automated animal audio classification. The animal audio datasets used are (i) birds and (ii) cat sounds, which are freely available. We exploit different clustering methods to reduce the spectrograms in the dataset to a number of centroids that are used to generate the dissimilarity space through the Siamese network. Once computed, we use the dissimilarity space to generate a vector space representation of each pattern, which is then fed into an support vector machine (SVM) to classify a spectrogram by its dissimilarity vector. Our study shows that the proposed approach based on dissimilarity space performs well on both classification problems without ad-hoc optimization of the clustering methods. Moreover, results show that the fusion of CNN-based approaches applied to the animal audio classification problem works better than the stand-alone CNNs.

Keywords: audio classification; dissimilarity space; siamese network; ensemble of classifiers; pattern recognition; animal audio

1. Introduction

Sound classification and recognition have been applied in different domains, e.g., speech recognition [1], music classification [2], environmental sound recognition, and biometric identification [3]. Traditionally, in pattern recognition problems, features have been extracted from the actual audio traces (e.g., Statistical Spectrum Descriptor and Rhythm Histogram [4]). However, by replacing audio traces by their visual representation, image classification techniques can be used to extract features on sound classification problems. The most commonly used visual representation of audio traces involves the display of their frequency spectrum as they vary in time, as in spectrograms [5] and Mel-frequency Cepstral Coefficients spectrograms [6]. A spectrogram can be described as a graph with two dimensions (time and frequency) plus a third dimension in terms of pixel intensity [7] that represents the signal amplitude in a specific frequency at a particular time step. Costa et al. [8,9] applied several classification and texture analysis techniques to music genre classification using such a method. In [9], the authors extracted grey level co-occurrence matrices (GLCMs) [10] from spectrograms, while in [8] they used the local binary pattern (LBP) [11], which is a popular texture descriptor. In [12], two other feature descriptors were extracted from audio images: local phase quantization (LPQ) and Gabor filters [13]. In 2017, Nanni et al. [2] demonstrated on multiple audio datasets how the fusion of acoustic features extracted from audio traces using state-of-the-art texture descriptors greatly improves the accuracy of acoustic and visual feature-based systems.

When deep learning became popular and Graphic Processing Units (GPUs) became more powerful at accessible costs, traditional pattern recognition changed, and attention focused even more on visual representations of acoustic traces. In the traditional machine learning framework, the optimization of the feature extraction step plays a key role, especially with the evolution of handcrafted features, which minimize the distance between patterns of the same class in the feature space while simultaneously attempting to maximize their distance from the patterns of other classes. Since deep classifiers learn the best features for describing patterns during the training process, these engineered features have diminished in significance, playing in the deep framework more of a supporting role when combined with features extracted from visual representations of acoustic traces that the deep classifiers determine are most informative. Another reason for the growing popularity of representing audio as images is the fact that the convolutional neural network (CNN), one of the most famous deep classifiers, requires images for its input. In their study, Humphrey and Bello [14,15] explored CNNs as an alternative approach to music classification problems, establishing the state-of-the-art in automatic chord detection and recognition. Nakashika et al. [16] converted spectrograms into GCLM maps to train CNNs for music genre classification, and Costa et al. [17] performed better than the state-of-the-art on the LMD dataset by fusing canonical approaches, e.g., LMP-trained SVMs with CNNs. Only a few studies, however, have focused on making these processes that were designed for image classification more specific for sound image recognition. In their study, Sigtia and Dixon [18] focused on adjusting CNN parameters and structures and showed how using Rectified Linear Units (ReLu) instead of stochastic gradient descent with the Hessian Free optimization and sigmoid units reduced training time. Wang et al. [19] presented an innovative CNN, which they named a *sparse coding CNN*, for sound event recognition and retrieval, which, when evaluated under noisy and clean conditions, achieved competitive and sometimes better performance than the majority of other approaches. In Oramas [20], a hybrid approach was presented that combined diverse modalities (album cover images, reviews, and audio tracks) for multi-label music genre classification by applying deep learning techniques appropriate for each modality, an approach that outperformed the single-modality methods. Finally, it should be mentioned that many methods in machine learning are also proposed for the human voice classification task: emotion recognition [21], English accent classification, and gender classification [22], to name a few.

Because deep classifiers have produced a patent improvement in music classification, researchers have begun to apply deep learning approaches to other sound recognition tasks, such as biodiversity assessment. Precise sound recognition systems can be of crucial importance in assessing and handling environmental threats like animal species loss and climate changes affecting wildlife fauna [23]. Birds, for instance, have been acknowledged as an indicator species for ecological research, and their monitoring has become increasingly important for biodiversity preservation [23], especially considering the minimal impact video and audio acquisition has on ecosystems. To date, many datasets are available to develop classifiers to identify and monitor different species, such as birds [24,25], whales [26], frogs [24], bats [25], and cats [27]. For instance, both Cap et al. [28] and Salamon et al. [29] have investigated the fusion of CNNs with other methods to classify animals. The former study combined CNNs with handcrafted features to classify marine animals [30] using the Fish and MBARI benthic animal dataset [31], while the latter fused deep learning with shallow learning for bird species identification based on 5428 bird flight calls from 43 species. In both cases, the fusion of CNNs with other canonical techniques outperformed the single approaches.

Existing approaches for animal audio classification can roughly be classified into two categories: fingerprinting and CNN approaches. Fingerprinting [32] relies on the compact representation of audio traces so that each one can be efficiently matched against other audio clips to compare for similarity and dissimilarity [33]. A sample of audio fingerprinting by CNN is shown in [34], where the authors used a Siamese neural network to produce semantic representations of the audio traces. However, fingerprinting is useful only in finding an exact match; the problem addressed in this work involves audio classification. As already noted, CNN-based approaches [35,36] train networks for animal audio classification starting from an image representation of the audio signal. Unfortunately, CNNs require a large number of training examples to be effective (larger than available in most animal audio datasets) and cannot generalize to new classes without retraining the network. The objective of this work is to solve these issues by proposing an approach based on Dissimilarity Spaces. Recently, Agrawal [37] proposed an approach that learns a distance model by training a Siamese neural network directly on dissimilarity values for brain image classification, and in [38] an approach is proposed for online signature verification using a Siamese neural network and a contrastive loss function. In the latter work, the authors claim that the main advantage a Siamese network offers over a canonical CNN is the ability to generalize: the Siamese network approach they developed was shown to verify the authenticity of the signature of a new user without being trained on any examples from this user.

In this work, the dissimilarity space is created using a Siamese Neural Network (SNN) trained on the entire training set to define a distance function among the samples. The training phase for SNN is aimed at maximizing the distance between patterns of different classes; the testing phase of the SNN is used to compare two spectrograms to obtain a measure of their dissimilarity. In theory, all the training samples can be selected as centroids of the dissimilarity space. Dimensionality reduction is obtained by selecting a smaller number (k) of prototypes via a clustering approach. The dissimilarity space is the space where each spectogram is represented by a its distance to each centroid/prototype: in this space, the SNN is used to compare the spectrogram to every centroid, obtaining the spectrogram's dissimilarity vector, which is the final descriptor. The classification task is performed by a support vector machine (SVM) trained using the dissimilarity descriptors generated from the training samples. The proposed system is evaluated on two different datasets for animal audio classification: domestic cat sounds [27] and bird sounds [23]. Results for the different clustering methods and different values of the hyperparameter (k) are reported.

In addition, an ensemble of SVMs trained on different dissimilarity spaces (by changing the value of k) are combined by sum rule, and its performance is compared with (i) some canonical CNN approaches and (ii) the fusion of the SVMs and the CNNs. Experiments demonstrate for the first time that the use of dissimilarity spaces based on SNN is a feasible representation for image data and can, when combined with a general purpose classifier, achieve high classification performance. Because the descriptors obtained in the dissimilarity space show high diversity with respect to the representations based on CNNs, their fusion can be exploited in an ensemble, as proven by the high classification accuracy obtained by the fusion of CNNs with our approach. The MATLAB code used in this study is freely available at https://github.com/LorisNanni.

2. Proposed Approach

The proposed method for spectrogram classification using dissimilarity space is based on several steps which are schematized in Figure 1. This figure is followed by the pseudo-code for each step (Algorithms 1 and 2). In order to define a similarity space, it is necessary to select a distance measure and a set of prototypes in the training phase. The distance measure $d(x,y)$ is learned by means of a SNN trained to maximize the similarity between couples of spectrograms in the same class, while minimizing the similarity for couples in different classes. The set of prototypes $P = p_1, ... p_k$ are obtained as the k centroids of the clusters generated by a supervised clustering procedure. The final step represents each training sample x in the dissimilarity space by a feature vector $f \in \Re^k$, where each component f_i is the distance between x and the prototype p_i: $f_i = d(x, p_i)$. These feature vectors are used to train a SVM for the final classification task. In the testing phase, each unlabeled spectrogram is first represented in the dissimilarity space by calculating its distance to all the prototypes, then the resulting feature vector is classified by SVM.

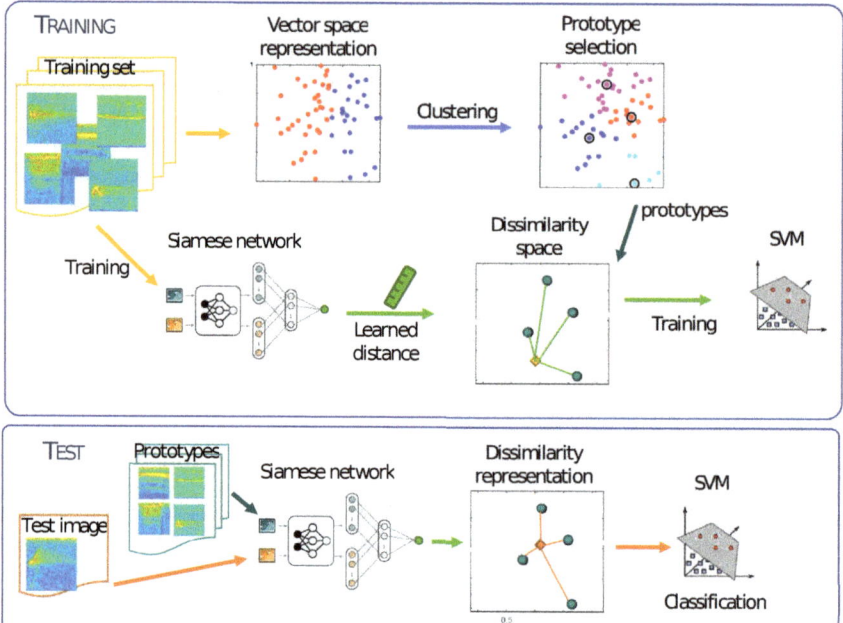

Figure 1. Proposed approach scheme.

Algorithm 1 Training phase

Input: Training images (*imgsTrain*), training labels (*labelTrain*), the number of training iterations (*trainIterations*), batch size (*trainBatchSize*), number of centroids (*k*), and the clustering technique (*type*).
Output: Trained SNN (*tSNN*), set of centroids (*C*), and trained SVM (*svm*).
1: $tSNN \leftarrow \text{TRAINSIAMESE}(imgsTrain, labelTrain, trainIterations, trainBatchSize)$
2: $P \leftarrow \text{CLUSTERING}(imgsTrain, labelTrain, k, type)$
3: $F \leftarrow \text{GETDISSSPACEPROJECTION}(imgsTrain, P, tSNN)$
4: $tSVM \leftarrow \text{TRAINSVM}(labelTrain, F)$

Algorithm 2 Testing phase

Input: Test images (*imgsTest*), trained SNN (*tSNN*), Set of centroids (*C*), Trained SVM (*tSVM*).
Output: Actual test labels (*labelTest*).
1: $F \leftarrow \text{GETDISSSPACEPROJECTION}(imgsTest, P, tSNN)$
2: $labelTest \leftarrow \text{PREDICTSVM}(F, tSVM)$

Each of the main functions used in the pseudo-code are described below.

2.1. Siamese Neural Network Training

The SNN, described in more detail in Section 3, is trained to compare a pair of spectrograms by returning a measure of their similarity. Algorithm 3 presents the pseudocode for this phase and corresponds with step 1 of Algorithm 1. The SNN architecture is defined in steps 2 and 3 of algorithm Algorithm 3. Steps 5–8 are repeated for each training iteration. Step 5 extracts randomly *batchSize* spectrograms pairs from the training set using the function GETSIAMESEBATCH. Step 6 feeds the pairs to the network and computes loss and gradients for gradient descent. Steps 7 and 8 use the gradients and loss to update the weights of the fully connected layer and the twin subnetworks.

Algorithm 3 Siamese training pseudocode

Input: Training image ($trainImgs$), training labels ($trainLabels$), batch size ($batchSize$), and iterations ($numberOfIterations$).
Output: Trained SNN ($tSNN$).
1: **function** TRAINSIAMESE
2:   $subnet \leftarrow$ NETWORK($[inputLayer, ..., FullyConnectedLayer]$)
3:   $fcWeights \leftarrow randomWeights$
4:   **for** $iteration \leftarrow$ from 1 to $numberOfIterations$ **do**
5:     $X1, X2, pairLabels \leftarrow$ GETSIAMESEBATCH($trainImgs, trainLabels, batchSize$)
6:     $gradients, loss \leftarrow$ EVALUATE($subnet, X1, X2, pairLabels$)
7:     UPDATE($subnet, gradients$)
8:     UPDATE($fcWeights, gradients$)
9:   **end for**
10: **return** $tSNN \leftarrow subnet, fcWeights$
11: **end function**

Note: in the case where the SNN fails to converge on the training set, training is rerun.

2.2. Prototype Selection

In this phase, k prototypes are extracted from the training set. In theory, every spectrogram in the training set could be selected as a prototype, but this would be too resource expensive and the dimensionality of the generated dissimilarity vectors would be too high. A better alternative is to employ clustering techniques to compute k centroids for each class. Clustering would significantly reduce the dimension of the resulting dissimilarity space and thus make the process more viable. Algorithm 4 presents the pseudo code for prototype selection, which provides a selection from among four clustering procedures, which are used separately to cluster the training samples belonging to each class.

Algorithm 4 Clustering pseudocode

Input: Training images ($imgsTrain$), training labels ($labelTrain$), number of clusters (k), and clustering technique ($type$).
Output: Centroids P.
1: **function** CLUSTERING
2:   $numClasses \leftarrow$ number of classes from $labelTrain$
3:   $kc \leftarrow k/numClasses$
4:   **for** $i \leftarrow$ from 1 to $numClasses$ **do**
5:     $images \leftarrow$ images of the class i from $imgsTrain$
6:     **switch** $type$ **do**
7:       **case** "k-means" $P_i \leftarrow$ KMEANS($imgs,kc$)
8:       **case** "k-medoids" $P_i \leftarrow$ KMEDOIDS($imgs,kc$)
9:       **case** "hierarchical" $P_i \leftarrow$ HIERARCHICAL($imgs,kc$)
10:      **case** "spectral" $P_i \leftarrow$ SPECTRAL($imgs,kc$)
11:     $P \leftarrow P \cup P_i$
12: **end for**
13: **return** P
14: **end function**

2.3. Projection in the Dissimilarity Space

Existent classification methods learn to classify patterns using their feature space. In this work, patterns are represented in a dissimilarity space in which every pattern x is represented by its similarity to a selected set of prototypes $P = p_1, ... p_k$ by a dissimilarity vector:

$$F(x) = [d(x, p_i), d(x, p_{i+1}), ..., d(x, p_k)], \qquad (1)$$

where the similarity among pattern $d(x, y)$ is obtained using a trained SNN. In order to project each image in the Dissimilarity space $\Re^k$, Algorithm 5 compares each input image (stored in X in step 3) with the k centroids (stored in P) using the trained SNN $tSNN$ with the PREDICTSIAMESE function (step 4). The resulting feature space F includes the projected features of all the input images.

Algorithm 5 Projection in the Dissimilarity space pseudocode

Input: Images (*imgs*), Centroids (*P*), number of centroids (*k*), and trained SNN (*tSNN*).
Output: Feature vectors (*F*).
 1: **function** GETDISSSPACEPROJECTION
 2:     **for** $j \leftarrow$ from 1 to SIZE(*imgs*) **do**
 3:         $X \leftarrow imgs[j]$
 4:         $F[j] \leftarrow$ PREDICTSIAMESE($tSNN, X, P$)
 5:     **end for**
 6:     **return** F
 7: **end function**

2.4. Support Vector Machine Training and Prediction

A Support Vector Machine (SVM) is a supervised learning model witch can be used to perform classification or regression. An SVM model represents each training example as a data point in space and is trained to construct one or more hyperplanes that divide the space in two, separating data points belonging to different classes (function TRAINSVM). The model will predict (function PREDICTSVM) the class of a new pattern mapped in the space according to the side of the hyperplane the data point falls into. The hyperplane found by an SVM is defined as follows:

$$D(x) = w * x - b, \qquad (2)$$

where $D(x)$ is the hyperplane, x is the data point vector, w is the hyperplane's normal vector, and the $\frac{b}{||w||}$ ratio is the hyperplane's distance from the origin. The optimal hyperplane is the one that maximizes the distance to the nearest data point of any class, defined as $\frac{2}{||w||}$, which is also called the *margin*. The i-th point x_i will be assigned to the first class when $D(x_i) \geq +1$ and to the second class when $D(x_i) \leq -1$. The points that lie on the margin line, defined by the equation $D(x_i) = \pm 1$, completely describe the solution to the problem and are called *support vectors*. An example of an optimal hyperplane with highlighted support vectors is shown in Figure 2.

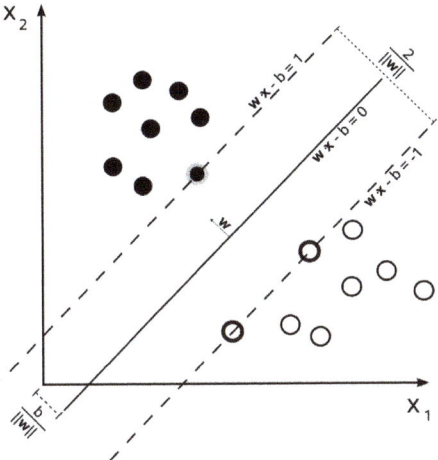

Figure 2. SVM's hyperplane.

Because SVMs use hyperplanes to discriminate data, they do not work well with data that is not linearly separable in its original space. This problem can be solved using kernel functions, which map data into a much higher dimensional space, presumably to make the separation easier in that space. To keep the computational complexity to an acceptable level, the kernel function of choice has to be computationally efficient.

Being binary classifiers, SVMs can only determine the separation surface between two classes of data; however, it is possible to apply SVMs to multi-label problems by training an ensemble of SVMs and combining them. In this work, the *One-Against-All* approach is used, where for each class an SVM is trained to discriminate between a given class and all the other classes put together. The pattern is then assigned to the class that gives the higher confidence score.

3. Siamese Neural Network

The Siamese Neural Network (SNN) is a class of neural network architectures that contains two or more twins, i.e., sub-networks with the same parameters and weights. SNNs are used in tasks involving similarity or in identifying correlations between different entities. SNN was first proposed by Bromley et al. [39] for performing signature verification. SNNs have since been used successfully in other application domains, such as face verification [40], image recognition [41], human fall detection [42], content-based audio representation [34], and sound search by vocal imitation [43]. The SNN architecture used in this work is similar to the one used in [43] and is represented in (Figure 3).

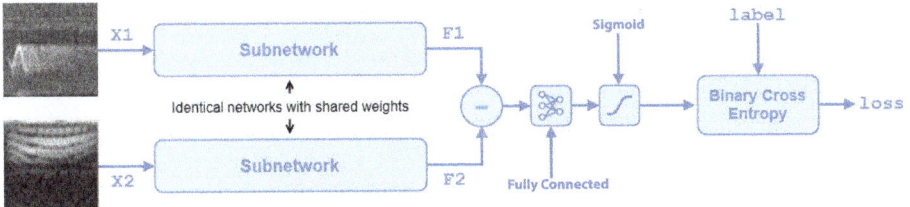

Figure 3. Siamese Neural Network architecture.

As shown in Figure 3, the SNN used in this work is composed of five blocks:

- *Two identical twin subnetworks*

    The twin subnetworks in our SNN are two Convolutional Neural Networks composed of 13 layers, as listed in Table 1.

    Table 1. Siamese subnetworks layers.

#	Layer	Filer Size	Number of Filters
1	Input Layer	224 × 224 images	
2	2D Convolution	10 × 10	64
3	ReLU		
4	Max Pooling	2 × 2	
5	2D Convolution	7 × 7	128
6	ReLU		
7	Max Pooling	2 × 2	
8	2D Convolution	4 × 4	128
9	ReLU		
10	Max Pooling	2 × 2	
11	2D Convolution	5 × 5	64
12	ReLU		
13	Fully Connected	Returns a 4096-dimensional vector	

    These subnetworks learn the features best representing the spectrograms in the input ($X1$ and $X2$), returning a 4096-dimensional feature vector for each ($F1$ and $F2$). The subnetworks share parameters and weights which are mirrored during the training.

- *Subtract block*

    The output vectors of the subnetworks are subtracted, resulting in a feature vector Y representing the features in which the images differ:

    $$Y = |F1 - F2| \qquad (3)$$

- *Fully Connected Layer*

    As in [37], the Fully Connected Layer (FCL) learns the distance model to calculate the dissimilarity. The output vector of the subtract block is fed to the FCL which returns a dissimilarity value for the pair of spectrograms in the input.

- *Sigmoid*

    The sigmoid function is a class of mathematical real functions having a characteristic S-shaped curve. We apply the sigmoid to the dissimilarity value returned by the FCL to convert it to a probability value in the range [0, 1], using the standard logistic function:

    $$S(x) = \frac{1}{1 + e^{-x}} \qquad (4)$$

- *Binary Cross Entropy*

    The Binary Cross Entropy (BCE) is a popular loss function, which, given the prediction of the model and the correct observation label (in our case, 1 if the two spectrograms belong to the same class, 0 otherwise) returns a measure of the performance of the model. Loss functions are used by learning algorithms to train the network by adjusting the weights. BCE is applied to the probability obtained from the sigmoid and computes the gradients of the loss function with respect to the weights of the network in order to adjust them. In a two-class problem, BCE can be calculated as:

$$BCE(y, p) = -(y \log(p) + (1 - y) \log(1 - p)), \tag{5}$$

where y is the binary value that indicates whether the class label c is correct for the observation o, p is the predicted probability that observation o is of class c, and log is the natural logarithm.

4. Clustering

Clustering is the task of organizing data in groups (Figure 4) so that patterns in the same cluster are more similar to each other than they are to patterns belonging to other clusters. Clustering is often used to find natural clusters in unlabeled data. Some clustering techniques calculate centroids during the process. A centroid is the mean vector of all the patterns in a cluster. Because it is a mean vector, it contains the most characterizing features of a cluster's patterns. Centroids are computed to reduce the dissimilarity space size without losing too much information. The greater the number of centroids used for each class, the more information that is retained. In this work, samples are divided into classes before clustering, and the clustering procedure is applied to each class separately. The remainder of this section describes the four clustering techniques used in this study.

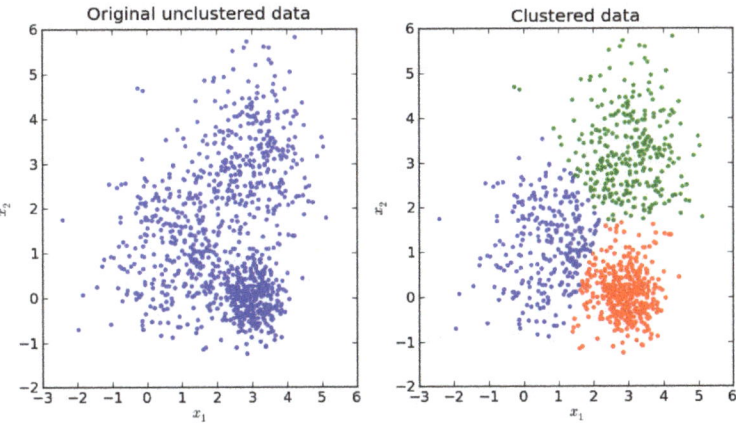

Figure 4. A sample of clusters found from unlabeled data: on the left the original 2D data, on the right clustered data, where different colors denote different clusters.

4.1. K-Means

K-means is a popular clustering algorithm that partitions a set of patterns into k clusters by assigning each observation to the cluster with the nearest centroid, or mean vector. There are several versions of this algorithm. In this study, the default implementation (with the Euclidean distance metric) in the MATLAB Statistics and Machine Learning Toolbox was applied. The standard k-means algorithm cycles through the following steps:

1. Choose k initial cluster centers (centroids) according to the k-means++ variation detailed below.
2. Compute point-to-cluster-centroid distances of all observations to each centroid.
3. Assign each observation to the cluster with the closest centroid.
4. Compute the average of the observations in each cluster to obtain k new centroids.
5. Repeat steps 2 through 4 until cluster assignments no longer change (i.e., until the algorithm converges) or until the maximum number of iterations is reached.

The k-means++ variation [44] employs a heuristic to find the initial centroids:

1. Choose one center uniformly at random from among the data points.
2. For each data point x, compute $d(x)$, the distance between x and the nearest center that has already been chosen.
3. Choose one new data point at random as a new center, using a weighted probability distribution where a point x is chosen with probability proportional to $d(x)^2$.
4. Repeat Steps 2 and 3 until k centers have been chosen.

4.2. K-Medoids

K-medoids is a clustering technique very similar to k-means. It partitions a set of observations into k clusters by minimizing the sum of distances between a pattern and the center of that pattern's cluster. The main difference between k-means and k-medoids is that, in the first case, the center of a cluster is its centroid, or mean, whereas, in the latter case, the center is a member, or medoid, of the cluster. A medoid is an observation in a cluster whose sum of distances from the other observations within the cluster is minimal. The basic algorithm for K-medoids loops through the following three steps:

1. Build-step: each k cluster is associated with a potential medoid. The first assignment can be performed in various ways; the standard MATLAB's implementations uses the k-means++ heuristic.
2. Swap-step: within each cluster, each point is tested as a potential medoid by checking whether the sum of the within-cluster distances gets smaller using that particular point as the medoid. If so, the point is defined as a new medoid. Every point is then assigned to the cluster with the closest medoid.
3. Repeat steps 1–4 until medoids no longer swap (i.e., until the algorithm converges) or until the maximum number of iterations is reached.

4.3. Hierarchical

Hierarchical clustering is a clustering technique that groups data by building a hierarchy of clusters. The hierarchy tree that is obtained is divided into n levels chosen for the application at hand. There are two main categories of hierarchical clustering:

- *Agglomerative*: each pattern starts in its own cluster; then, moving up the hierarchy, each cluster in one level is obtained by merging two clusters in the previous level.
- *Divisive*: all patterns start in one cluster; then, by moving down the hierarchy, each pair of clusters is obtained by splitting a single cluster in the previous level.

In this work, the default MATLAB implementation of hierarchical clustering is used, which is the agglomerative type. The MATLAB algorithm loops through the following three steps:

1. Find the similarity or dissimilarity between every pair of objects in the dataset using a distance metric.
2. Group the objects into a binary hierarchical cluster tree by linking objects in pairs based on their distance. As objects are paired into binary clusters, the newly formed clusters are grouped into larger clusters until a hierarchical tree is formed.
3. Determine where to cut the hierarchical tree into clusters. Here, MATLAB's *cluster* function is used to prune branches off the bottom of the hierarchical tree and to assign all the objects below each cut to a single cluster. In this way, k clusters are obtained.

After applying this algorithm, centroids, as the mean vectors of each cluster, are computed.

4.4. Spectral

The spectral clustering technique splits data into groups using the data's undirected similarity graph represented by a similarity matrix (also called an *adjacency matrix*). In the similarity graph, every·node is an observation, and two nodes are connected by an edge if their similarity is larger then a certain threshold, which is often 0. The algorithm uses four mathematical expressions:

- *Similarity Matrix*: a square symmetrical matrix that represents the similarity graph. Letting M be the similarity matrix, each cell value m_{ij} is the similarity value of two connected nodes in the graph, which, in turn, represent the spectrogram pairs (s_i, s_j).
- *Degree Matrix*: a diagonal matrix obtained by summing the similarity matrix rows. The degree matrix is defined by the equation

$$D_g(i,i) = \sum_j m_{ij},$$

where D_g is the degree matrix, and m_{ij} is a value of the similarity matrix.
- *Laplacian Matrix*: another way of representing the similarity graph that is defined as

$$L = D_g - M.$$

Here are the steps required by the spectral algorithm:

- For each spectrogram in the dataset, define a local neighborhood. There are different ways such a neighborhood can be defined. The MATLAB implementation defaults to the nearest-neighbor method. Once the neighborhood is defined, compute the pairwise similarities of each spectrogram in the neighborhood using some distance metric.
- Calculate the Laplacian matrix L.
- Create a matrix V containing columns $v_1, ..., v_k$, where the columns are the k eigenvectors that correspond to the k smallest eigenvalues of the Laplacian matrix. The eigenvalues of the matrix are also called *spectrum*, hence the algorithm's name.
- Treating each row of V as a pattern, perform k-means clustering or k-medoids clustering.
- Assign the original spectrograms in the dataset to the same clusters as their corresponding rows in V.

5. Experimental Results

The approach proposed in this paper is tested, along with some comparison canonical approaches, using a stratified ten-fold cross validation protocol and the classification accuracy as the performance indicator. Tests were performed on two datasets:

- BIRDZ, which was also used as a control and as a real-world audio dataset in [23]. The real-world tracks were obtained from the Xeno-canto Archive (http://www.xeno-canto.org/) and cover 11 widespread North American bird species. Thus, the dataset contains 11 classes: (1) Blue Jay, (2) Song Sparrow, (3) Marsh Wren, (4) Common Yellowthroat, (5) Chipping Sparrow, (6) American Yellow Warbler, (7) Great Blue Heron, (8) American Crow, (9) Cedar Waxwing, (10) House Finch, and (11) Indigo Bunting. BIRDZ is composed of five types of spectrograms: constant frequency, frequency modulated whistles, broadband pulses, broadband with varying frequency components, and strong harmonics, for a total of 2762 bird acoustic events with 339 detected "unknown" events corresponding to noise and other unknown species vocalizations. Including the "unknown class", BIRDZ has 3101 samples for 12 classes.
- CAT, which was first presented in [27,45]. This dataset is composed of 10 balanced classes with about 300 samples per class: (1) Resting, (2) Warning, (3) Angry, (4) Defence, (5) Fighting, (6)·Happy, (7) Hunting mind, (8) Mating, (9) Mother call, and (10) Paining. The samples have an average duration of about 4s and were collected by the author from different sources: Kaggle, Youtube and Flickr. CAT has a total of 2962 samples.

In Tables 2 and 3, the performance of the four tested clustering algorithms is reported using different values of *kc* (i.e., the number of clusters per class). As a baseline for comparison, the classification accuracy is also reported for the following well-known CNN models, each fine-tuned on the problem (for 30 epochs, using a batch size of 30, and a learning rate of 0.0001, no freezing):

- Googlenet [46], VGG16 and VGG19 [47], all pretrained on ImageNet [48];
- GoogleNetP365, a GoogleNet model pretrained on Places365 [49].

Moreover, in Tables 2 and 3, the accuracy obtained by the following fusion approaches are reported:

- KAll, fusion by sum rule of the four SVMs trained using the dissimilarity space built with all tested values for *kc* = 15, 30, 45, 60;
- ALL, fusion by average rule of the four approaches KAll (one for each clustering method);
- eCNN, fusion by sum rule of the four CNNs;
- ALL+eCNN, fusion by sum rule between ALL and eCNN;
- ALL+GoogleNet, fusion by sum rule between ALL and GoogleNet;
- ALL+GoogleNetP365, fusion by sum rule between ALL and GoogleNetP365.

Table 2. Classification accuracy on the BIRDZ dataset.

BIRDZ	k-Means	K-Medoids	Hierarchical	Spectral
kc = 15	91.85	92.06	91.85	92.09
kc = 30	91.81	92.05	91.94	92.05
kc = 45	92.03	91.69	91.90	92.24
kc = 60	91.61	91.77	91.39	91.79
KAll	92.71	92.59	92.63	92.95
ALL			92.97	
GoogleNet			92.41	
VGG16			95.30	
VGG19			95.19	
GoogleNetP365			92.94	
eCNN			95.81	
ALL+eCNN			**95.95**	
ALL+GoogleNet			95.64	
ALL+GoogleNetP365			94.74	

Table 3. Classification accuracy on the CAT dataset.

CAT	k-Means	K-Medoids	Hierarchical	Spectral
$kc = 15$	65.83	78.75	75.12	65.69
$kc = 30$	56.88	78.47	77.66	71.29
$kc = 45$	72.81	72.31	66.34	67.36
$kc = 60$	80.37	78.54	80.37	73.73
KAll	80.61	81.59	81.29	81.69
ALL		82.41		
GoogleNet		82.98		
VGG16		84.07		
VGG19		83.05		
GoogleNetP365		85.15		
eCNN		87.36		
ALL+eCNN		**87.76**		
ALL+GoogleNet		85.02		
ALL+GoogleNetP365		87.49		

From the results reported in Tables 2 and 3, the following conclusions can be drawn:

1. KAll outperforms each stand alone method based on a single value of kc;
2. ALL outperforms each KAll in both datasets;
3. Performance of ALL is similar to that obtained by GoogleNet;
4. The ensemble ALL based on our dissimilarity space is a feasible representation for spectograms and achieves a performance that is comparable to the CNNs.
5. In both datasets, the best performance is obtained by ALL+eCNN, (even though the improvement with respect to eCNN is negligible).
6. ALL+GoogleNet strongly outperforms ALL and Googlenet; this light ensemble, which uses only one CNN, is our recommended method.

The proposed approach based on the representation of animal sound in a dissimilarity space has two main advantages: (1) it produces a compact representation on the signal (ranging from 15 to 60, depending on the number of clusters for the single space, to 150 for the KAll ensemble); (2) it generates a high diversity of classification results with respect to the baseline CNNs, which can be exploited to improve the performance in an ensemble method (i.e., ALL+GoogleNet).

In Table 4, the ensembles proposed in this work are shown to achieve a performance on the two datasets that is similar to some of the state-of-the-art approaches reported in the literature. Two results are taken from [27], and are labeled [27] and [27]-*CNN*.

Unfortunately, most published papers in the field of acoustic animal classification focus only on a single dataset. The authors of this paper are aware that evaluating the proposed approach on two different datasets instead of focusing on just one limits the strength of the conclusions drawn. Be that as it may, the experiments reported here prove the robustness of the proposed approach, which obtains good classification accuracy on two different problems without any ad-hoc parameter optimization and according to a clear and unambiguous testing protocol. As a result, the performances reported in this paper can be used for baseline comparisons with other audio classification methods developed in the future.

Table 4. Literature results.

Descriptor	BIRDZ	CAT
[50]	96.3	—
[2]	95.1	—
[23]	93.6	—
[45]	—	87.7
[27]	—	91.1
[27] −CNN	—	90.8
[51]	96.7 *	—

* Note that the results in [51] are based on a feature selection approach where the number of selected features is the hyperparameters selected on that dataset; the approach presented here has no hyperparameters selected on a given dataset.

6. Conclusions

In this work, a method using dissimilarity space is presented that achieves competitive results in automated audio classification of animal sounds (bird and cat sounds). Different types of clustering techniques to obtain centroids for dissimilarity space generation were tested and compared. A set of SVMs was trained on the dissimilarity spaces generated using four clustering techniques and different numbers of centroids. These SVMs were then combined by sum rule to obtain a high performing ensemble.

Moreover, it is shown that the method presented here can be fused with other state-of-the-art approaches to improve classification accuracy. The proposed ensemble of SVMs was fused with other state-of-the-art approaches. The fusions improved performance on the two audio classification problems and were shown to outperform the standalone approaches.

In the future, this study will be further developed by including other sound classification problems, e.g., those cited in [26,37], in order to obtain a more comprehensive validation of the proposed approach. The plan is also to test the proposed method on some image classification problems using additional supervised and unsupervised clustering techniques.

Author Contributions: L.N. conceived of the presented idea., A.R. carried out the implementation. L.N., A.L. performed the experiments. A.L. and S.B. wrote the manuscript with input from all authors. S.B. provided some resources. All authors have read and agreed to the published version of the manuscript.

Funding: This research received no external funding.

Acknowledgments: The authors thank NVIDIA Corporation for supporting this work by donating a Titan Xp GPU.

Conflicts of Interest: The authors declare no conflict of interest.

References

1. Padmanabhan, J.; Premkumar, M.J.J. Machine learning in automatic speech recognition: A survey. *IETE Tech. Rev.* **2015**, *32*, 240–251. [CrossRef]
2. Nanni, L.; Costa, Y.M.G.; Lucio, D.R.; Silla, C.N., Jr.; Brahnam, S. Combining visual and acoustic features for audio classification tasks. *Pattern Recognit. Lett.* **2017**, *88*, 49–56. [CrossRef]
3. Sahoo, S.K.; Choubisa, T.; Prasanna, S.R.M. Multimodal biometric person authentication: A review. *IETE Tech. Rev.* **2012**, *29*, 54–75. [CrossRef]
4. Lidy, T.; Rauber, A. *Evaluation of Feature Extractors and Psycho-Acoustic Transformations for Music Genre Classification*; ISMIR: Washington, DC, USA, 2005; pp. 34–41.
5. Wyse, L. Audio spectrogram representations for processing with convolutional neural networks. *arXiv* **2017**, arXiv:1706.09559.

6. Rubin, J.; Abreu, R.; Ganguli, A.; Nelaturi, S.; Matei, I.; Sricharan, K. Classifying heart sound recordings using deep convolutional neural networks and mel-frequency cepstral coefficients. In Proceedings of the 2016 Computing in Cardiology Conference, Vancouver, BC, Canada, 11–14 September 2016; pp. 813–816.
7. Nanni, L.; Costa, Y.; Brahnam, S. Set of Texture Descriptors for Music Genre Classification. In *WSCG 2014: Communication Papers Proceedings: 22nd WSCG International Conference on Computer Graphics, Visualization and Computer Vision*; UNION Agency: Plzen, Czech Republic, 2014.
8. Costa, Y.M.G.; Oliveira, L.S.; Koerich, A.L.; Gouyon, F.; Martins, J.G. Music genre classification using LBP textural features. *Signal Process.* **2012**, *92*, 2723–2737. [CrossRef]
9. YCosta, M.G.; Oliveira, L.S.; Koericb, A.L.; Gouyon, F. Music genre recognition using spectrograms. In Proceedings of the 18th International Conference on Systems, Signals and Image Processing, Sarajevo, Bosnia-Herzegovina, 16–18 June 2011; pp. 1–4.
10. Haralick, R.M. Statistical and structural approaches to texture. *Proc. IEEE* **1979**, *67*, 786–804. [CrossRef]
11. Ojala, T.; Pietikainen, M.; Maenpaa, T. Multiresolution gray-scale and rotation invariant texture classification with local binary patterns. *IEEE Trans. Pattern Anal. Mach. Intell.* **2002**, *24*, 971–987. [CrossRef]
12. Costa, Y.; Oliveira, L.; Koerich, A.; Gouyon, F. Music genre recognition using gabor filters and lpq texture descriptors. In *Iberoamerican Congress on Pattern Recognition*; Springer: Berlin/Heidelberg, Germany, 2013; pp. 67–74.
13. Ojansivu, V.; Heikkilä, J. Blur insensitive texture classification using local phase quantization. In *Lecture Notes in Computer Science (Including Subser. Lect. Notes Artif. Intell. Lect. Notes Bioinformatics)*; Springer: Berlin, Germany, 2008; pp. 236–243._27. [CrossRef]
14. Humphrey, E.J.; Bello, J.P. Rethinking automatic chord recognition with convolutional neural networks. In Proceedings of the 11th International Conference on Machine Learning and Applications, Boca Raton, FL, USA, 12–15 December 2012; pp. 357–362.
15. Humphrey, E.J.; Bello, J.P.; LeCun, Y. Moving beyond feature design: Deep architectures and automatic feature learning in music informatics. In Proceedings of the 13th International Society for Music Information Retrieval Conference ISMIR, Porto, Portugal, 8–12 October 2012; pp. 403–408.
16. Nakashika, T.; Garcia, C.; Takiguchi, T. Local-feature-map integration using convolutional neural networks for music genre classification. In Proceedings of the Thirteenth Annual Conference of the International Speech Communication Association, Portland, OR, USA, 9–13 September 2012.
17. Costa, Y.M.G.; Oliveira, L.S.; Silla, C.N., Jr. An evaluation of convolutional neural networks for music classification using spectrograms. *Appl. Soft Comput.* **2017**, *52*, 28–38. [CrossRef]
18. Sigtia, S.; Dixon, S. Improved music feature learning with deep neural networks. In Proceedings of the 2014 IEEE International Conference on Acoustics, Speech and Signal Processing, Florence, Italy, 4–9 May 2014; pp. 6959–6963.
19. Wang, C.-Y.; Santoso, A.; Mathulaprangsan, S.; Chiang, C.-C.; Wu, C.-H.; Wang, J.-C. Recognition and retrieval of sound events using sparse coding convolutional neural network. In Proceedings of the 2017 IEEE International Conference on Multimedia and Expo, Hong Kong, China, 10–14 July 2017; pp. 589–594.
20. Oramas, S.; Nieto, O.; Barbieri, F.; Serra, X. Multi-label music genre classification from audio, text, and images using deep features. *arXiv* **2017**, arXiv:1707.04916.
21. Badshah, A.M.; Ahmad, J.; Rahim, N.; Baik, S.W. Speech Emotion Recognition from Spectrograms with Deep Convolutional Neural Network. International Conference on Platform Technology and Service (PlatCon), Busan, Korea, 13–15 February 2017; pp. 1–5.
22. Zeng, Y.; Mao, H.; Peng, D.; Yi, Z. Spectrogram based multi-task audio classification. *Multimed. Tools Appl.* **2019**, *78*, 3705–3722. [CrossRef]
23. Zhao, Z.; Zhang, S.; Xu, Z.; Bellisario, K.; Dai, N.; Omrani, H.; Pijanowski, B.C. Automated bird acoustic event detection and robust species classification. *Ecol. Inform.* **2017**, *39*, 99–108. [CrossRef]
24. Acevedo, M.A.; Corrada-Bravo, C.J.; Corrada-Bravo, H.; Villanueva-Rivera, L.J.; Aide, T.M. Automated classification of bird and amphibian calls using machine learning: A comparison of methods. *Ecol. Inform.* **2009**, *4*, 206–214. [CrossRef]
25. Cullinan, V.I.; Matzner, S.; Duberstein, C.A. Classification of birds and bats using flight tracks. *Ecol. Inform.* **2015**, *27*, 55–63. [CrossRef]
26. Fristrup, K.M.; Watkins, W.A. *Marine Animal Sound Classification*; No. WHOI-94-13; Woods Hole Oceanographic Institution: Falmouth, MA, USA, 1993.

27. Pandeya, Y.; Kim, D.; Lee, J. Domestic Cat Sound Classification Using Learned Features from Deep Neural Nets. *Appl. Sci.* **2018**, *8*, 1949. [CrossRef]
28. Cao, Z.; Principe, J.C.; Ouyang, B.; Dalgleish, F.; Vuorenkoski, A. Marine animal classification using combined CNN and hand-designed image features. In Proceedings of the Oceans 2015-MTS/IEEE Washington, Washington, DC, USA, 19–22 October 2015; pp. 1–6.
29. Salamon, J.; Bello, J.P.; Farnsworth, A.; Kelling, S. Fusing shallow and deep learning for bioacoustic bird species classification. In Proceedings of the 2017 IEEE International Conference on Acoustics, Speech and Signal Processing, New Orleans, LA, USA, 5–9 March 2017; pp. 141–145.
30. Nanni, L.; Brahnam, S.; Lumini, A.; Barrier, T. Ensemble of local phase quantization variants with ternary encoding. In *Local Binary Patterns: New Variants and Applications*; Springer: Berlin/Heidelberg, Germany, 2014; doi:10.1007/978-3-642-39289-4_8. [CrossRef]
31. Edgington, D.R.; Cline, D.E.; Davis, D.; Kerkez, I.; Mariette, J. Detecting, tracking and classifying animals in underwater video. In Proceedings of the Oceans 2006, Boston, MA, USA, 18–21 September 2006; pp. 1–5.
32. Wang, A. *An Industrial Strength Audio Search Algorithm*; ISMIR: Washington, DC, USA, 2003; Volume 2003, pp. 7–13.
33. Haitsma, J.; Kalker, T. *A Highly Robust Audio Fingerprinting System*; ISMIR: Washington, DC, USA, 2002; Volume 2002, pp. 107–115.
34. Manocha, P.; Badlani, R.; Kumar, A.; Shah, A.; Elizalde, B.; Raj, B. Content-based Representations of audio using Siamese neural networks. In Proceedings of the 2018 IEEE International Conference on Acoustics, Speech and Signal Processing (ICASSP), Calgary, AB, Canada, 15–20 April 2018; IEEE: Piscataway, NJ, USA, 2018; pp. 3136–3140.
35. Şaşmaz, E.; Tek, F.B. Animal Sound Classification Using A Convolutional Neural Network. In Proceedings of the 2018 3rd International Conference on Computer Science and Engineering (UBMK), Sarajevo, Bosnia-Herzegovina, 20–23 September 2018; pp. 625–629.
36. Oikarinen, T.; Srinivasan, K.; Meisner, O.; Hyman, J.B.; Parmar, S.; Fanucci-Kiss, A.; Desimone, R.; Landman, R.; Feng, G., Deep convolutional network for animal sound classification and source attribution using dual audio recordings. *J. Acoust. Soc. Am.* **2019**, *145*, 654–662. [CrossRef] [PubMed]
37. Agrawal, A. Dissimilarity learning via Siamese network predicts brain imaging data. *arXiv* **2019**, arXiv:1907.02591.
38. Sekhar, C.; Mukherjee, P.; Guru, D.S.; Pulabaigari, V. OSVNet: Convolutional Siamese Network for Writer Independent Online Signature Verification. In Proceedings of the International Conference on Document Analysis and Recognition (ICDAR), Sydney, Australia, 20–25 September 2019. [CrossRef]
39. Bromley, J.; Guyon, I.; LeCun, Y.; Säckinger, E.; Shah, R. Signature verification using a "siamese" time delay neural network. *Adv. Neural Inf. Process. Syst.* **1994**, *7*, 737–744.
40. Chopra, S.; Hadsell, R.; LeCun, Y. Learning a similarity metric discriminatively, with application to face verification. In Proceedings of the Computer Vision and Pattern Recognition (CVPR), San Diego, CA, USA, 20–25 June 2005; pp. 539–546.
41. Koch, G.; Zemel, R.; Salakhutdinov, R. Siamese neural networks for one-shot image recognition. In Proceedings of the 32nd International Conference on Machine Learning (ICML), Lille, France, 6–11 July 2015.
42. Droghini, D.; Vesperini, F.; Principi, E.; Squartini, S.; Piazza, F. Few-shot siamese neural networks employing audio features for human-fall detection. In Proceedings of the International Conference on Pattern Recognition and Artificial Intelligence (PRAI 2018). Association for Computing Machinery, New York, NY, USA, 15–17 August 2018; pp. 63–69.
43. Zhang, Y.; Pardo, B.; Duan, Z. Siamese style convolutional neural networks for sound search by vocal imitation. *IEEE/ACM Trans. Audio, Speech, Lang. Process.* **2018**, *27*, 429–441. [CrossRef]
44. David, A.; Vassilvitskii, S. K-means++: The Advantages of Careful Seeding. In Proceedings of the SODA '07: Proceedings of the Eighteenth Annual ACM-SIAM Symposium on Discrete Algorithms, New Orleans, LA, USA, 7–9 January 2007; pp. 1027–1035.
45. Pandeya, Y.R.; Lee, J. Domestic cat sound classification using transfer learning. *Int. J. Fuzzy Log. Intell. Syst.* **2018**, *18*, 154–160. [CrossRef]

46. Szegedy, C.; Liu, W.; Jia, Y.; Sermanet, P.; Reed, S.; Anguelov, D.; Rabinovich, A. Going deeper with convolutions. In Proceedings of the IEEE Conference on Computer Vision and Pattern Recognition, Boston, MA, USA, 7–12 June 2015; pp. 1–9.
47. Simonyan, K.; Zisserman, A. Very deep convolutional networks for large-scale image recognition. *arXiv* **2014**, arXiv:1409.1556.
48. Deng, J.; Dong, W.; Socher, R.; Li, L.J.; Li, K.; Fei-Fei, L. Imagenet: A large-scale hierarchical image database. In Proceedings of the 2009 IEEE conference on computer vision and pattern recognition, Miami, FL, USA, 20–25 June 2009; pp. 248–255.
49. Zhou, B.; Lapedriza, A.; Xiao, J.; Torralba, A.; Oliva, A. Learning deep features for scene recognition using places database. In *Proceedings of the 27th International Conference on Neural Information Processing Systems (NIPS'14) 2014*; MIT Press: Cambridge, MA, USA, 2014; Volume 1, pp. 487–495.
50. Nanni, L.; Costa, Y.M.G.; Lumini, A.; Kim, M.Y.; Baek, S.R. Combining visual and acoustic features for music genre classification. *Expert Syst. Appl.* **2016**, *45*, 108–117. [CrossRef]
51. Zhang, S.; Zhao, Z.; Xu, Z.; Bellisario, K.; Pijanowski, B.C. Automatic Bird Vocalization Identification Based on Fusion of Spectral Pattern and Texture Features. In Proceedings of the 2018 IEEE International Conference on Acoustics, Speech and Signal Processing, Calgary, AB, Canada, 15–20 April 2018; pp. 271–275.

© 2020 by the authors. Licensee MDPI, Basel, Switzerland. This article is an open access article distributed under the terms and conditions of the Creative Commons Attribution (CC BY) license (http://creativecommons.org/licenses/by/4.0/).

Article

Texture Segmentation: An Objective Comparison between Five Traditional Algorithms and a Deep-Learning U-Net Architecture

Cefa Karabağ [1], Jo Verhoeven [2,3], Naomi Rachel Miller [2] and Constantino Carlos Reyes-Aldasoro [1,*]

[1] Department of Electrical and Electronic Engineering, Research Centre for Biomedical Engineering, School of Mathematics, Computer Science and Engineering, City, University of London, London EC1V 0HB, UK; cefa.karabag.1@city.ac.uk

[2] School of Health Sciences, Division of Language & Communication Science, Phonetics Laboratory, University of London, London EC1R 1UW, UK; johan.verhoeven.1@city.ac.uk or jo.verhoeven@uantwerpen.be (J.V.); naomi-rachel.miller@city.ac.uk (N.R.M.)

[3] Department of Linguistics CLIPS, University of Antwerp, 2000 Antwerp, Belgium

* Correspondence: reyes@city.ac.uk

Received: 30 July 2019; Accepted: 10 September 2019; Published: 17 September 2019

Abstract: This paper compares a series of traditional and deep learning methodologies for the segmentation of textures. Six well-known texture composites first published by Randen and Husøy were used to compare traditional segmentation techniques (co-occurrence, filtering, local binary patterns, watershed, multiresolution sub-band filtering) against a deep-learning approach based on the U-Net architecture. For the latter, the effects of depth of the network, number of epochs and different optimisation algorithms were investigated. Overall, the best results were provided by the deep-learning approach. However, the best results were distributed within the parameters, and many configurations provided results well below the traditional techniques.

Keywords: texture; segmentation; deep learning

1. Introduction

Texture, and more specifically textural characteristics in images, has been widely studied in the past decades as texture is one of the most important features present in images and can be used for feature extraction [1–8] and classification and segmentation [9–14]. The areas of study where texture is present range from crystallographic texture [15], stratigraphy [16,17], food science of potatoes [18] or apples [19], patterned fabrics [20] to natural stone industry [21]. In medical imaging, there is a large volume of research which exploits the use of texture for different purposes, like segmentation or classification in most acquisition modalities like magnetic resonance imaging (MRI) [22–26], ultrasound [27,28], computed tomography (CT) [29–31], microscopy [32,33] and histology [34]. There are numerous approaches to texture: Haralick's co-occurrence matrix [4,5] on the spatial domain, Gabor filters [35–37] and ordered pyramids [8] on the spectral domain, wavelets [38,39] or Markov random fields [3,40].

In recent years, advances in artificial intelligence have revolutionised image processing tasks. Several deep learning approaches [41–43] have achieved outstanding results in difficult tasks such as those of the ImageNet Large Scale Visual Recognition Challenge (ILSVRC) [44]. Convolutional Neural Networks (CNNs) are well suited to analyse textures as their repetitive patterns can be learned and identified by filter banks [45]. The U-Net architecture proposed by Ronneberger [46] has become a very widely used tool for segmentation and analysis reaching thousands of citations in the few years since it was published. U-Nets have been used widely, for instance, for road extraction [47], singing voice separation [48], automatic brain tumour detection and segmentation [49] and cell counting, detection, and morphometry [50]. The success of these deep learning approaches in very different areas invites for their application on texture analysis.

In this work, a U-Net architecture for the segmentation of textures is implemented and objectively compared against several popular traditional segmentation strategies. The traditional algorithms (co-occurrence matrices [5], watershed [51], local binary patterns (LBP) [52,53], filtering [54] and multiresolution sub-band filtering (MSBF) [8]) were selected as these have been previously published using the texture composites proposed by Randen [55] and thus an objective numerical comparison is possible.

To perform an objective comparison, six well-known texture composites from the Brodatz [56] album, first published by Randen and Husøy [54], are segmented with U-Nets of different configurations and parameters and the results compared against previously published results. The effects of the configuration of the networks, namely, number of epochs, depth of the network in the number of layers, and type of optimisation algorithm are assessed. All the programming was performed in Matlab® (The Mathworks™, Natick, MA, USA) and the code is freely available through GitHub (https://github.com/reyesaldasoro/Texture-Segmentation).

2. Materials and Methods

2.1. Texture Composite Images

Six composite texture images were segmented in this work (Figure 1). The first five composites are images of 256 × 256 pixels and consist of five different textures whilst the last one is 512 × 512 pixels and is formed with 16 different textures. The masks with which these were formed are shown in Figure 2. It should be highlighted that these textures have been histogram equalised prior to the arrangement and thus they cannot be distinguished by the general intensity of each region. It is frequent that comparisons are made over textures that are not equalised (e.g., [57] Figure 3, [45] Figure 2) and thus the segmentation is not only based on the texture but the average intensity of the regions. Furthermore, whilst some textures are easy to distinguish, there are some that are quite challenging, for instance, the difference between the central and bottom regions in Figure 1c or the top left corners of Figure 1d,e.

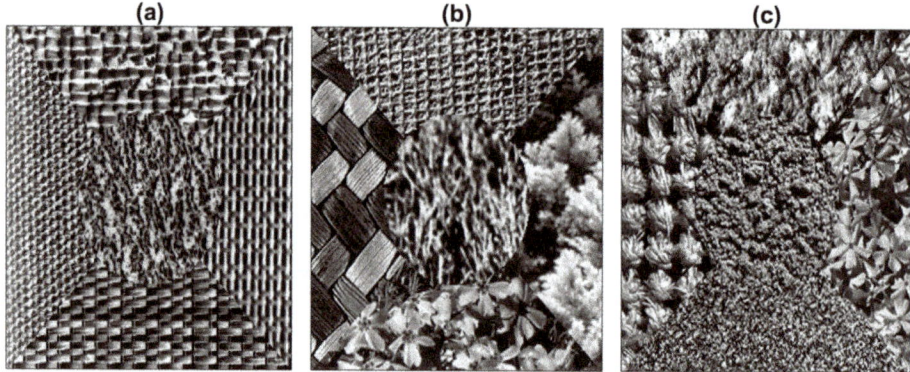

Figure 1. *Cont.*

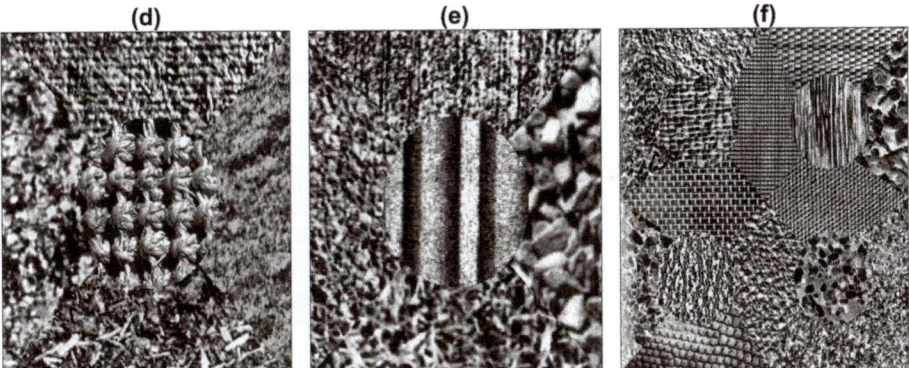

Figure 1. Six composite texture images. (**a–e**) Texture arrangements with five textures. (**f**) Texture arrangement with sixteen textures. Notice first, that individual textures have been histogram equalised and thus each region cannot be distinguished by the intensity, and second, some textures are easier to distinguish (e.g., (**a**)) than others (e.g., (**d**)).

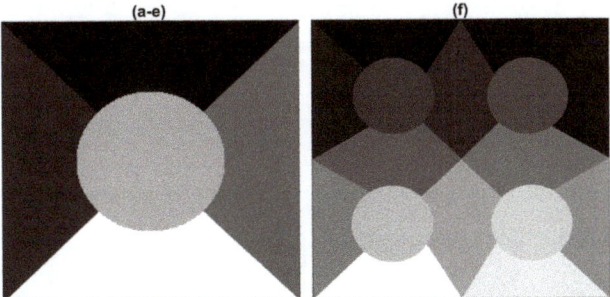

Figure 2. (**Left**) Mask corresponding to texture arrangements of Figure 1a–e. (**Right**) Mask corresponding to texture arrangements of Figure 1f.

2.2. Training Data

The training data in [54] is provided separately and is shown in Figure 3 for the first five composites and in Figure 4 for the last case. For the purpose of training the U-Nets, the training images were tessellated into sub-regions of 32 × 32 pixels each.

Pairs of textures and labels were constructed simultaneously in the following way: two training images were selected. Sub-regions of each image were selected and for every pair of the sub-regions, half of each was selected and placed together so that a new 32 × 32 patch with both textures was created with a corresponding 32 × 32 patch with the classes. The patches were created with diagonal, vertical and horizontal pairs. The training images were traversed horizontally and vertically without overlap creating numerous training pairs. A montage of the texture pairs and labels corresponding to Figure 1a is illustrated in Figure 5. All pairs between classes were considered i.e., 1-2, 1-3, 1-4, 1-5, 2-1, 2-3, ..., 5-3, 5-4. In total, 2940 patches were created for the five composites with five textures and 35,280 were created for the composite with sixteen textures.

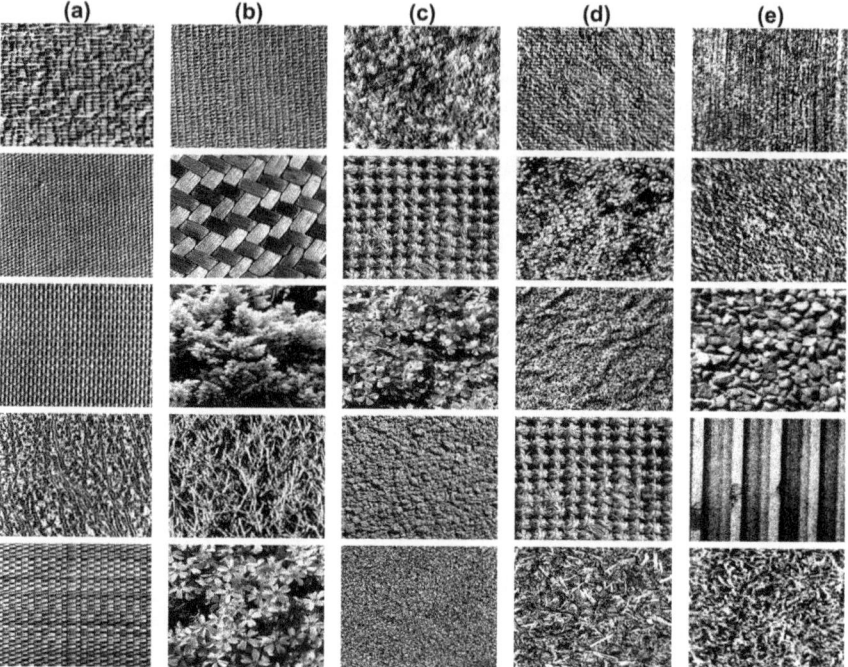

Figure 3. Training images corresponding to the texture arrangements of Figure 1a–e.

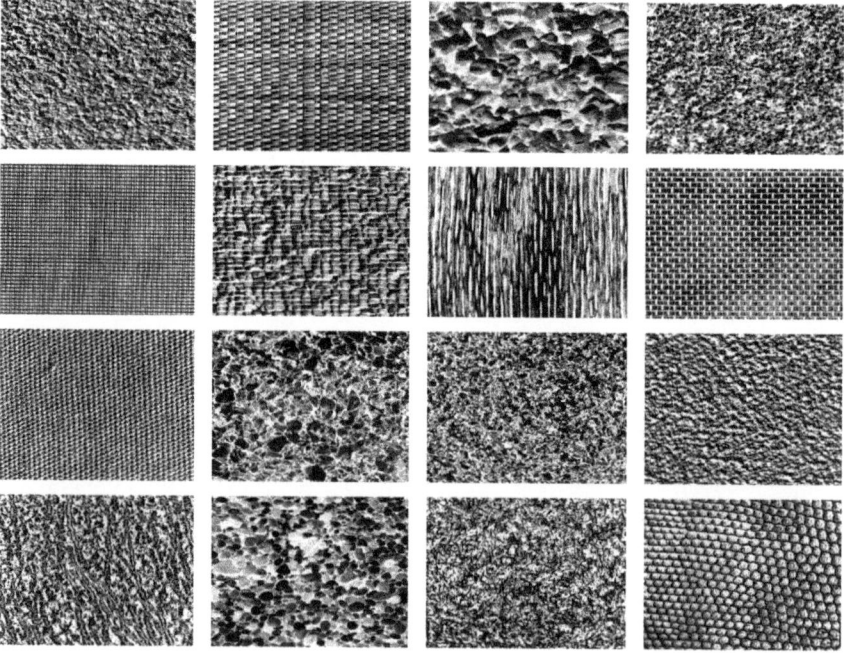

Figure 4. Training images corresponding to the texture arrangements of Figure 1f.

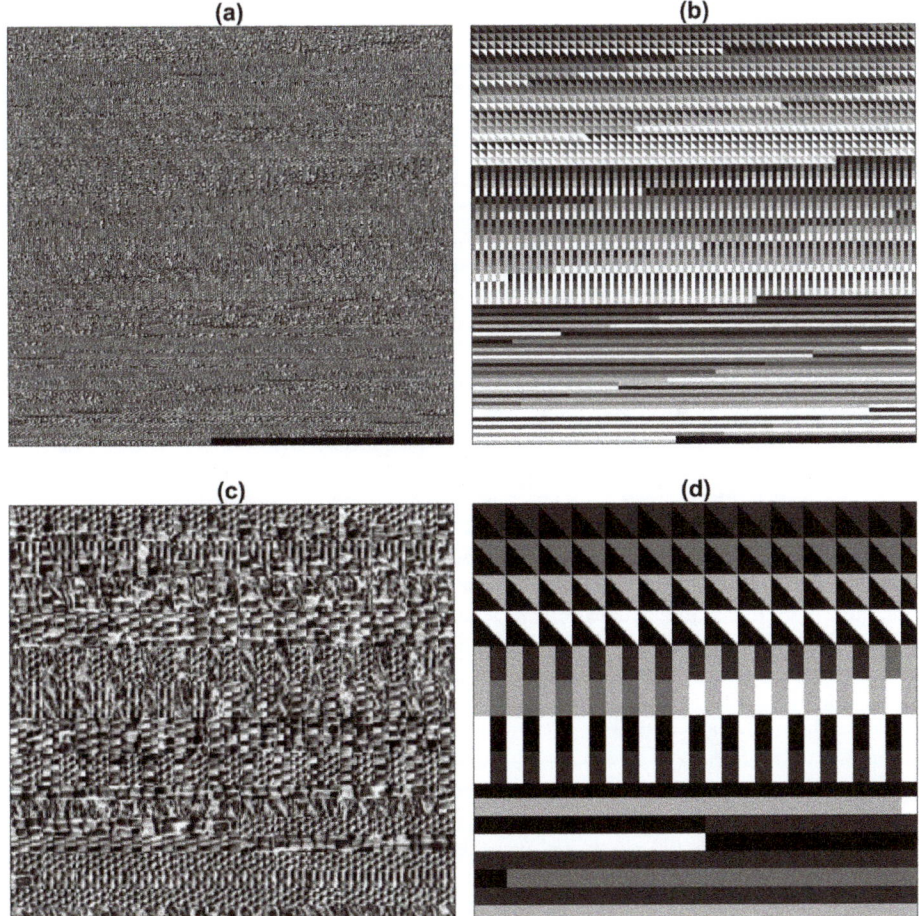

Figure 5. Montages of the texture pairs created to train the deep learning networks. Training images shown in Figures 3 and 4 were tessellated and arranged in diagonal, vertical and horizontal pairs. (**a**) Texture pairs. (**b**) Labels. (**c**) Detail of the texture pairs. (**d**) Detail of the labels.

2.3. Traditional Texture Segmentation Algorithms

For this paper, we compared the results of the following texture segmentation algorithms: co-occurrence matrices [5], watershed [51], local binary patterns (LBP) [52,53], filtering [54] and multiresolution sub-band filtering (MSBF) [8] against a U-Net architecture [46].

The traditional algorithms have been thoroughly described in the literature; however, for completeness, a short explanation of how features are extracted with each algorithm will follow. For a discussion of traditional texture techniques, the reader is referred to any of the following reviews [58–60].

Co-occurrence matrices are constructed from a quantised version of a grey level image so that if an image is quantised to 8 levels, the co-occurrence matrix will have 8 rows and columns. The values of each location of the matrix will depend on the number of times that a pair of grey levels jointly occur at a neighbouring distance (e.g., 1 pixel away) with a certain orientation (e.g., horizontally). In this way, a co-occurrence matrix is able to measure local grey level dependence: textural coarseness and directionality. For example, in coarse images, the grey level of the pixels change slightly with distance, while for fine textures the levels change rapidly. From this matrix, different features like entropy, uniformity, maximum probability, contrast, difference moment, inverse difference moment and correlation can be calculated [5]. Once the features have been calculated, classifiers can be applied directly, or further processing like the watershed transforms can be applied.

Watershed transforms are based on a topographical analogy of a landscape. Should water fall in this landscape, it would find the path through which it could reach a region of minimum altitude, i.e., a basin, sometimes called lake or sea. For each point in the landscape (or pixel of the image) there is a path towards one and only one basin. Thus, the landscape can be partitioned into catchment basins or regions of influence of the regional minima and the boundaries between the basins (e.g., points of inflection) are called the watershed lines. [61]. The watershed transform can be applied to features extracted from the co-occurrence matrix [51]. The basins produced can further be iteratively merged to segment textured regions.

Local binary patterns (LBP) [52], explore the relations between neighbouring pixels. These methods concentrate on the relative intensity relations between the pixels in a small neighbourhood and not in their absolute intensity values or the spatial relationship of the whole data. The underneath assumption is that texture is not properly described by the Fourier spectrum and traditional frequency filters. The texture analysis is based on the relationship of the pixels of a 3×3 neighbourhood. A Texture Unit is first calculated by differentiating the grey level of a central pixel with the grey level of its neighbours. The difference is measured if the neighbour is greater or lower than the central pixel. Two advantages of LBP is that there is no need of quantising images and there is a certain immunity to low frequency artefacts. In a more recent paper, Ojala [53] presented another variation to the LBP by considering the sign of the difference of the grey-level differences histograms. Under the new consideration, LBP is a particular case of the new operator called p_8. This operator is considered as a probability distribution of grey levels, when $p(g_0, g_1)$ denotes the co-occurrence probabilities, they use $p(g_0, g_1 - g_0)$ as a joint distribution.

Filtering, in the context of image processing, consists of a process that will modify the pixel values. There are spatial filters, which are applied directly to the values of the images (e.g., average neighbouring pixels to blur an image) and filters which are applied after a transformation of the data has been performed. Thus a filter in the frequency or Fourier domain will be applied after the image has been converted through the Fourier transform. The filters in the Fourier domain are sometimes named after the frequencies that are to be allowed to pass through them: low pass, band pass and high pass filters. Since textures can vary in their spectral distribution in the frequency domain, a set of sub-band filters can help in their discrimination. One common frequency filtering approach is that of Gabor multichannel filter banks [2,10,62–64].

The partitioning of the Fourier space can be achieved in different ways, Gabor being only one. A multiresolution approach, based on finite prolate spheroidal sequences is described in [8]. The Fourier space is divided into frequencies and orientations, which are further subdivided in a multiresolution approach. Each filter then produces a feature; different textures are captured by different filters. In addition, a feature selection strategy can improve the texture segmentation.

2.4. U-Net Configuration

The basic U-Net architecture was formed with the following layers: *Input, Convolutional, ReLu, Max Pooling, Transposed Convolutional, Convolutional, Softmax* and *Pixel Classification*. Two levels of depth were investigated by repeating the downsampling and upsampling blocks in the following configurations:

15 layers:
Input,
Convolutional, ReLu, Max Pooling,

Convolutional, ReLu, Max Pooling,
Convolutional, ReLu,
Transposed Convolutional, Convolutional,
Transposed Convolutional, Convolutional,
Softmax,
Pixel Classification

20 layers:
Input,
Convolutional, ReLu, Max Pooling,
Convolutional, ReLu, Max Pooling,
Convolutional, ReLu, Max Pooling,
Convolutional, ReLu,
Transposed Convolutional, Convolutional,
Transposed Convolutional, Convolutional,
Transposed Convolutional, Convolutional,
Softmax,
Pixel Classification.

The image input layer was configured for the 32 × 32 patches. The convolutional layers consisted of 64 filters of size 3 and padding of 1. The pooling size was 2 with stride of 2. The transposed convolutional had a filter size of 4, stride of 2 and cropping of 1. The numbers of epochs evaluated were 10, 20, 50, 100. The following optimisation algorithms were analysed: stochastic gradient descent (sgdm), Adam (Adam) [65] and Root Mean Square Propagation (RMSprop). One last investigation was performed by training the 20 layer network two separate times to investigate the variability of the process.

2.5. Misclassification

For the purposes of assessing the algorithms, a pixel-based assessment will be considered. Each pixel whose class is correctly determined by the segmentation algorithm will be counted as **Correct**, every pixel which the algorithm assigns a different class will be considered as **Incorrect**. Notice that since there is no foreground/background distinction but rather correct or incorrect, both **True Positive** (TP) and **True Negative** (TN) are included as correct, and **False Positive** (FP) and **False Negative** (FN) are included in the incorrect. Thus, the **misclassification** in percentage, or classification error, will be calculated as number of incorrect pixels divided by the total number of pixels of the image $m = 100 * (FP + FN)/(TP + TN + FP + FN)$. The accuracy can be calculated as the complement $a = 100 * (TP + TN)/(TP + TN + FP + FN)$.

3. Results

For each image, the networks were trained with the 3 different optimisation algorithms, 3 layer configurations and 4 epoch numbers, for a total of 36 different combinations. Thus for the 6 composite images there were 216 results. The misclassification of each segmentation was measured against the ground truth as the percentage of pixels classified incorrectly. These results are summarised in Table 1.

The best results for each image were selected and compared against traditional methodologies and are shown in Table 2. The results are illustrated graphically in two ways. Figure 6 shows segmented the classes overlaid as different colours over the original textured images. Figure 7 shows correctly segmented pixels in white and the misclassified pixels in black.

Table 1. Comparative misclassification (%) results of the different U-Net configurations. (Bold and underline denotes the best result for each image).

Method			Figures					
Layers	Optimisation Algorithm	Epochs	a	b	c	d	e	f
15	sgdm	10	6.8	21.5	40.8	31.2	27.2	20.9
20	sgdm	10	33.0	59.0	74.3	79.1	77.3	41.9
20	sgdm	10	71.9	62.9	74.3	78.8	72.1	39.0
15	Adam	10	3.2	10.4	7.9	<u>**7.1**</u>	17.8	19.3
20	Adam	10	7.4	15.5	46.5	25.0	45.1	94.2
20	Adam	10	6.4	15.5	36.0	21.1	26.7	32.9
15	RMSprop	10	5.1	<u>**8.9**</u>	14.0	18.3	12.1	17.6
20	RMSprop	10	5.3	42.4	45.3	59.9	56.2	27.7
20	RMSprop	10	20.2	37.4	47.0	43.7	44.2	26.1
15	sgdm	20	3.8	23.1	17.5	15.9	14.1	19.8
20	sgdm	20	27.3	60.5	74.8	69.3	73.9	27.4
20	sgdm	20	23.8	51.0	63.6	66.8	56.5	26.7
15	Adam	20	3.7	11.6	7.5	7.4	9.5	71.7
20	Adam	20	6.1	13.3	28.7	18.5	40.8	32.2
20	Adam	20	5.6	17.9	27.4	22.5	39.3	94.0
15	RMSprop	20	3.8	11.7	14.5	19.2	11.7	17.9
20	RMSprop	20	6.1	42.2	54.7	47.5	42.6	22.3
20	RMSprop	20	19.1	30.3	44.7	51.7	37.1	26.9
15	sgdm	50	3.2	15.3	9.2	7.7	13.8	19.6
20	sgdm	50	18.2	32.2	60.3	42.8	30.2	28.9
20	sgdm	50	9.4	55.2	56.0	16.0	32.4	32.4
15	Adam	50	3.4	10.4	9.8	9.9	39.1	22.6
20	Adam	50	8.3	80.3	19.8	82.3	79.6	34.8
20	Adam	50	7.2	9.6	41.4	10.0	27.6	23.6
15	RMSprop	50	3.4	18.7	10.0	8.3	11.2	<u>**17.5**</u>
20	RMSprop	50	5.6	33.2	25.7	34.8	34.4	22.4
20	RMSprop	50	5.4	22.8	45.3	20.0	34.7	29.2
15	sgdm	100	3.9	10.6	7.9	7.7	<u>**7.7**</u>	21.4
20	sgdm	100	9.6	22.1	39.4	39.7	30.3	23.8
20	sgdm	100	13.7	17.1	52.8	26.3	37.1	30.5
15	Adam	100	2.7	16.6	80.3	7.2	18.2	21.9
20	Adam	100	**2.6**	38.9	79.9	80.1	31.1	25.7
20	Adam	100	3.4	80.0	79.7	80.9	80.3	28.6
15	RMSprop	100	4.8	11.2	<u>**7.2**</u>	8.1	9.5	18.1
20	RMSprop	100	7.1	66.0	46.0	28.6	30.9	24.0
20	RMSprop	100	5.6	29.5	26.9	18.5	29.3	22.9
Max			71.9	80.3	80.3	82.3	80.3	94.1
Mean			10.4	30.7	39.4	33.7	35.6	30.7
Min			2.6	8.9	7.2	7.1	7.7	17.5

Table 2. Comparative misclassification (%) results with co-occurrence [5], best filtering result from Randen [54], p_8 and LBP [53], Watershed [51], Multiresolution sub-band filtering (MSBF) [8] and U-Net [46]. (Bold is the best for each image).

Method	Figures						
	a	b	c	d	e	f	Average
Co-occurrence [5]	9.9	27.0	26.1	51.1	35.7	49.6	33.23
Best in Randen [55]	7.2	18.9	20.6	16.8	17.2	34.7	19.23
p_8 [52]	7.4	12.8	15.9	18.4	16.6	27.7	16.46
LBP [52]	6.0	18.0	12.1	9.7	11.4	**17.0**	12.36
Watershed [51]	7.1	10.7	12.4	11.6	14.9	20.0	12.78
MSBF [8]	2.8	14.8	8.4	7.3	**4.3**	17.9	9.25
U-Net [46]	**2.6**	**8.9**	**7.2**	**7.1**	7.7	17.5	**8.50**

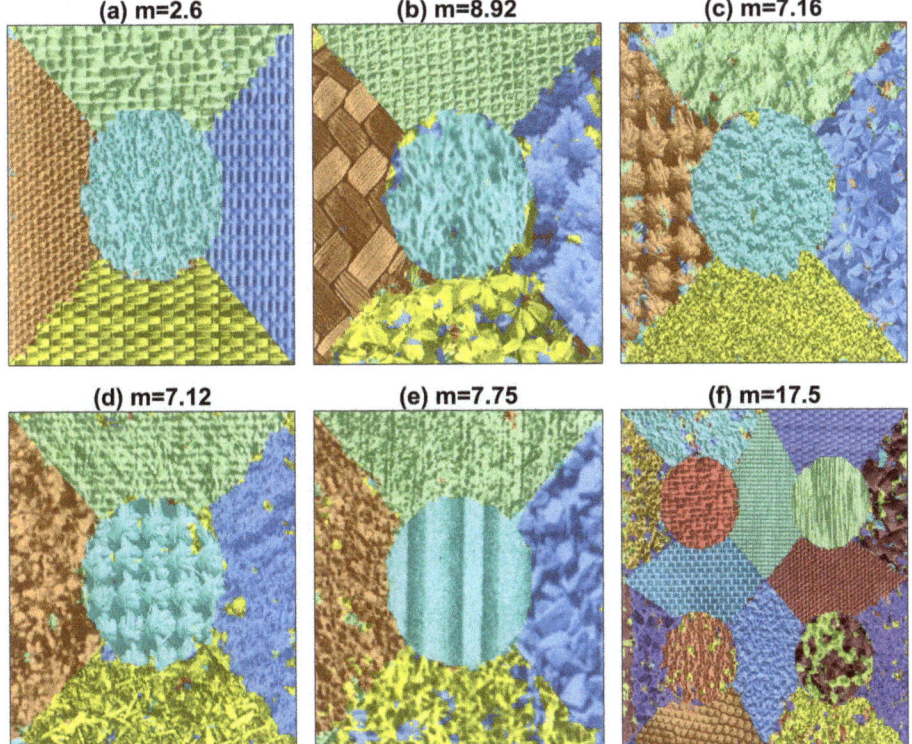

Figure 6. (a–f) Results of the segmentation with U-Nets for the six texture arrangments. The misclassification (%) is shown in each case. The classes are shown as overlaid colours.

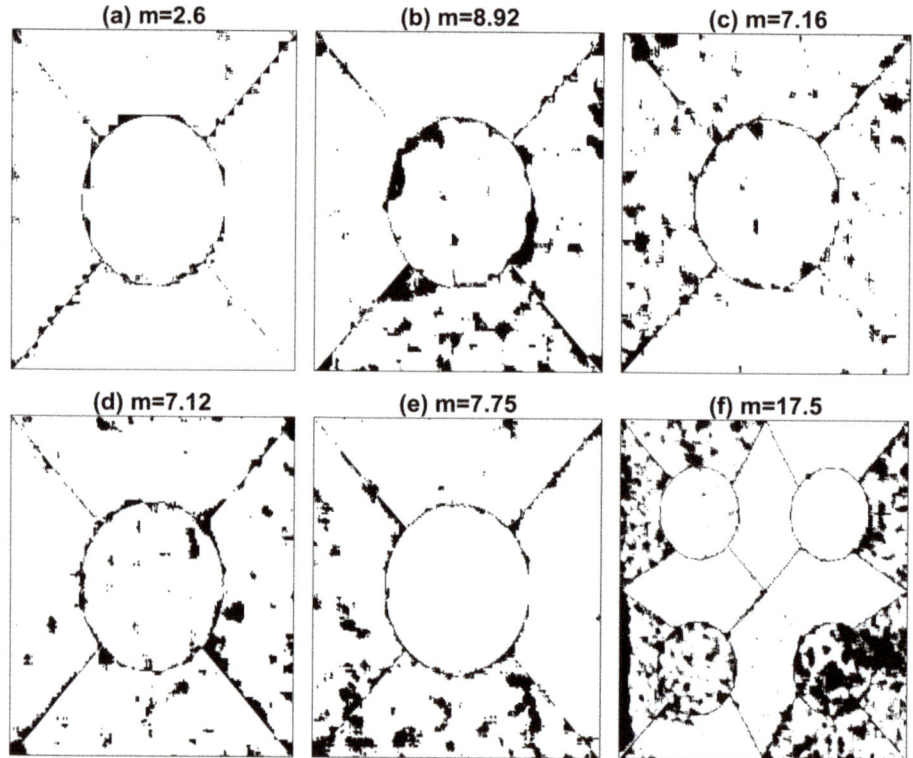

Figure 7. (a–f) Results of the segmentation with U-Nets for the six texture arrangments. The misclassification (%) is shown in each case. Pixels that are correctly classified appear in white.

4. Discussion

The results provided by the U-Net algorithm provided interesting results in terms of the actual misclassification results against traditional algorithms, and the variability of the U-Net cases. The segmentation results provided by the U-Nets were better in four of the six images. In some cases, the results were very close to the second best option (a: 2.8/2.6, d: 7.3/7.1) and in two cases (e,f) traditional algorithms provided better results (e: 4.3/7.7, f: 17.0/17.5). The average for all the six composites was best for U-Nets, however, given the fact that the difference with the second best is relatively small (0.75), and that traditional algorithms provided better results in 1/3 cases shows that care should be taken when selecting algorithms. This is similar to the conclusion of Randen who stated that "No single approach did perform best or very close to the best for all images" [55].

In terms of the U-Net configuration there are several interesting observations. First, there was a great variability in the results produced by the different U-Net configurations. It was surprising that the maximum value of the misclassification in some cases was extremely high, 80% in the cases of 5 textures and 94% in the case of 16 textures, those cases are equivalent of selecting a single class for all textures. Second, three of the best results were obtained with 100 epochs, 2 with 10 epochs, and 1 with 50, which is counter-intuitive as it would be expected that longer training times would provide better results. Third, three of the best results were provided by RMSprop optimisation, two by Adam and one by sgdm. Fourth, and perhaps the most surprising result was that the results provided by the two 20 layer configurations were very different. In a few cases the result were equal (e.g., image c, sgdm, 10 epochs; image b, Adam, 10 epochs) but in others the variation was huge (e.g., image b, Adam, 50 epochs).

In terms of texture, it can be highlighted that not all textures are the same, the five textures of image (a) are far easier to distinguish and correctly segment than those of image (b) and image (f). The U-Net was capable of segmenting these textures with accuracy comparable or better than traditional techniques. As mentioned previously, the fact that the textures have been histogram equalised removes the discrimination of the regions by their average intensities. More complex architectures, e.g., Siamese Networks [57] could provide better results, but it is important to use a standard benchmark such as that provided by Randen [55].

There are many other configuration parameters that could be varied; *learning rate, batch size, variations of the training data, different number of layers*, but for the purpose of this work, the results show first, the capability of deep learning architectures for segmentation of textured images and second, in some cases better results than traditional methodologies. However, the configuration of the network is not trivial and variations of some parameters can provide sub-optimal results. The experiments conducted in this work did not provide conclusive evidence for the selection of any of the parameters evaluated. Furthermore, training of the networks requires considerable resources. The training times for the images with 5 textures took around 5 hours and for the image with 16 textures around 96 hours on a Apple (Cuppertino, CA, USA) Mac Pro (Late 2013) with a 3.7 GHz Quad-Core and 32 GB Memory with Dual AMD FirePro D300 graphics processors.

Therefore, it can be concluded that U-Net convolutional neural networks can be used for texture segmentation and provide results that are comparable or better than traditional texture algorithms. Furthermore, these results encourage the application of deep learning to other areas. If we assume that different textures are characterised by patterns, i.e., repetitions of certain sequences or particular variation of intensities, then any data which is characterised by patterns could be analysed. For instance, phonemes in human speech have different patterns, which when combined form words. Thus one line of an image with different textures would have similar characteristics as the intensity variation of a phrase with different phonemes. Moreover, voice signals, which are one-dimensional can be converted into two-dimensional spectrograms [66] with time on one axis and frequency in another axis. In these cases, the spectrograms can be analysed for texture directly.

Author Contributions: Conceptualization: C.K., J.V., N.R.M. and C.C.R.-A.; Methodology C.K. and C.C.R.-A., writing, reviewing and editing, C.K., J.V., N.R.M. and C.C.R.-A.; funding acquisition J.V. and C.C.R.-A.

Funding: This work was funded by the Leverhulme Trust, Research Project Grant RPG-2017-054. C.K. is partially funded by the School of Mathematics, Computer Science and Engineering at City, University of London.

Conflicts of Interest: The authors declare no conflict of interest.

References

1. Bigun, J. Multidimensional Orientation Estimation with Applications to Texture Analysis and Optical Flow. *IEEE Trans. Pattern Anal. Mach. Intell.* **1991**, *13*, 775–790. [CrossRef]
2. Bovik, A.C.; Clark, M.; Geisler, W.S. Multichannel Texture Analysis Using Localized Spatial Filters. *IEEE Trans. Pattern Anal. Mach. Intell.* **1990**, *12*, 55 – 73. [CrossRef]
3. Cross, G.R.; Jain, A.K. Markov Random Field Texture Models. *IEEE Trans. Pattern Anal. Mach. Intell.* **1983**, *5*, 25–39. [CrossRef] [PubMed]
4. Haralick, R.M. Statistical and Structural Approaches to Texture. *Proc. IEEE* **1979**, *67*, 786–804. [CrossRef]
5. Haralick, R.M.; Shanmugam, K.; Dinstein, I. Textural Features for Image Classification. *IEEE Trans. Syst. Man Cybern.* **1973**, *3*, 610–621. [CrossRef]
6. Tamura, H.; Mori, S.; Yamawaki, T. Texture Features Corresponding to Visual Perception. *IEEE Trans. Syst. Man Cybern.* **1978**, *8*, 460–473. [CrossRef]
7. Tuceryan, M.; Jain, A.K. Texture Analysis. In *Handbook of Pattern Recognition and Computer Vision*, 2nd ed.; Chen, C.H., Pau, L.F., Wang, P.S.P., Eds.; World Scientific Publishing: Singapore, 1998; pp. 207–248.
8. Reyes-Aldasoro, C.C.; Bhalerao, A. The Bhattacharyya Space for Feature Selection and Its Application to Texture Segmentation. *Pattern Recognit.* **2006**, *39*, 812–826. [CrossRef]

9. Bouman, C.; Liu, B. Multiple Resolution Segmentation of Textured Images. *IEEE Trans. Pattern Anal. Mach. Intell.* **1991**, *13*, 99–113. [CrossRef]
10. Jain, A.K.; Farrokhnia, F. Unsupervised Texture Segmentation using Gabor Filters. *Pattern Recognit.* **1991**, *24*, 1167–1186. [CrossRef]
11. Kadyrov, A.; Talepbour, A.; Petrou, M. Texture Classification with Thousand of Features. In Proceedings of the 13th British Machine Vision Conference (BMVC), Cardiff, UK, 2–5 September 2002; pp. 656–665.
12. Kervrann, C.; Heitz, F. A Markov Random Field Model-based Approach to Unsupervised Texture Segmentation using Local and Global Spatial Statistics. *IEEE Trans. Image Process.* **1995**, *4*, 856–862. [CrossRef]
13. Unser, M. Texture Classification and Segmentation Using Wavelet Frames. *IEEE Trans. Image Process.* **1995**, *4*, 1549–1560. [CrossRef] [PubMed]
14. Weszka, J.; Dyer, C.; Rosenfeld, A. A Comparative Study of Texture Measures for Terrain Classification. *IEEE Trans. Syst. Man Cybern.* **1976**, *6*, 269–285. [CrossRef]
15. Tai, C.; Baba-Kishi, K. Microtexture Studies of PST and PZT Ceramics and PZT Thin Film by Electron Backscatter Diffraction Patterns. *Text. Microstruct.* **2002**, *35*, 71–86. [CrossRef]
16. Carrillat, A.; Randen, T.; Snneland, L.; Elvebakk, G. Seismic Stratigraphic Mapping of Carbonate Mounds using 3D Texture Attributes. In Proceedings of the 64th EAGE Conference & Exhibition, Florence, Italy, 27–30 May 2002.
17. Randen, T.; Monsen, E.; Abrahamsen, A.; Hansen, J.O.; Schlaf, J.; Snneland, L. Three-dimensional Texture Attributes for Seismic Data Analysis. In Proceedings of the 70th SEG Annual Meeting, Calgary, AB, Canada, 6–11 August 2000.
18. Thybo, A.K.; Martens, M. Analysis of Sensory Assessors in Texture Profiling of Potatoes by Multivariate Modelling. *Food Qual. Prefer.* **2000**, *11*, 283–288. [CrossRef]
19. Létal, J.; Jirák, D.; Šuderlová, L.; Hájek, M. MRI 'Texture' Analysis of MR Images of Apples during Ripening and Storage. *LWT Food Sci. Technol.* **2003**, *36*, 719–727. [CrossRef]
20. Lizarraga-Morales, R.A.; Sanchez-Yanez, R.E.; Baeza-Serrato, R. Defect Detection on Patterned Fabrics using Texture Periodicity and the Coordinated Clusters Representation. *Text. Res. J.* **2017**, *87*, 1869–1882. [CrossRef]
21. Bianconi, F.; González, E.; Fernández, A.; Saetta, S.A. Automatic Classification of Granite Tiles through Colour and Texture Features. *Expert Syst. Appl.* **2012**, *39*, 11212–11218. [CrossRef]
22. Kovalev, V.A.; Petrou, M.; Bondar, Y.S. Texture Anisotropy in 3D Images. *IEEE Trans. Image Process.* **1999**, *8*, 346–360. [CrossRef]
23. Reyes-Aldasoro, C.C.; Bhalerao, A. Volumetric Texture Description and Discriminant Feature Selection for MRI. In Proceedings of the Information Processing in Medical Imaging, Ambleside, UK, 20–25 July 2003; Taylor, C., Noble, A., Eds.; pp. 282–293.
24. Lerski, R.; Straughan, K.; Schad, L.R.; Boyce, D.; Bluml, S.; Zuna, I. MR Image Texture Analysis—An Approach to tissue Characterization. *Magn. Resonance Imaging* **1993**, *11*, 873–887. [CrossRef]
25. Schad, L.R.; Bluml, S.; Zuna, I. MR Tissue Characterization of Intracranial Tumors by means of Texture Analysis. *Magn. Resonance Imaging* **1993**, *11*, 889–896. [CrossRef]
26. Reyes Aldasoro, C.C.; Bhalerao, A. Volumetric Texture Segmentation by Discriminant Feature Selection and Multiresolution Classification. *IEEE Trans. Med. Imaging* **2007**, *26*, 1–14. [CrossRef] [PubMed]
27. Zhan, Y.; Shen, D. Automated Segmentation of 3D US Prostate Images Using Statistical Texture-Based Matching Method. In Proceedings of the Medical Image Computing and Computer-Assisted Intervention (MICCAI), Montréal, QC, Canada, 15–18 November 2003; pp. 688–696.
28. Xie, J.; Jiang, Y.; tat Tsui, H. Segmentation of Kidney from Ultrasound Images based on Texture and Shape Priors. *IEEE Trans. Med. Imaging* **2005**, *24*, 45–57. [CrossRef] [PubMed]
29. Hoffman, E.A.; Reinhardt, J.M.; Sonka, M.; Simon, B.A.; Guo, J.; Saba, O.; Chon, D.; Samrah, S.; Shikata, H.; Tschirren, J.; et al. Characterization of the Interstitial Lung Diseases via Density-Based and Texture-Based Analysis of Computed Tomography Images of Lung Structure and Function. *Acad. Radiol.* **2003**, *10*, 1104–1118. [CrossRef]
30. Segovia-Martínez, M.; Petrou, M.; Kovalev, V.A.; Perner, P. Quantifying Level of Brain Atrophy Using Texture Anisotropy in CT Data. In Proceedings of the Medical Image Understanding and Analysis, Oxford, UK, 19–20 July 1999; pp. 173–176.

31. Ganeshan, B.; Goh, V.; Mandeville, H.C.; Ng, Q.S.; Hoskin, P.J.; Miles, K.A. Non–Small Cell Lung Cancer: Histopathologic Correlates for Texture Parameters at CT. *Radiology* **2013**, *266*, 326–336. [CrossRef] [PubMed]
32. Sabino, D.M.U.; da Fontoura Costa, L.; Gil Rizzatti, E.; Antonio Zago, M. A Texture Approach to Leukocyte Recognition. *Real-Time Imaging* **2004**, *10*, 205–216. [CrossRef]
33. Wang, X.; He, W.; Metaxas, D.; Mathew, R.; White, E. Cell Segmentation and Tracking using Texture-Adaptive Snakes. In Proceedings of the 2007 4th IEEE International Symposium on Biomedical Imaging: From Nano to Macro, Washington, DC, USA, 12–16 April 2007; pp. 101–104. [CrossRef]
34. Kather, J.N.; Weis, C.A.; Bianconi, F.; Melchers, S.M.; Schad, L.R.; Gaiser, T.; Marx, A.; Zollner, F. Multi-class Texture Analysis in Colorectal Cancer Histology. *Sci. Rep.* **2016**, *6*, 27988. [CrossRef] [PubMed]
35. Dunn, D.; Higgins, W.; Wakeley, J. Texture Segmentation using 2-D Gabor Elementary Functions. *IEEE Trans. Pattern Anal. Mach. Intell.* **1994**, *16*, 130–149. [CrossRef]
36. Bigun, J.; du Buf, J.M.H. N-Folded Symmetries by Complex Moments in Gabor Space and Their Application to Unsupervised Texture Segmentation. *IEEE Trans. Pattern Anal. Mach. Intell.* **1994**, *16*, 80–87. [CrossRef]
37. Bianconi, F.; Fernández, A. Evaluation of the Effects of Gabor Filter Parameters on Texture Classification. *Pattern Recognit.* **2007**, *40*, 3325–3335. [CrossRef]
38. Rajpoot, N.M. Texture Classification Using Discriminant Wavelet Packet Subbands. In Proceedings of the 45th IEEE Midwest Symposium on Circuits and Systems (MWSCAS 2002), Tulsa, OK, USA, 4–7 August 2002.
39. Chang, T.; Kuo, C.C.J. Texture Analysis and Classification with Tree-Structured Wavelet Transform. *IEEE Trans. Image Process.* **1993**, *2*, 429–441. [CrossRef]
40. Chellapa, R.; Jain, A. *Markov Random Fields*; Academic Press: Boston, MA, USA, 1993.
41. Krizhevsky, A.; Sutskever, I.; Hinton, G.E. ImageNet Classification with Deep Convolutional Neural Networks. In Proceedings of the 25th International Conference on Neural Information Processing Systems—Volume 1 (NIPS'12), Lake Tahoe, NV, USA, 3–6 December 2012; Curran Associates Inc.: New York, NY, USA, 2012; pp. 1097–1105.
42. Zeiler, M.D.; Fergus, R. Visualizing and Understanding Convolutional Networks. In Proceedings of the Computer Vision—ECCV 2014, Zurich, Switzerland, 6–12 September 2014; Fleet, D., Pajdla, T., Schiele, B., Tuytelaars, T., Eds.; Springer International Publishing: Berlin, Germany, 2014; pp. 818–833.
43. Simonyan, K.; Zisserman, A. Very Deep Convolutional Networks for Large-Scale Image Recognition. *arXiv* **2014**, arXiv:1409.1556.
44. Russakovsky, O.; Deng, J.; Su, H.; Krause, J.; Satheesh, S.; Ma, S.; Huang, Z.; Karpathy, A.; Khosla, A.; Bernstein, M.; et al. ImageNet Large Scale Visual Recognition Challenge. *Int. J. Comput. Vis.* **2015**, *115*, 211–252. [CrossRef]
45. Andrearczyk, V.; Whelan, P.F., Chapter 4—Deep Learning in Texture Analysis and Its Application to Tissue Image Classification. In *Biomedical Texture Analysis*; Depeursinge, A., Al-Kadi, O.S., Mitchell, J.R., Eds.; Academic Press: Cambridge, MA, USA, 2017; pp. 95–129. [CrossRef]
46. Ronneberger, O.; Fischer, P.; Brox, T. U-Net: Convolutional Networks for Biomedical Image Segmentation. In Proceedings of the Medical Image Computing and Computer-Assisted Intervention—MICCAI 2015, Munich, Germany, 5–9 October 2015; Navab, N., Hornegger, J., Wells, W.M., Frangi, A.F., Eds.; Springer International Publishing: Berlin, Germany, 2015; Volume 9350, pp. 234–241.
47. Zhang, Z.; Liu, Q.; Wang, Y. Road Extraction by Deep Residual U-Net. *IEEE Geosci. Remote Sens. Lett.* **2018**, *15*, 749–753. [CrossRef]
48. Jansson, A.; Humphrey, E.J.; Montecchio, N.; Bittner, R.M.; Kumar, A.; Weyde, T. Singing Voice Separation with Deep U-Net Convolutional Networks. In Proceedings of the 18th International Society for Music Information Retrieval Conference, Suzhou, China, 23–27 October 2017.
49. Dong, H.; Yang, G.; Liu, F.; Mo, Y.; Guo, Y. Automatic Brain Tumor Detection and Segmentation Using U-Net Based Fully Convolutional Networks. In Proceedings of the Annual Conference on Medical Image Understanding and Analysis, Edinburgh, UK, 11–13 July 2017; Valdés Hernández, M., González-Castro, V., Eds.; Springer International Publishing: Berlin, Germany, 2017; Volume 723, pp. 506–517.
50. Falk, T.; Mai, D.; Bensch, R.; Çiçek, Ö.; Abdulkadir, A.; Marrakchi, Y.; Böhm, A.; Deubner, J.; Jäckel, Z.; Seiwald, K.; et al. U-Net: Deep Learning for Cell Counting, Detection, and Morphometry. *Nat. Methods* **2019**, *16*, 67. [CrossRef] [PubMed]
51. Malpica, N.; Ortuño, J.E.; Santos, A. A Multichannel Watershed-based Algorithm for Supervised Texture Segmentation. *Pattern Recognit. Lett.* **2003**, *24*, 1545–1554. [CrossRef]

52. Ojala, T.; Pietikäinen, M.; Harwood, D. A Comparative Study of Texture Measures with Classification based on Feature Distributions. *Pattern Recognit.* **1996**, *29*, 51–59. [CrossRef]
53. Ojala, T.; Valkealahti, K.; Oja, E.; Pietikäinen, M. Texture Discrimination with Multidimensional Distributions of Signed Gray Level Differences. *Pattern Recognit.* **2001**, *34*, 727–739. [CrossRef]
54. Randen, T.; Husøy, J.H. Filtering for Texture Classification: A Comparative Study. *IEEE Trans. Pattern Anal. Mach. Intell.* **1999**, *21*, 291–310. [CrossRef]
55. Randen, T.; Husøy, J.H. Texture Segmentation using Filters with Optimized Energy Separation. *IEEE Trans. Image Process.* **1999**, *8*, 571–582. [CrossRef]
56. Brodatz, P. *Textures: A Photographic Album for Artists and Designers*; Dover: New York, NY, USA, 1996.
57. Yamada, R.; Ide, H.; Yudistira, N.; Kurita, T. Texture Segmentation using Siamese Network and Hierarchical Region Merging. In Proceedings of the 2018 24th International Conference on Pattern Recognition (ICPR), Beijing, China, 20–24 August 2018; pp. 2735–2740. [CrossRef]
58. Petrou, M.; Garcia-Sevilla, P. *Image Processing: Dealing with Texture*; John Wiley & Sons: Hoboken, NJ, USA, 2006.
59. Reyes-Aldasoro, C.C.; Bhalerao, A.H. Volumetric Texture Analysis in Biomedical Imaging. In *Biomedical Diagnostics and Clinical Technologies: Applying High-Performance Cluster and Grid Computing*; Pereira, M., Freire, M., Eds.; IGI Global: Hershey, PA, USA, 2011; pp. 200–248.
60. Mirmehdi, M.; Xie, X.; Suri, J. *Handbook of Texture Analysis*; Imperial College Press: London, UK, 2009.
61. Vincent, L.; Soille, P. Watersheds in digital spaces: An efficient algorithm based on immersion simulations. *IEEE Trans. Pattern Anal. Mach. Intell.* **1991**, *13*, 583–598. [CrossRef]
62. Gabor, D. Theory of Communication. *J. IEE* **1946**, *93*, 429–457. [CrossRef]
63. Knutsson, H.; Granlund, G.H. Texture Analysis Using Two-Dimensional Quadrature Filters. In Proceedings of the IEEE Computer Society Workshop on Computer Architecture for Pattern Analysis and Image Database Management—CAPAIDM, Pasadena, CA, USA, 12–14 October 1983; pp. 206–213.
64. Randen, T.; Husy, J.H. Multichannel filtering for image texture segmentation. *Opt. Eng.* **1994**, *33*, 2617–2625. [CrossRef]
65. Kingma, D.P.; Ba, J. Adam: A Method for Stochastic Optimization. *arXiv* **2014**, arXiv:1412.6980.
66. Verhoeven, J.; Miller, N.R.; Daems, L.; Reyes-Aldasoro, C.C. Visualisation and Analysis of Speech Production with Electropalatography. *J. Imaging* **2019**, *5*, 40. [CrossRef]

© 2019 by the authors. Licensee MDPI, Basel, Switzerland. This article is an open access article distributed under the terms and conditions of the Creative Commons Attribution (CC BY) license (http://creativecommons.org/licenses/by/4.0/).

Review

Recent Advances in Saliency Estimation for Omnidirectional Images, Image Groups, and Video Sequences

Marco Buzzelli

Department of Informatics, Systems and Communication, University of Milano-Bicocca, Viale Sarca 336, 20126 Milan, Italy; marco.buzzelli@unimib.it

Received: 9 July 2020; Accepted: 24 July 2020 ; Published: 27 July 2020

Abstract: We present a review of methods for automatic estimation of visual saliency: the perceptual property that makes specific elements in a scene stand out and grab the attention of the viewer. We focus on domains that are especially recent and relevant, as they make saliency estimation particularly useful and/or effective: omnidirectional images, image groups for co-saliency, and video sequences. For each domain, we perform a selection of recent methods, we highlight their commonalities and differences, and describe their unique approaches. We also report and analyze the datasets involved in the development of such methods, in order to reveal additional peculiarities of each domain, such as the representation used for the ground truth saliency information (scanpaths, saliency maps, or salient object regions). We define domain-specific evaluation measures, and provide quantitative comparisons on the basis of common datasets and evaluation criteria, highlighting the different impact of existing approaches on each domain. We conclude by synthesizing the emerging directions for research in the specialized literature, which include novel representations for omnidirectional images, inter- and intra- image saliency decomposition for co-saliency, and saliency shift for video saliency estimation.

Keywords: co-saliency; omnidirectional images; video saliency; visual saliency estimation

---

1. Introduction

Visual saliency is defined as a property of a scene in relation to an observer. This follows from a commonly-accepted interpretation [1–3] that defines it as the set of subjective and perceptual attributes that make certain items stand out from their surroundings, and therefore grab the viewer's attention.

In the vision system of human beings and other animals, two components typically contribute to the overall saliency: bottom-up and top-down factors [4]. Bottom-up saliency is driven by low-level activations in the vision system, based for example on pre-attentive computational mechanisms in the primary visual cortex [5], and does not depend on specific tasks and objectives. Conversely, top-down saliency is defined as being goal-directed [6], and as such it is highly dependent on the intrinsic biases of the observer, and correlated to the semantics of the depicted elements. Scientific literature reviews for automatic visual saliency estimation often adopt these two categories to classify existing methods [2]. For example, deep learning solutions are rightfully labeled as top-down approaches due to their intrinsic ability to extract and exploit semantic pieces of information [7], whereas hand-crafted methods tend to rely on lower-level features such as contrasting patterns, and are therefore categorized as bottom-up solutions. In practice, though, multiple interacting factors (both top-down and bottom-up) are considered to determine which parts of the scenes are further processed by the attentional process of the biological vision system [8].

Properly modeling visual saliency means emulating the widest set of factors that influence the evaluation of saliency as performed by a human being. This goal has been pursued by many authors, both in the neuroscience community and, more recently, in the computer vision and image processing communities. Due to the different levels of involved complexity, bottom-up saliency estimation methods are generally faster than top-down methods [9], and thus useful in applications where real-time feedback is considered more important than reaching higher accuracy. For example, in a live augmented reality scenario, fast saliency estimation would locate image regions deemed important for further localized computer vision analysis, and would provide precious information to avoid covering areas of potential interest with the rendering of augmentation elements. Conversely, top-down saliency estimation methods tend to be more robust, at the cost of a higher demand for computational resources. These are therefore typically employed in applications with looser time constraints, and which benefit from semantic interpretation. For example, a system for storing and organizing personal photos could exploit saliency estimation to detect objects of interests based on visual composition, and their reoccurring presence in multiple photos. In general, visual saliency estimation has been successfully employed in multiple tasks, such as image retargeting [10], video summarization [11], and photo-collage creation [12]. It has also been adopted as an intermediate pre-processing step for other computer-vision problems, such as scene recognition [13], object detection [14] and segmentation [15]. Since the advent and diffusion of deep-learning, many of these problems have been reformulated in an end-to-end fashion that does not rely on explicitly estimating the salient component, as proven by state of the art solutions in each field [16–18]. There exist, however, problems that remain directly related to the evaluation of saliency information such as advertisement assessment [19], and domains where its explicit computation is particularly relevant, for example in reducing the computational effort for analysis of large quantities of data (such as video sequences, or high-resolution panoramic images exploited in the virtual reality domain).

By analyzing the recent scientific literature on saliency estimation, in fact, specific topics emerged as persistently reoccurring amidst works dedicated to saliency on regular images, due to a combination of the excellent results already reached by the scientific community, and the paradigm shift in solving certain problems without explicitly modeling general-purpose saliency. Such trending topics are, namely, saliency in omnidirectional images, and multiple-input scenarios, which include co-saliency and video saliency estimation. Although visual saliency has been studied in other fields as well (such as light field and hyper-spectral imaging) most of the current domain-specific research happens to converge on the three mentioned topics, while the literature does not offer enough material to produce a valuable review of recent solutions related to other less widespread domains. Our goal is therefore to highlight the recent trends of research in these fields, providing a concise yet exhaustive insight into each analyzed method, and summarizing the similarities and differences across different solutions. The investigated domains are either very recent, or have lived a particularly dynamic evolution. As a consequence, different methods are typically evaluated and/or optimized on different datasets, making comparative evaluations extremely challenging. Nonetheless, we conduct an analysis on the joint occurrence of methods and datasets, and we benchmark solutions that are directly comparable as they were evaluated in equivalent conditions.

Accompanying the development of research into visual saliency estimation through the years, the scientific literature has periodically offered different benchmarks and surveys, typically concerning general-purpose saliency. Borji et al. (2012) [20] provide an in-depth comparison of 35 state of the art methods for saliency estimation, over both synthetic and natural images. A second work by Borji et al. (2015) [9] conduct a similar benchmark including newly developed solutions, the most recent of which, however, was released in the year 2014. A more recent review is presented by Wang et al. (2019) [21], offering an in-depth survey over methods for salient object detection specifically based on deep learning approaches. Concerning domain-specific analyses, Cong et al. (2019-I) [22] cover methods for saliency detection that rely on so-called "comprehensive information", such as depth cues, inter-image correspondence (equivalent to co-saliency), and multiple frames. Zhang et al. (2018) [23] review the concepts, applications, and challenges intrinsic into co-saliency detection, whereas Riche et al. (2016) [24] focus on video saliency estimation approaches based on a bottom-up interpretation. With the current survey, our goal is to inform on up-to-date developments in the fields of domain-specific visual saliency estimation. To the best of our knowledge, there are currently no surveys that specifically focus on saliency estimation in omnidirectional images, which is the most recent domain-specific development in the field.

The main contributions of this paper are the following:

- We highlight domains that naturally emerged from a literature review as being particularly timely and relevant.
- Through a systematic analysis of the methods in each domain, we show their commonalities and differences.
- We provide clear information regarding the targeted ground truth representation, as well as the output that each method can explicitly generate.
- We conduct, where deemed fair, a quantitative comparison of the selected methods, and provide some insights on the basis of such comparison.
- We report an in-depth analysis of the most common datasets for the analyzed domains, including the representation used for the ground truth saliency information.
- We present the commonly used evaluation measures, which can be either domain-specific or general-purpose.
- We conclude by synthesizing the emerging directions for research in the specialized literature.

The rest of the paper is structured as follows: Section 2 presents the systematic approach that led to the selection of works in this review. Section 3 introduces the three domains of interest and their peculiarities, followed by the description of different interpretations and representations commonly adopted for visual saliency, and an overview of existing metrics and measures used to assess saliency estimation algorithms. The subsequent sections present methods, datasets, and measures for each domain of interest: Section 4 focuses on omnidirectional images, Section 5 relates to co-saliency estimation, and finally, Section 6 presents developments in the field of video saliency.

2. Methodology for Literature Review

The selection of literature works included in this review paper has been determined through a systematic approach, which is described in the following.

The initial prompt was to observe and highlight the current trends in visual saliency estimation. With this objective, we performed a keyword-based search on the academic search engine Google Scholar, using the terms: "visual saliency", "saliency estimation", "salient object detection". Given the time-sensitive nature of our goal, we restricted the results to works published no earlier than 2017. For each resulting paper, we retrieved the following information:

- Title
- Year of publication
- Author list
- Venue (the specific journal or conference)
- Abstract
- Number of citations

For the years 2018 and 2017, we restricted the number of results to those having collected at least one citation at the time of the review, intending to focus on the dissemination of works that are considered relevant by the scientific community. Based on the title and abstract analysis, then, we excluded some further results:

- Works that do not fit in the field of visual saliency (retrieved due to the unreliability of keyword-based search alone);

- Works that focus on extremely narrow tasks (e.g., saliency estimation for skin lesions, or for comic strips).

Works related to datasets and surveys have also been isolated and used as a reference for the corresponding sections. We annotated the remaining results in terms of domains of application, and the most recurring themes emerged as being: saliency estimation for omnidirectional images, co-saliency estimation, and saliency estimation for video sequences. We therefore focused on these domains to provide the scientific community with an analysis of relevant and recent developments. For each of the selected works, domain-specific and cross-domain characteristics have been collected through careful study of the corresponding manuscripts.

The final selection of recent and relevant methods for saliency estimation has then been used as the starting point to identify the associated evaluation measures and the associated datasets. Evaluation measures have been classified as either general-purpose (presented in Section 3.3), or domain-specific (presented in the corresponding Sections 4–6).

In virtue of the importance of data in training and assessing methods for visual saliency estimation, we dedicated for each domain an in-depth analysis of the corresponding datasets. A matrix describing the joint occurrences of datasets and methods has been defined and presented for all three domains. At this stage, no explicit constraint on the release date has been imposed: the rationale is that if a dataset is still widely adopted as a benchmark for new methods, it is to be considered relevant and worth mentioning. Multiple instances of the same dataset being reported with different names have been identified and merged. Conversely, whenever two or more saliency estimation methods refer to the same dataset in different versions, this piece of information has been annotated and reported. Finally, all datasets that were identified during the preliminary, keyword-based, search have been found to be already present in the current selection. Detailed characteristics of the identified datasets have been presented.

3. Visual Saliency Estimation

In this section, we describe the recently emerged domains for visual saliency, and provide background information about the different types of saliency representation, as well as commonly used evaluation measures.

3.1. Domains

The scientific literature on saliency estimation has witnessed the emergence of domain-specific solutions, covering a wide range of topics that go beyond the traditional regular-image input. Specifically, recent developments have shown several works in the domains of omnidirectional images, image groups for co-saliency estimation, and video sequences, as exemplified in Figure 1. Other fields of application, such as light field [25] and hyper-spectral imaging [26,27] have also caught the attention of saliency-related research. Visual saliency estimation is, however, still at its early stages in such domains, and a full review of related methods is therefore left as a future development.

Figure 1. Visualization of the three domains of interest for visual saliency estimation: omnidirectional images (**a**), image groups for co-saliency (**b**), and clips for video saliency (**c**).

Another relevant domain of research is depth-assisted visual saliency estimation, which explores the advantages of predicting saliency on so-called RGB-D images [28]. In this case, the additional knowledge associated to the distance between the camera and the depicted elements can improve the separation of subjects from the background, providing a precious piece of information for better saliency estimation. Despite the clear relevance of the topic, we chose not to explicitly discuss this domain since such a wide field deserves a whole dedicated survey paper. Nonetheless, we found that depth-assisted saliency estimation is sometimes included in the analyzed domains of omnidirectional images [29], co-saliency [30], and video saliency [31]. We will, therefore, reference and discuss only these works in the corresponding sections.

Omnidirectional images (ODIs) are panoramic representations of a scene, covering a 360° solid angle from a single viewpoint, typically employed in passive virtual reality. Virtual reality is the experience of a simulated world, which can be navigated by the user to varying degrees of freedom [32]. In a passive virtual reality scenario, the spatial movements are predefined, and the virtual environment is precomputed in a sequence of omnidirectional images. During fruition, the user only determines the direction of viewing, i.e., the subpart of the ODI to visualize at any given time. Image cropping for thumbnail selection is particularly valuable when operating on large omnidirectional images, depicting wide sceneries in high-resolution [33]. Storing and transmitting these ODIs can then benefit from perceptually-aware compression, i.e., reducing the represented detail over areas that are considered "less-interesting" [34].

Image **co-saliency** refers to the problem of estimating the saliency from a group of images that depict the same subject. The rationale behind this approach is to provide the saliency estimation model with additional information, and thus partially compensate for the ill-posed nature of the problem. Depending on the chosen level of abstraction, image groups for co-saliency estimation could either represent exactly the same instance from multiple points of view, or different instances of the same category, possibly characterized by slight variations in appearance.

Video saliency is the task of performing saliency estimation on a sequence of frames. By considering the time component, in fact, estimation of visual saliency acquires additional value in terms of understanding how people react to, and learn from images [35]. If we exclude the naive frame-by-frame approach, multiple-frame analysis helps a given model gain a global view of the input, in a fashion similar to what happens with co-saliency estimation. In addition, the annotated sequences are expected to exhibit different patterns compared to single image saliency, as the vision of each single video frame is both limited in time and highly influenced by the previous frames.

Cross-talk between domains is of course highly present, with approaches aiming at video saliency estimation in omnidirectional images [36], as well as co-saliency estimation in video sequences [37].

3.2. Saliency Representation

Ground truth for visual saliency estimation is typically collected and distributed in one of three possible representations: scanpaths, saliency maps, and salient object regions. These are visually shown in Figure 2.

  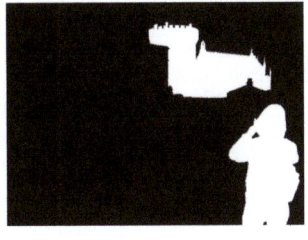

(a) (b) (c)

Figure 2. Different representations used for visual saliency ground truth: fixation points relative to scanpaths (**a**), continuous saliency maps directly related to fixations (**b**), and sharp salient object regions (**c**).

3.2.1. Fixations and Scanpaths

Human eyes have been shown to explore a given scene in saccades, which are rapid movements from a point of interest to another. Between saccades, a temporary pause, called a fixation, is spent in the area of the point of interest [38]. The ordered sequence of fixations is called scanpath [39], or gaze trajectory [33], and it is the first and most direct way to represent the salient areas of an image. In some cases, such as omnidirectional images explored with virtual reality displays, gaze trajectory is complemented with head trajectory [33,40,41], tracking the movement of the whole head of the viewing subjects.

3.2.2. Fixation-Related Saliency Maps

Scanpaths can be processed by discretizing fixations coordinates to pixel coordinates. The result is a scattered map of pixel saliency, which is typically convolved with a bidimensional Gaussian kernel [41] in order to create a proper saliency map through kernel density estimation [42]. This type of representation removes, by definition, the temporal relationship between fixations, which can be considered non-necessary for specific tasks, such as thumbnail selection [33]. Mixed representations have been proposed, such as saliency volumes [35], giving the possibility to produce both scanpaths and saliency maps.

3.2.3. Salient Object Regions

Another commonly used representation for saliency information consists of pixel-precise binary segmentation maps. This type of annotation can be generated by pre-segmenting each element of the scene and subsequently selecting one, or more, segments that overlap with the largest amount of fixations [43] or explicit selections [44]. Alternatively, one or multiple users can be asked to directly provide a hand-drawn segmentation of the area they consider most relevant [45]. In the case of multiple proposals, these are then reduced to one annotation with a predefined aggregation strategy.

Different representations are more suited to different applications. For example, temporally-aware scanpaths can be useful to determine the optimal path of a virtual camera in an omnidirectional video [46]. Continuous saliency maps can be employed for saliency-aware image compression, specifically tuning non-uniform bit allocation as a function of the estimated local saliency [47], while a sharp salient object estimation is typically used for automatic or semi-automatic object segmentation in photo-manipulation tools [48]. There is no hard evidence that explicitly optimizing for one representation may help improving performance on others, and methods tend to be developed for clusters of datasets sharing the same type of salient representation, with a few isolated exceptions [35,49].

3.3. Evaluation Measures

We provide a selection of the evaluation measures most commonly used by the saliency estimation methods analyzed in this paper. An exhaustive review of evaluation measures for saliency models is provided by Riche et al. [50]. Domain-specific measures will also be presented, when existing, in each of the subsequent sections, regarding omnidirectional images, co-saliency, and video saliency estimation.

In the following, we categorize the selected evaluation measures on the basis of the involved ground truth representation. We will refer to the predicted saliency map as P, and to the corresponding ground truth as G. Some formulations will rely on the sum-normalized versions P' and G'. X will be the total number of image pixels, and F the total number of fixations.

3.3.1. Measures for Fixations, Scanpaths, and Saliency Maps

The **Pearson Correlation Coefficient** (CC) [51] is a measure of the linear correlation between prediction P and ground truth G considered as two statistical variables:

$$\text{CC} = \frac{\text{cov}(P, G)}{\sigma_P \sigma_G} \tag{1}$$

where $\text{cov}(\cdot, \cdot)$ is the co-variance, σ_P and σ_G are the standard deviation values for, respectively, the predicted saliency data and ground truth saliency data.

The **Normalized Scanpath Saliency** (NSS) [52] is used to compare a densely-estimated saliency map with a fixation-based ground truth. Specifically, it is the average of the estimated saliency values P in the locations indicated by eye fixations f:

$$\text{NSS} = \frac{1}{F} \sum_{f=1}^{F} \frac{P(f) - \mu_P}{\sigma_P} \tag{2}$$

Note that the saliency estimation map is normalized to have zero mean and unitary standard deviation through corresponding statistics μ_P and σ_P. In this scenario, a null NSS indicates a correspondence between estimation and ground truth equivalent to random chance. Conversely, very high or very low NSS suggests a high correspondence or anti-correspondence.

The **Kullback–Leibler** (KL) [53] divergence between two saliency maps considered as probability density functions, is computed as:

$$\text{KL} = \sum_{x=1}^{X} G'(x) \cdot \log \left(\frac{G'(x)}{P'(x) + \gamma} + \gamma \right) \tag{3}$$

where γ is a protection constant.

The **SIMilarity measure** (SIM) [54], also called histogram intersection, compares two different saliency maps when viewed as normalized distributions:

$$\text{SIM} = \sum_{x=1}^{X} \min \left(P'(x), G'(x) \right) \tag{4}$$

The **Earth Mover's Distance** (EMD) [55] quantifies the minimal cost to transform probability distribution P into G:

$$\text{EMD} = (\min_{f_{ij}} \sum_{i,j} f_{ij} d_{ij}) + | \sum_{i} G_i - \sum_{j} P_j | \max_{i,j} d_{ij} \tag{5}$$

with:

$$\sum_{i,j} f_{ij} = \min(\sum_{i} G_i - \sum_{j} P_j) \tag{6}$$

where d_{ij} represents the difference between bin i in G and bin j in P.

3.3.2. Measures for Salient Object Regions

The **Precision/Recall curve** is computed by varying a binarization threshold on the continuous saliency estimation map P, and computing at each level the precision PR and recall RE components:

$$\text{PR} = \frac{\text{TP}}{\text{TP} + \text{FP}} \tag{7}$$

$$\text{RE} = \frac{\text{TP}}{\text{TP} + \text{FN}} \tag{8}$$

where TP is the number of True Positive pixels, FP False Positives, and FN False Negatives, obtained by comparing predicted saliency P with ground truth map G.

The **F-measure** (F_β) [56] corresponds to the weighted harmonic mean between precision PR and recall RE:

$$F_\beta = \frac{(1+\beta^2)\, \text{PR} \cdot \text{RE}}{\beta^2 \text{PR} + \text{RE}} \qquad (9)$$

In this case, the continuous-valued saliency estimation can be binarized with different techniques before effectively computing precision and recall. Furthermore, it is common practice [9] to give more weight to the precision component (considered more important than recall for the saliency estimation task), by setting parameter β^2 to 0.3.

The **Mean Absolute Error** (MAE) is computed directly on the prediction, without any threshold, as:

$$\text{MAE} = \frac{1}{X} \sum_{x=1}^{X} |P(x) - G(x)| \qquad (10)$$

The **Structure measure** (S_α) [57], inspired by the structure similarity (SSIM) from image quality assessment, is the weighted mean between region-aware structural similarity S_r and object-aware structural similarity S_o:

$$S_\alpha = \alpha \cdot S_o + (1 - \alpha) \cdot S_r \qquad (11)$$

where S_r covers the object-part similarity with the ground truth, while S_o accounts for the global similarity based on sharp estimation contrast and uniform distribution.

The **enhanced-alignment measure** (Q) [58] captures both pixel-level matching and image-level statistics as:

$$Q = \frac{1}{X} \sum_{x=1}^{X} \frac{1}{4}(1 + \xi(x))^2 \qquad (12)$$

where:

$$\xi = \frac{2\varphi_G \circ \varphi_P}{\varphi_G \circ \varphi_G + \varphi_P \circ \varphi_P} \qquad (13)$$

Bias matrix $\varphi_{\{G,P\}}$ is the distance between each value of the binary map (G or P) and its global mean, and the two matrices are compared through the Hadamard product ($\circ$).

3.3.3. Representation-Independent Measures

Area Under Curve (AUC) is the area under the Receiver Operating Characteristic (ROC) curve. The latter is computed by varying the binarization threshold and plotting False Positive Rate (FPR) against True Positive Rate (TPR):

$$\text{TPR} = \text{RE} = \frac{\text{TP}}{\text{TP} + \text{FN}} \qquad (14)$$

$$\text{FPR} = \frac{\text{FP}}{\text{FP} + \text{TN}} \qquad (15)$$

Variants of the general concept of AUC take into consideration data distribution at various levels, in order to normalize the evaluation of estimated saliency. These include AUC-Judd [54], AUC-Borji [20], AUC-Zhao [59] and AUC-Li [60]. The AUC measure has been used to evaluate saliency estimation under different representations: from fixations, scanpaths, and saliency maps [34,36,61–66] to salient object regions [67–70].

4. Omnidirectional Images

Omnidirectional images, also known as 360° images, or panoramic images, present a set of domain-specific peculiarities. An omnidirectional image is digitally stored in equirectangular format, projecting a spherical surface into a planar and rectangular one. Any such projection inevitably introduces distortions in the representation, as a direct consequence of the *Theorema Egregium* [71,72], therefore saliency estimation methods for regular images would behave sub-optimally without a proper adaptation. For this reason, several methods specifically aimed at omnidirectional images focus on producing an alternative projection or transformation, that fully exploits existing approaches for classical image saliency estimation [35,36,73].

When a user explores an omnidirectional image, he/she normally uses a head-mounted display to freely navigate the scene. In this case, only a portion of it is shown at any given time, under a so-called Normal Field of View (NFoV), which introduces less-noticeable distortions. The starting point of view, a non-ODI thumbnail, and a suggested exploration pattern can all be optimized for the best user experience by exploiting saliency estimation.

Saliency maps related to several omnidirectional datasets have been observed to exhibit a bias in fixations close to the equator line of view [33,34,41]. This bias has been exploited by different methods [62,63] to produce more accurate estimations.

4.1. Methods for Omnidirectional Images

Table 1 presents a synthetic overview of recent methods for saliency estimation in omnidirectional images. All analyzed methods target a ground truth in the form of fixation saliency maps. They all produce a continuous saliency map output related to fixation data, whereas only a limited subset also explicitly predicts scanpath trajectories [34,35].

Table 1. Characteristics of recent methods for visual saliency estimation in omnidirectional images. The "Target" column indicates the nature of the ground truth used to train or develop the methods, while "Output" describes what data they are explicitly able to generate (FM = Fixation maps, SP = Scanpaths).

Method	2D-to-ODI Adaptation	Custom Representation	Backbone CNN	Deep Learning	Hand-Crafted	Target	Output
Battisti 2019 [29]			(none)		✓	FM	FM
Sitzmann 2018 [33]	✓		(none)		✓	FM	FM
Monroy 2018 [61]	✓		VGG_CNN_M	✓		FM	FM
Ling 2018 [62]			(none)		✓	FM	FM
Lebreton 2018 [63]	✓		(none)		✓	FM	FM
Cheng 2018 [36]		✓	ResNet-50/VGG-16	✓		FM	FM
Fang 2018 [64]			(none)		✓	FM	FM
De Abreu 2017 [34]	✓		(none)		✓	FM	FM, SP
Assens 2017 [35]		✓	VGG-16	✓		FM, SP	FM, SP
Maugey 2017 [73]	✓	✓	(none)		✓	FM	FM

Part of the research in this field consists of evaluating the transferability of existing methods originally designed for classical images [33,34,61,63,73]. Sitzmann et al. [33] initially collected the SVR (Saliency in Virtual Reality) dataset. Through observations on the extensive and diverse set of acquisitions, they acquired knowledge about fixation bias, which they used to improve upon existing saliency estimation solutions when applied in the field of omnidirectional images. They applied the developed method to a wide range of use cases, including automatic montage and summarization of videos, thumbnail extraction, and video compression. De Abreu et al. [34] gathered data only relative to the whole head movement, instead of tracking the viewers' eyes, when collecting their own dataset. The authors first proposed a method to convert this information into saliency maps. They then observed a fixation bias as well, which is addressed using the proposed Fused Saliency Maps (FSM) method, operating on existing saliency estimation solutions. Monroy et al. [61] presented an architectural extension that can be applied to any existing neural network for saliency estimation, in order to fine-tune it to the specific domain of omnidirectional images. The underlying idea is the extraction of six undistorted patches of the panoramic view, their independent evaluation, and subsequent fusion.

As previously mentioned, some methods devised specific representations of the input data, that allow for full exploitation of the domain, but without suffering from its intrinsic disadvantages (namely, image distortions) [36,39,73]. Assens et al. [35] propose a novel representation, called saliency volume, to extract saliency information that can be adapted to different forms: from the time-independent saliency maps, to the ordered scanpaths (extracted through specific sampling strategies), to a hybrid representation, which consists in temporally weighted saliency maps. Cheng et al. [36], who also collected the Wild-360 dataset, presented a weakly-supervised training for a spatial-temporal neural network architecture. They also proposed working on a six-face cube projection, in order to mitigate the heavy distortions of equirectangular projection, and implemented so-called cube padding to hide the discontinuities of representation to the neural network processing. Maugey et al. [73] proposed an aggregation technique for the application of existing saliency estimation methods to different map projections. They mitigate the discontinuities introduced at the edge of 2D representations by performing a double cube projection, the results of which are eventually merged. They also proposed the automation of a navigation pattern that maximizes exposition to estimated salient areas.

Despite the clear dominance of machine learning approaches to saliency estimation (and, specifically, deep learning approaches), a good deal of recent methods for omnidirectional saliency are based on hand-crafted design and combination of visual features [29,33,34,62–64,73]. Ling et al. [62] defined a hand-crafted approach to saliency estimation for omnidirectional images. Their color-dictionary sparse representation (CDSR) is applied in conjunction with multi-patch analysis to simulate human color perception. Fixation bias is also taken into consideration for the specific characteristics of the domain. Lebreton et al. [63] extended existing solutions to the estimation of saliency in omnidirectional images, namely Boolean Map Saliency (BMS) and Graph-Based Visual Saliency (GBVS). They then defined a novel framework, called Projected Saliency, to adapt existing estimation models with a simple mechanism, which allowed extensive analysis of features interaction in computational saliency models. Fang et al. [64] developed a hand-crafted solution based on the fusion of feature contrast and boundary connectivity, leaning on the figure-ground law from Gestalt Theory. Boundary connectivity is designed to describe the spatial layout of the image region with an upper and a lower boundary. Feature contrast is based on luminance and color features from the CIE Lab color space. Battisti et al. [29] presented a hand-crafted approach based on low-level image descriptors, such as edges and texture features. They also exploit a depth description of the image itself, to produce a more robust estimation of image saliency, which is evaluated using metrics such as the Kullback–Leibler divergence, and the correlation coefficient.

Methods based on deep learning [35,36,61] are built on the basis of existing Convolutional Neural Network (CNN) backbones, such as the VGG_CNN_M [74] and VGG-16 [75] architectures (from the Visual Geometry Group), and the residual-learning-based ResNet-50 [76] architecture.

4.2. Datasets for Omnidirectional Images

Table 2 presents a synthetic overview describing the adoption of different datasets by different methods for visual saliency estimation in omnidirectional images. The most frequently adopted dataset is the one published with the Salient360! challenge [41], in some cases based on an old version of the same set [40]. The iSUN dataset [77] (interactive Scene UNderstanding) was used by Assens et al. [35] to pre-train their solution, but does not involve omnidirectional images. The MIT dataset [78] from Massachusetts Institute of Technology was adopted for evaluation by Maugey et al. [73], but does not contain saliency ground truth information.

A detailed description of all the relevant datasets is consequently presented in Table 3. These can be differentiated first and foremost by the stimuli characteristics, ranging from image resolution (when stored in equirectangular format), to duration of the exposition to the stimulus itself. The display device is typically either a head-mounted display such as Oculus Discovery Kit 2 (DK2), or a classical computer screen. In the latter case, the image is visualized in Normal Field of View, allowing the user to navigate the whole panorama with the use of mouse and keyboard.

Table 2. Dataset/method matrix for visual saliency estimation in omnidirectional images.

Dataset	Method	Battisti 2019 [29]	Sitzmann 2018 [33]	Monroy 2018 [61]	Ling 2018 [62]	Lebreton 2018 [63]	Cheng 2018 [36]	Fang 2018 [64]	De Abreu 2017 [34]	Assens 2017 [35]	Maugey 2017 [73]
Salient360! (2017) [40] (2018) [41]		✓						✓		✓	
SVR [33]			✓	✓	✓	✓					
Wild-360 [36]							✓				
LAY [34]									✓	✓	
iSUN [77]											
MIT [78]											✓

Table 3. Datasets for visual saliency in omnidirectional images and related characteristics (FM = Fixation maps, SP = Scanpaths).

	Stimuli					
Dataset	Video/Image	CGI/Real	Resolution (Pixels)	Duration (Seconds)	Conditions	
Salient360! [41]	Mixed	Mixed	3000 × 1500 ÷ 18,332 × 9166	25	Seated	
SVR [33]	Image	CGI	8192 × 4096	30	Seated, standing	
Wild-360 [36]	Video	Real	1920 × 960	~20 (length)	N/A	
LAY [34]	Image	Real	4096 × 2048	10, 20	N/A	

	Cardinalities		Responses		Devices	
Dataset	Input Data	Users	Acquisitions	FM/SP	Head/Eyes	Display	Eye Tracker	Project Page
Salient360! [41]	85 images/19 videos	63	≤32 maps/paths per stimulus	FM, SP	Head, eyes	Oculus DK2	SMI	[79]
SVR [33]	22 images	169	1980 maps/paths	FM, SP	Head, eyes	Oculus DK2, PC screen	Pupil-labs, Tobii EyeX	[80]
Wild-360 [36]	85 videos (29 ann.)	30	12,926 maps	FM	(manual)	PC screen	(none)	[81]
LAY [34]	21 images	32	704 paths	SP	Head	Oculus DK2	(none)	[82]

245

All analyzed datasets provide a ground truth in terms of either scanpaths (for eyes and head movement) or fixations-related saliency maps, i.e., without precisely-annotated salient object regions, possibly due to the intrinsic difficulty in segmenting equirectangular projection images.

The Salient360! [41] dataset was created for the Grand Challenge "Salient360!" organized in conjunction with ICME 2017 (International Conference on Multimedia and Expo). The dataset has been updated through the years [40], with the last edition also including a set of omnidirectional video clips. It is supplied with a script toolbox for the evaluation of predicted scanpaths and saliency maps.

The SVR [33] (Saliency in Virtual Reality) dataset is a collection of both head and eye orientation data (scanpaths), coming from the observation of 22 stereoscopic omnidirectional images. The environmental condition of the stimuli include combinations of users being seated or standing, with or without a head-mounted display. In all conditions, an eye tracking device was used.

Wild-360 [36] is an exclusively video-based dataset for omnidirectional saliency. The original clips were retrieved from YouTube using specific keywords such as "nature", "wildlife", and "animals", in order to collect a dataset with heterogeneous and dynamic contents. The video sequences were manually annotated by multiple users, without any head- or eye- tracking device.

LAY [34] (Look Around You) was built with the objective of developing saliency estimation methods without the support for an eye tracking device for data collection. Specifically, the head orientation of the viewers (called Viewport Center Trajectory) is used as a proxy ground truth for the generation of saliency maps. Different experiments have been conducted by varying the viewing time of each stimulus.

4.3. Evaluation of Saliency for Omnidirectional Images

Methods for saliency estimation in omnidirectional images are evaluated with a variety of measures, most of which are common to visual saliency in traditional images, such as the Pearson correlation coefficient CC ([29,33,36,61–64]), and the area under the ROC curve AUC ([34,36,61–64]).

The Salient360! benchmark[41] introduced, among other criteria, an evaluation based on the Kullback–Leibler divergence (KL). Although not specifically designed for omnidirectional images, this has been widely adopted as an evaluation measure[29,61–64] thanks to Salient360! being the de-facto reference for saliency in omnidirectional images.

Regarding domain-specific evaluation, the same benchmark also introduced the evaluation of scanpaths based on the comparison metric by Jarodzka et al. [83] properly adjusted to incorporate orthodromic distances in 360° instead of Euclidean distances. The original metric is based on a comparison between each fixation from the prediction with all the fixations from the provided ground truth. Such comparison is applied on the basis of multiple elements, namely the spatial proximity of starting points, the difference in direction and magnitude of the saccades, and the temporal proximity of saccade midpoints.

Based on the dataset/method matrix in Table 2, the Salient360! dataset is the best candidate benchmark to compare the largest subset of selected methods. Results are presented in Table 4 according to four different metrics reported in the corresponding publications. The VGG-based model by Assens et al. [35] has been excluded as it does not report performance on metrics comparable with other methods. Ling et al. [62] generates in absolute terms the best results across all considered measures, while the second-best is the model by Fang et al. [64], according to three measures out of four. Both solutions are based on hand-crafted algorithms with a specific focus on emulating the color perception in human vision. Omnidirectional images, therefore, would appear to represent a domain where manually-defined criteria still outperform machine learning, possibly due to the stronger positional bias, and to image distortions that are uncommon in large datasets used for neural network pre-training [84].

Table 4. Quantitative comparison of selected methods for saliency estimation in omnidirectional images, on the Salient360! challenge dataset. Best results are highlighted in boldface.

Method	CC ↑	KL ↓	NSS ↑	AUC ↑
Monroy 2018 [61]	0.536	0.487	0.757	0.702
Ling 2018 [62]	**0.550**	**0.477**	**0.939**	**0.736**
Lebreton 2018 [63]	0.527	0.698	0.851	0.714
Fang 2018 [64]	0.538	0.508	0.910	**0.736**

5. Co-Saliency

The concept of co-saliency was first introduced by Toshev et al. [85] to address the problem of image matching, exploiting local point feature correspondence and region segmentation. By its original definition, therefore, co-saliency estimation refers to determining the common element from two or more instances of exactly the same subject. A more general interpretation would extend the concept to groups of images depicting different instances of the same category [86–88] (e.g., many images of different lions). Regardless of the specific definition, the presence of multiple images can provide a useful constraint in the otherwise ill-posed problem of saliency estimation, thanks to the assumption that all images (or a subgroup [89]) contain the same salient element.

Co-saliency estimation is often encountered along with other related tasks, namely co-segmentation [90] and co-localization [91]. While the output of a method for co-saliency is a continuous map, representing the probability of each pixel being salient, the output of a method for co-segmentation is typically a binary mask, that precisely separates the foreground from the background. Following a similar abstraction, co-localization refers to generating a bounding-box over common elements in multiple images.

5.1. Methods for Co-Saliency

Table 5 presents a selection of recent methods for co-saliency estimation that were well received by the scientific community. All presented methods target a binary salient object region ground truth. The output of these methods is a continuous-valued saliency map, which is, however, optimized to be as sharp as possible. In some cases [37,68,69,92], the methods also produce a segmentation-oriented binary mask.

Table 5. Characteristics of the analyzed methods for co-saliency estimation. The "Target" column indicates the nature of the ground truth used to train or develop the methods, while "Output" describes what data they are explicitly able to generate (OR = Object Regions, SM = Sharp saliency Maps, BM = Binary Masks).

Method	Early Fusion	Late Fusion	Backbone CNN	Deep Learning	Deep Features	Hand-Crafted	Target	Output
Cong 2019-II [93]	✓	✓	(none)			✓	OR	SM
Zhang 2019 [67]		✓	FCN (VGG-16)	✓			OR	SM
Jerripothula 2018 [92]		✓	(none)			✓	OR	SM, BM
Hsu 2018 [94]		✓	ResNet-50 + FCN (VGG-16)	✓			OR	SM
Tsai 2018 [68]	✓	✓	CNN-S		✓	✓	OR	SM, BM
Jeong 2018 [69]	✓	✓	DeepLab (VGG-16)	✓			OR	SM, BM
Zheng 2018 [95]		✓	FCN-32s (VGG-16)	✓			OR	SM
Cong 2018 [96]	✓	✓	VGG-16		✓	✓	OR	SM
Wang 2017-I [37]	✓	✓	(none)			✓	OR	SM, BM
Wei 2017 [70]	✓		FCN (VGG-16)	✓			OR	SM
Yao 2017 [89]	✓		(none)		✓	✓	OR	SM

The co-saliency domain involves, by definition, the analysis of multiple images. How these are handled can help in differentiating among different methods for co-saliency estimation. Early-fusion techniques [70,89] initially extract a global representation of all the images in the input group, capturing relationships between different images. Conversely, late-fusion techniques [67,92,94,95] are designed to estimate single-image saliency from each input individually, and reciprocally update them in a second phase, based on the extracted information.

Joining the efforts of early and late fusion techniques, are methods that exploit both approaches by extracting so-called "intra-image saliency" (i.e., from each individual image) as well as "inter-image saliency" (as the correspondence among multiple images), to eventually combine them [37,68,69,96]. Cong et al. (2018)[96] proposed computing intra-saliency maps exploiting the depth information associated with each image, and calculating the inter-saliency maps based on multi-constraint feature matching to improve the overall performance. A cross-label-propagation scheme was adopted to optimize and refine both maps in a cross-way, eventually integrated into a final co-saliency map. In a subsequent work, Cong et al. (2019-II) [93] formulated the inter-image correspondence as a hierarchical sparsity reconstruction framework. They addressed image-pairs correspondences through a set of pairwise dictionaries, and global image group characteristics through a ranking-scheme-based common dictionary. A three-term energy function refinement model is introduced in order to improve the intra-image smoothness and inter-image consistency. Wang et al. [37] extended the concept of co-saliency from images to videos, and as such operate on multiple input video sequences. They took into consideration both inter-video foreground correspondences and intra-video saliency stimuli, with the objective of ignoring background distraction elements and concurrently emphasizing salient foreground regions. Tsai et al. [68] observed that the auxiliary task of co-segmentation improves object boundaries in co-saliency detection, and proposed a joint optimization of the two tasks by solving an energy minimization problem over a graph. The resulting model iteratively transfers useful information in both directions, to improve the prediction of both domains. The solution by Jeong et al. [69] produces an initial set of co-saliency maps, which are then refined on object boundaries. The authors then introduced a seed-propagation step over an integrated multilayer graph, aimed at detecting regions missed by lower-level descriptors. Such descriptors are pooled both within-segment and within-group, in order to handle input images having different sizes.

Another possible criterion to discriminate among different approaches, is the distinction between deep-learning solutions, and those based on hand-crafted design and traditional techniques. Methods in the deep learning group [67,70,94,95] typically benefit from end-to-end learning, therefore optimizing the final objective of co-saliency estimation regardless of the adopted early-fusion or late-fusion approach. Many are based on the Fully-Convolutional Network (FCN) by Long et al. [97] or DeepLab by Chen et al. [18], both leveraging the VGG backbone [75]. Other adopted neural architectures include the "Slow" CNN-S model by Chatfield et al. [74]. Zhang et al. [67] presented a coarse-to-fine framework for co-saliency detection: they first generate an initial proposal using a mask-guided fully convolutional network, based on the high-level feature response maps of a pre-trained VGG network [75]. They then defined a multi-scale label smoothing model to refine the prediction, optimizing both the label smoothness of pixels and superpixels. Wei et al. [70] presented an end-to-end co-saliency estimation neural network. The model adopts an early-fusion approach by extracting high-level descriptions of the input images, and capturing the group-wise interaction information for group images. It was proven to be able to learn the collaborative relationships between single-image features and group-wise features. Hsu et al. [94] presented an original unsupervised approach to co-saliency estimation, addressed in a graphical model based on two losses: the single-image saliency (SIS) loss, acting as the unary term, and the Co-occurrence (COOC) loss, acting as the pairwise term. The authors also presented two refining extensions, namely boundary preservation and map sharpening. Zheng et al. [95] presented FASS: a feature-adaptive semi-supervised framework for co-saliency estimation. The proposed solution addresses and exploits the difference in efficacy of image features, by a joint formulation of element-level feature selection and view-level feature weighting. It optimizes co-saliency label prorogation over both labeled and unlabeled image regions.

The purely hand-crafted methods include the aforementioned video co-saliency solution by Wang et al. [37], and the more recent work by Jerripothula et al. [92]. Specifically, the latter focuses on predicting the saliency map for one selected key image, and subsequently extending the prediction to other images in the group. The authors proposed fusing individual saliency maps using the "dense correspondence" technique, and evaluating a no-reference concentration-based saliency quality to decide whether the fused saliency map improves upon the original one.

Finally, crossing the gap between deep learning solutions, and purely hand-crafted ones, are all those traditional methods that exploit the extraction of high-level deep features from a pre-trained model, as a descriptor to be used in combination with other pieces of information for co-saliency estimation [68,89,96]. A notable example is represented by Yao et al. [89], who generalized the problem of co-saliency estimation to the case where multiple object categories are present in the input image group. The task has been therefore decomposed into two sub-problems: automatically identifying subgroups of images, based on multi-view spectral rotation co-clustering, and subsequently extracting the co-saliency information from such groups.

5.2. Datasets for Co-Saliency

Table 6 presents the combination of methods and datasets used in the corresponding experiments for training and evaluation. The most frequently adopted datasets are iCoseg [98] and various versions of the MSRC from Microsoft Research [86]. The latter is particularly old, the first version going back to 2005 as it was originally collected for a different purpose than saliency estimation. Different updates of the dataset have been released through the years, and the specific version is indicated in Table 6 by specifying the number of input image groups.

The number of image groups is also one of the discriminating elements reported in Table 7 along with other cardinality-related information. The stimuli are described in terms of data and content type. For most reported datasets, the resolution is extremely heterogeneous across images, and it is therefore reported as a minimum-maximum side pair. The "same subject" column indicates whether each image group depicts exactly the same instance of the subject from different points of view, or multiple instances of the same category. All the reported co-saliency datasets provide a binary salient object region annotation, i.e., none have been collected with the aid of eye tracking devices for scanpath acquisition, relying instead on manual annotation of the contours of salient objects.

RGBD Coseg183 [30] is a dataset developed for those co-saliency estimation methods that exploit the depth information associated with the input RGB image. It is partially composed of images from the RGBD Scenes Dataset [99], which were acquired using a prototype PrimeSense RGB-D camera and a firewire camera from Point Grey Research.

RGBD Cosal150 [96] is a selection of images and depth maps originally coming from the RGBD NJU-1985 dataset [100] (Nanjing University), which are augmented with co-saliency pixel-level annotations. The depth information in the original dataset comes from mixed sources: either from the Kinect device used for acquisition, or inferred through an optical-flow-based method [101]. This dataset has been presented in the previously discussed method by Cong et al. (2018).

iCoseg [98] was collected using the "Group" functionality in the Flickr photography platform, in order to collect groups of images belonging to the same category (and sometimes, the same photographer), which includes various wild animals, popular landmarks, and sports teams. The authors also made available for public download the developed interface that was used to interactively annotate the dataset.

MSRC [86] (from Microsoft Research) is the oldest dataset commonly used for training and evaluation of co-saliency algorithms, although originally collected for applications related to image classification. Multiple versions of the dataset exist, with the number of image groups ranging from 7 to 23. Table 7 reports information regarding the 14-groups version of the dataset.

Table 6. Dataset/method matrix for co-saliency estimation. The number in parentheses indicates the version of the MSRC dataset identified by the number of image groups.

Dataset Method	Cong 2019-II [93]	Zhang 2019 [67]	Jerripothula 2018 [92]	Hsu 2018 [94]	Tsai 2018 [68]	Jeong 2018 [69]	Zheng 2018 [95]	Cong 2018 [96]	Wang 2017-I [37]	Wei 2017 [70]	Yao 2017 [89]
RGBD Coseg183 [30]	✓							✓			
RGBD Cosal150 [96]	✓							✓			
iCoseg [98]		✓	✓(14)	✓(7)	✓	✓(8)	✓(7)			✓(23)	✓(7)
MSRC [86]		✓(8)	✓(14)	✓(7)	✓(14)	✓(8)	✓(7)				
Cosal2015 [87]		✓	✓								
Coseg-Rep [88]											
Internet Images [102]					✓						
Image-Pair [103]									✓		
Safari [104]									✓		
Vicosegment [37]											

Table 7. Selected datasets for co-saliency estimation with corresponding characteristics (OR = Object Regions).

Dataset	Stimuli					Cardinalities			Responses	Project Page
	Video/ Image	CGI/ Real	Resolution (Pixels)	Same Subject	Groups	Images Per Group	Total Images			
RGBD Coseg183 [30]	Image	Real	640 × 480	Yes	16	6 ÷ 18	183		OR	[105]
RGBD Cosal150 [96]	Image	Mixed	303 ÷ 1177	Mixed	21	2 ÷ 20	150		OR	[106]
iCoseg [98]	Image	Real	333 ÷ 500	Mixed	38	4 ÷ 41	643		OR	[107]
MSRC [86]	Image	Real	320 × 213	No	14	24 ÷ 32	418		OR	[108]
Cosal2015 [87]	Image	Real	93 ÷ 3008	No	50	26 ÷ 52	2015		OR	[109]
Coseg-Rep [88]	Image	Real	137 ÷ 1280	No	22+1	9 ÷ 49 (+116)	572		OR	[110]
Internet Images [102]	Image	Real	107 ÷ 340	No	3	561 ÷ 1306	2746		OR	[111]
Image-Pair [103]	Image	Real	66 ÷ 500	Mixed	105	2	210		OR	[112]
Safari [104]	Video	Real	270 ÷ 640	Yes	9	20 ÷ 50	415		OR	[113]
Vicosegment [37]	Video	Real	216 ÷ 480	Yes	10+38	18 ÷ 40	743		OR	[114]

Authors of the Cosal2015 [87] dataset gathered images in challenging scenarios from the YouTube video set [115] and the ILSVRC2014 detection set [84] (ImageNet Large Scale Visual Recognition Competition), observing that images belonging to the same group often involve similar backgrounds, leading to potentially wrong co-saliency estimations. The dataset has been annotated by 20 different users, whereas most of the other reported datasets involve one human annotation per image.

Coseg-Rep [88] is a dataset for co-segmentation and co-sketch, the objective being to automatically infer a common pattern from instances of the same subject. It is composed of 22 categories of different flowers and animals, plus a special "repetitive" category, which contains images with repeating patterns aimed at inter-image co-segmentation and co-saliency.

Internet Images [102], also known as Internet Datasets, is composed of only three image groups (car, horse, and airplane), characterized however by high cardinality inside each group. It presents a total of 15,270 images, out of which 2746 are provided with a segmentation ground truth that was acquired using both the LabelMe annotation toolbox [116] and Amazon Mechanical Turk.

The Safari dataset [104] is a video-based collection of annotated sequences for object co-segmentation, partially built upon the existing MOViCS dataset [117] (Multi-Object Video Co-Segmentation). It is composed of nine videos of five animal classes: for each class, there is one video sequence containing only that specific class, plus one or more videos of the class in conjunction with other classes.

Vicosegment [37] is another, more recent, video dataset for co-segmentation and co-saliency. It is composed of 10 category groups containing similar foreground objects, and a total of 38 videos with cardinality ranging between 18 frames and 40 frames. This dataset was presented in conjunction with the already presented method by Wang et al. based on inter-video foreground correspondence and intra-video saliency stimuli.

The Image-Pair [103] dataset contains groups of only two images, depicting (at least) one common object on two different background scenes. It is composed of image pairs collected from the dataset from Hochbaum et al. [118], the Caltech-256 Object Categories database [119], and the PASCAL Visual Object Challenge dataset [120].

5.3. Evaluation of Co-Saliency

Although not specifically designed for co-saliency estimation with image groups, the Average Precision score AP is often applied for evaluation in this specific domain ([67–69,89,94,95]). It is proportional to the area under the Precision/Recall curve, generated as defined in Section 3.3.

Other measures commonly used for co-saliency evaluation are the F_β ([37,67–70,89,94–96]) and the area under the ROC curve AUC ([67–70]).

The dataset/method matrix for co-saliency estimation presented in Table 6 suggests using either iCoseg [98] or MSRC [86] as a comparison benchmark. We decided to focus on iCoseg, due to the extreme variability of MSRC versions adopted by different methods. Results are reported in Table 8: the overall best performance is reached by Zheng et al. [95], followed by Zhang et al. [67] and Hsu et al. [94] for F_β and Average Precision (AP).

Table 8. Quantitative comparison of selected methods for co-saliency estimation on the iCoseg dataset. Best results are highlighted in boldface.

Method	F_β ↑	AP ↑	AUC ↑
Zhang 2019 [67]	0.855	0.906	0.974
Hsu 2018 [94]	0.850	0.911	-
Tsai 2018 [68]	0.820	0.878	0.968
Jeong 2018 [69]	0.823	0.896	**0.979**
Zheng 2018 [95]	**0.873** *	**0.920** *	-
Yao 2017 [89]	0.810	0.868	-

* Values inferred from graphs in the corresponding publication.

All these solutions are VGG-16-based neural networks, adopting a late-fusion approach. This common pattern can be justified as semantic interpretation is particularly relevant in a domain that requires finding common elements across different images. At the same time, the recent inter-saliency/intra-saliency paradigm, although promising in the context of the corresponding publications, is possibly not yet mature enough. In this specific evaluation setup, in fact, the work by Yao et al. [89] presents the lowest performance. It should be noted, however, that the corresponding solution performs the selection of image groups in a completely unsupervised manner, while all other methods rely on existing annotated clusters.

6. Video Saliency

Saliency estimation in video sequences presents a specific set of advantages as well as original challenges. In the same spirit as co-saliency, the availability of multiple images (i.e., frames) imposes useful constraints on the ill-posed problem of saliency estimation. Unlike co-saliency datasets, video saliency ones are sometimes collected with the use of an eye tracking device, instead of manually segmenting the elements of interest in each frame. One effect of this approach is the high variability in the ground truth saliency maps across different frames: Li et al. [43] and Fan et al. [49] recently proposed to explicitly consider the phenomenon of saliency shift, where the viewer's attention can briefly change due to distracting elements, or even transfer indefinitely to a whole different salient object. Furthermore, as noted by Ullah et al. [121], saliency estimation in videos can prove to be particularly difficult when the salient object is in motion, it is small, it changes shape, and it is embedded in a context where the whole camera is moving.

6.1. Methods for Video Saliency

Table 9 enumerates recent solutions for saliency estimation in video sequences, along with additional pieces of information. Particular attention should be paid in differentiating methods that target salient object region annotations, and those who target fixations-related saliency maps [65,66]. Specifically for the former category, some of the described solutions are tested against datasets that were originally annotated for video segmentation [122–124], and in some cases the method itself is described as addressing "saliency-based video segmentation" [121,125,126], showing once again the correlation between such tasks.

Table 9. Methods for video saliency. The temporal window is indicated in relation to the underlying technique: OF (Optical Flow), LSTM (Long-Short Term Memory), CNN (Convolutional Neural Network), GRU (Gated Recurrent Unit). The "Target" column indicates the nature of the ground truth used to train or develop the methods, while "Output" describes what data they explicitly generate (OR = Object Regions, SM = Sharp Maps, BM = Binary Masks, FM = Fixation Maps).

Method	Saliency Shift	Temporal Window	Optical Flow	Backbone CNN	Deep Learning	Hand-Crafted	Target	Output
Fan 2019 [49]	✓	∞ (LSTM)		ResNet-50	✓		OR	SM
Li 2019 [127]		2 (OF)	✓	ResNet-34/101	✓		OR	SM
Yan 2019 [128]		4 (CNN) + ∞(GRU)	✓	ResNet-50	✓		OR	SM, BM
Cong 2019-III [129]		2 (OF) + ∞ (energy)	✓	(none)		✓	OR	SM
Hu 2018 [125]		2 (OF) + ∞ (diffusion)	✓	(none)		✓	OR	SM, BM
Zhou 2018 [130]		3	✓	(none)		✓	OR	SM
Ullah 2018 [121]		2 (OF)	✓	(none)		✓	OR	SM, BM
Wang 2017-II [126]		5 (OF)	✓	(none)		✓	OR	SM, BM
Chen 2017 [131]		4 ÷ 20 (diffusion)	✓	(none)		✓	OR	SM
Min 2019 [65]		32 (CNN)		S3D (inception)	✓		FM	FM
Gorji 2018 [66]	✓	∞ (LSTM)		VGG-16	✓		FM	FM

An inherent characteristic of video-based processing is the temporal window, i.e., the number of frames that are jointly analyzed in order to exploit the time dimension. Methods indicated with ∞ are not constrained with an explicit limit in the temporal window, although the influence of other frames to the current one typically decreases with their distance. Other criteria useful in discriminating among different methods include the type of representation involved (such as optical flow), and the type of model involved. In this case, for deep learning methods, the backbone CNN is also reported.

In computer vision, optical flow can be defined as a displacement vector field that describes, for each pixel in each frame, the direction and intensity movement from the previous frame (or frames). Solutions for video saliency estimation based on optical flow [18,21,66,121,125,127,128,130] demonstrate that explicitly and compactly representing the time-wise variations provide a valuable piece of information for accurate detection of salient objects in video sequences. Cong et al. (2019-III) [129] designed a single-frame saliency model based on sparsity-based reconstruction, and an inter-frame saliency map based on progressive sparsity-based propagation. The two maps are then incorporated in a global consistency energy formulation to achieve spatio-temporal smoothness. Hu et al. [125] framed the problem at hand as an "unsupervised video segmentation" task. They exploited edge-aware features and the optical flow representation to develop a novel diffusion technique based on a neighborhood graph. With this approach, they were able to eventually produce a generic object segmentation based on the propagation of estimated saliency information. Zhou et al. [130] developed a three-step framework. A set of localized estimation models, generated through a random forest regressor, can be first used to create a temporary saliency map. This is then improved through a spatio-temporal refinement step, based on appearance and motion information. The resulting map is finally used to provide saliency cues for the following frame estimation. Ullah et al. [121] presented an approach for so-called "unconstrained video segmentation". They first generate an initial set of saliency regions through a novel saliency measure. They then compute a homography over optical flow information to retrieve motion cues that are robust to background motion. The two pieces of information can be eventually combined, expanded and refined. Wang et al. [126] developed a super-pixel-based technique that initially produces a prior map for pixel-wise labeling, exploiting a geodesic distance. They then formulated the task as an energy minimization problem, operating on foreground-background models and dynamic location models as unary terms, as well as label smoothness potentials as pairwise terms. Chen et al. [131] designed a method for video saliency detection based on spatio-temporal fusion and low-rank coherency guided diffusion. They first segment the input video into batches, where motion clues are internally diffused. Interbatch saliency priors are then taken into account for a low-level saliency fusion. These clues are eventually used to guide a saliency diffusion step.

Similarly to what has been observed with co-saliency and saliency in omnidirectional images, recent methods in the domain of video saliency are also equally spread among hand-crafted solutions [121,125,126,130,131], and those based on a deep-learning approach [49,65,66,127,128]. Belonging to the latter category, Fan et al. [49] collected and annotated the DAVSOD dataset (Densely Annotated Video Salient Object Detection), and proposed a neural-network-based approach to video saliency detection that explicitly addresses the problem of "saliency-shift" (the phenomenon where human attention switches from one element to another during the stimulus). Their solution is based on convolutional LSTM (Long-short term memory) modules. Li et al. [127] designed a multi-task neural network for salient object detection in video sequences. The first task, accomplished by the first sub-network, consists of still-image saliency estimation. The second task aims at motion saliency detection based on optical flow images. The two sub-networks were trained end-to-end with the integration of specifically-designed motion-guided attention modules. Yan et al. [128] proposed a solution for video saliency estimation that does not rely on densely-annotated video sequences. They first developed a technique to generate pixel-level pseudo- ground truths from sparsely annotated video frames, based on a neural network operating on optical flow images. They then trained a neural model composed of a spatial refinement network and a spatio-temporal module on their artificially-augmented training data.

As mentioned, some solutions target a different representation of video saliency information, namely fixation-related saliency maps. Gorji et al. [66] focused on the concept of attentional push:

a family of saliency cues that include following the gaze of depicted subjects, accounting for the salient element leaving the scene, and for abrupt scene changes in general. They exploited these concepts to augment a static saliency estimation with the objective of minimizing the relative entropy between estimated and expected fixation patterns. Min et al. [65] presented TASED-Net: a Temporally-Aggregating Spatial Encoder-Decoder neural architecture based on the S3D [132] model (and, consequently, on the Inception model [133]), that produces an estimation of saliency for a single frame based on a finite number of previous frames. In order to produce a continuous saliency estimation, the developed network can be applied in a temporal-sliding-window fashion over the whole input sequence.

6.2. Datasets for Video Saliency

Table 10 illustrates the datasets that were involved in the experiments of each analyzed method for video saliency estimation, with the objective of highlighting the relevant benchmarks for recent developments in this field. We separate the datasets related to methods that target different types of ground truth data, highlighting how UCFSports [134] is in fact used by solutions belonging to both worlds. Regarding methods aimed at salient object regions, it can be observed that the most frequently-adopted datasets are FBMS [122] (Freiburg–Berkeley Motion Segmentation) and SegTrackV2 [123]. Despite not being very recent (both were released in the year 2013), they are described in-depth in the following, due to their high relevance. Conversely, datasets that are particularly old, and which have been tested against only by one or a few methods, are no further analyzed.

Table 11 therefore presents detailed information for the selected datasets, reporting information on both the stimuli and the user responses. As indicated, some saliency datasets that are specific for the video domain are exclusively annotated with salient object regions [43,122–124]. Others are collected with an eye tracking device, thus producing saliency maps based on user fixations [134,135]. Finally, the very recent DAVSOD [49] provides both types of annotation, thus highlighting the existing relationship between these different representations.

DAVSOD [49] (Densely Annotated Video Salient Object Detection) is built upon the stimuli from the DHF1K [135] (Dynamic Human Fixation 1000) eye tracking dataset, manually trimmed into short video clips. The scenes are enriched with additional annotations, which include: timestamp of the shift in visual attention, category labeling into 7 classes and 70 sub-classes, and conversion of the fixation records into hand-drawn object segmentation masks, performed per-frame by multiple annotators.

FBMS [122] (Freiburg–Berkeley Motion Segmentation) is a dataset composed from existing sources (Brox et al. [136] and the Hopkins 155 [137]) as well as new sequences, for a total of 59 video clips. The videos have been specifically collected aiming at high variation in image resolution and motion types, and have been manually annotated every 20th frame, thus providing a sparse ground truth.

SegTrack [138] and SegTrackV2 [123] are among the most tested-against datasets for video saliency estimation, despite being originally addressed at video segmentation. Both versions were collected with particular attention at equally representing challenging aspects, namely: color overlap between foreground and background, inter-frame motion, and changing target shape. The second version of the dataset introduces additional sequences and annotations.

VOS [43] (Video Object Segmentation) is composed of videos collected from internet sources as well as personal collections, divided into an easy and a difficult subset. One keyframe every 15 frames has been segmented at the object-level by a pool of four subjects. A different set of subjects have been asked to free-view the videos, in order to collect their eye tracking data, which are eventually used to unambiguously define and annotate the salient objects.

Table 10. Dataset/method matrix for video saliency estimation. The number in parentheses identifies the version of the dataset through the corresponding cardinality of video sequences.

Dataset Method	Fan 2019 [49]	Li 2019 [127]	Yan 2019 [128]	Cong 2019-III [129]	Hu 2018 [125]	Zhou 2018 [130]	Ullah 2018 [121]	Wang 2017-II [126]	Chen 2017 [131]	Min 2019 [65]	Gorji 2018 [66]
DAVSOD [49]	✓										
FBMS [122]	✓(30)		✓(59)		✓(59)			✓(59)	✓(26)		
ViSal [139]	✓	✓		✓							
MCL [140]	✓	✓		✓							
SegTrack v1 [138]	✓(13)				✓(14)	✓(14)	✓(14)	✓(14)			
SegTrack v2 [123]									✓(10)		
UVSD [141]	✓		✓								
VOS [43]	✓										
DAVIS [124]		✓	✓	✓	✓	✓					
I2R [142]							✓				
Wallflower [143]							✓				
MOViCS [117]							✓				
DS [144]									✓		
UCFSports [134,145]							✓		✓	✓	✓
DHF1K [135]										✓	
Hollywood-2 [134,146]										✓	✓
DIEM [147]											✓

Table 11. Selected datasets for video saliency estimation, with related features (SP = Scanpaths, FX = Fixations only, FM = Fixation maps, OR = Object Regions).

Dataset	Stimuli Characteristics			Devices	Users	
	Resolution (pixels)	FPS	Display	Eye Tracker	Users (Fixations)	Users (Objects)
DAVSOD [49]	640 × 360	30	N/A	SMI RED 250	17 (1/video)	20
FBMS [122]	288 ÷ 960	30	N/A	(none)	(none)	N/A
SegTrack v2 [123]	<640 × 360	N/A	N/A	(none)	(none)	N/A
VOS [43]	800 × 448	30	1680 × 1050	SMI RED 500	23	4
DAVIS [124]	1920 × 1080	24	N/A	(none)	(none)	N/A
UCFSports [134]	<720 × 480	10	22" 1280 × 1024	SMI iView X HiSpeed 1250	16	(none)
DHF1K [135]	640 × 360	30	19" 1440 × 900	SMI RED 250	17 (1/video)	(none)

Dataset	Cardinalities			Responses	
	Total Videos	Total Frames	Frames Per Video	SP/FX/FM/OR	Project Page
DAVSOD [49]	187	23,938	~128	FX, FM, OR	[148]
FBMS [122]	59	13,860 (720 annotated)	~235 (~12 annotated)	OR	[149]
SegTrack v2 [123]	14	1066	~76	OR	[150]
VOS [43]	200	114,421 (7467 annotated)	~722 (~37 annotated)	OR	[151]
DAVIS [124]	50	3455	~69	OR	[152]
UCFSports [134]	150	9578	~64	SP	[153]
DHF1K [135]	1000	582,605	~583	FX, FM	[154]

DAVIS [124] (Densely Annotated VIdeo Segmentation) comprises high-resolution short sequences that are manually annotated for pixel-accurate segmentation. Each clip depicts up to two spatially-connected objects, aiming at constraining the problem to a controlled and limited domain. Finally, all sequences are labeled with multiple attributes covering challenging aspects such as clutter, blur, appearance change, and many others.

UCFsports [134] was built upon the pre-existing large scale video dataset of the same name by Rodriguez et al. [145] from the University of Central Florida, originally published for human action recognition. This collection is composed of high-resolution recordings from television shows, covering nine sport action classes. Nineteen human subjects were divided into three groups and tasked with different objectives, namely: action recognition, context recognition, and free-viewing. The same procedure has been applied to build the Hollywood-2 saliency dataset, on top of the existing data from the dataset by Marszalek et al. [146].

DHF1K [135] (Dynamic Human Fixation 1000) has been collected with YouTube videos retrieved through 200 key search terms, following the principles of large scale and high quality, diverse content, varied motion patterns, and various objects. Seventeen subjects were tasked with free-viewing 10 sessions of non-overlapping videos presented in random order. Furthermore, five subjects were asked to provide an additional piece of annotation regarding the number of objects in each sequence.

6.3. Evaluation of Video Saliency

The analyzed methods for video saliency estimation introduce two domain-specific evaluation measures, namely the Temporal stability ($\mathcal{T}$) [125], and the Per-frame pixel error rate (ϵ) [121]. Both are based on a salient object region ground truth. Temporal stability $\mathcal{T}$ is computed as the distance between the descriptors of the segmentation boundaries between two successive frames, in terms of shape context descriptors [155]. Per-frame pixel error rate ϵ is computed as:

$$\epsilon = \frac{\text{XOR}(th(P), G)}{N} \tag{16}$$

where $th(P)$ is a binary (thresholded) version of the predicted saliency map, G is the reference ground truth, and N is the total number of frames in the input sequence.

Other general-purpose measures often used to evaluate saliency estimation in the video domain include F_β ([49,126–128,130,131]), the Precision/Recall curve ([126–128,130,131]), and MAE ([49,126,127,130]).

The landscape defined by the dataset/method matrix for video saliency estimation in Table 10 is particularly scattered. We report in Table 12 quantitative results for the frequently adopted SegTrack v2 dataset, and for the DAVIS dataset. These two datasets are comparable in terms of video length and type of annotations, with DAVIS being composed of about three times as many sequences, at a higher resolution.

Table 12. Quantitative comparison of selected methods for video saliency estimation on the SegTrack v2 and DAVIS datasets. Best results are highlighted in boldface.

Method	SegTrack v2		DAVIS	
	max$F_\beta\uparrow$	MAE$\downarrow$	max$F_\beta\uparrow$	MAE$\downarrow$
Fan 2019 [49]	0.801	0.0230	-	-
Li 2019 [127]	-	-	**0.902**	**0.0220**
Yan 2019 [128]	-	-	0.859	-
Cong 2019-III [129]	-	-	0.765	0.0588
Zhou 2018 [130]	**0.899** *	0.0807 *	0.747 *	0.0636 *
Wang 2017-II [126]	0.890 *	**0.0489** *	-	-
Chen 2017 [131]	0.810 *	-	-	-

* Values inferred from graphs in the corresponding publication.

We did not include an analysis of FBMS due to the wider variability of versions (subsets of video sequences) used by different methods. Drawing any conclusions in the reported scenario is particularly challenging: on the SegTrack v2 dataset, the hand-crafted method by Wang et al. [126] appears to be the most well-balanced solution according to F_β and MAE, while on DAVIS the best results are obtained by Li et al. [127], which is a deep-learning model. At the same time, the best-F_β method on SegTrack, developed by Zhou et al. [130], reports worse performance on other metrics and datasets. Fan et al. [49], which is based on the recently-introduced concept of saliency shift, reaches the best result in terms of MAE, at the cost of penalizing F_β-based evaluation. It is therefore ultimately not clear whether one type of solution should be preferred against another, for saliency estimation in video sequences.

7. Conclusions

We presented a survey on visual saliency estimation, by focusing on recent developments in domains that are not restricted to the traditional single-image input. Adequately modeling the process of visual saliency has been shown to be particularly useful and/or effective in specific cases, such as omnidirectional images employed in virtual reality scenarios, image groups depicting the same subject for co-saliency estimation, and finally video sequences for video saliency estimation.

Omnidirectional images, in particular, are the most recently-introduced domain for saliency. Many different methods in the analyzed literature approached the problem by developing novel representations of the input data, in a form that does not introduce, or that prevents, image distortions which might negatively impact the saliency estimation process. An evaluation of methods that are directly comparable showed that hand-crafted solutions present excellent results in this particular domain. Co-saliency estimation exploits the concept of image groups to partially constrain the ambiguous concept of visual saliency. Recent methods in this domain are focusing on the independent estimation of intra-image saliency (the traditional concept of image saliency) and inter-image saliency (finding common elements among images in the same group), and their subsequent combination. Direct comparison showed the apparent superiority of deep learning solutions for this specific domain. Video sequences offer yet another example of leveraging multiple pieces of input data to facilitate the saliency estimation process. The nature of ground truth data is inherently different from that of the traditional domain, as the viewer's attention can move to different elements in the short or long term. This phenomenon is called "saliency shift", and it has been explicitly addressed by recent methods in the field.

The ground truth information for visual saliency can be collected in different forms and levels of abstraction: scanpaths (directly related to eye-gaze trajectories), continuous saliency maps, and binary salient object regions. The datasets involved in recent methods for saliency estimation have been described, among other criteria, in terms of their ground truth nature. Datasets composed of omnidirectional images are provided with either scanpaths or saliency maps, i.e., no binary segmentation masks are provided. Conversely, all analyzed datasets for co-saliency are manually annotated in terms of binary salient objects, without the use of eye tracking devices. Finally, the domain of video saliency offers a heterogeneous scenario, with many datasets offering ground truth data at all levels of abstraction.

As a general observation that covers all analyzed domains, it is worth noting that a well-balanced distribution persists, between traditional hand-crafted algorithms and deep learning methods, among recent solutions for the problem of visual saliency estimation.

In conclusion, this work complements existing state of the art analyses that mainly focuses on regular images. We integrated such studies with a review on saliency estimation for omnidirectional images, image groups, and video sequences. A natural extension of this work is to develop a thorough analysis of emerging topics such as light field saliency and hyper-spectral saliency, as well as widely-explored domains such as depth-assisted visual saliency estimation.

Funding: The research leading to these results has received funding from TEINVEIN: TEcnologie INnovative per i VEicoli Intelligenti, CUP (Codice Unico Progetto - Unique Project Code): E96D17000110009 - Call "Accordi per la Ricerca e l'Innovazione", cofunded by POR FESR 2014-2020 (Programma Operativo Regionale, Fondo Europeo di Sviluppo Regionale—Regional Operational Programme, European Regional Development Fund).

Conflicts of Interest: The author declares no conflict of interest.

References

1. Itti, L. Visual salience. *Scholarpedia* **2007**, *2*, 3327. [CrossRef]
2. Bianco, S.; Buzzelli, M.; Ciocca, G.; Schettini, R. Neural architecture search for image saliency fusion. *Inf. Fusion* **2020**, *57*, 89–101. [CrossRef]
3. Kruthiventi, S.S.; Ayush, K.; Babu, R.V. Deepfix: A fully convolutional neural network for predicting human eye fixations. *IEEE Trans. Image Process.* **2017**, *26*, 4446–4456. [CrossRef] [PubMed]
4. Niebur, E. Saliency map. *Scholarpedia* **2007**, *2*, 2675. [CrossRef]
5. Li, Z. A saliency map in primary visual cortex. *Trends Cogn. Sci.* **2002**, *6*, 9–16. [CrossRef]
6. Hamker, F. The role of feedback connections in task-driven visual search. In *Connectionist Models in Cognitive Neuroscience*; Springer: Berlin/Heidelberg, Germany, 1999; pp. 252–261.
7. Bianco, S.; Buzzelli, M.; Schettini, R. Multiscale fully convolutional network for image saliency. *J. Electron. Imaging* **2018**, *27*, 051221. [CrossRef]
8. Bylinskii, Z.; Recasens, A.; Borji, A.; Oliva, A.; Torralba, A.; Durand, F. Where Should Saliency Models Look Next? In Proceedings of the European Conference on Computer Vision, Amsterdam, The Netherlands, 11–14 October 2016; Springer: Berlin/Heidelberg, Germany, 2016; pp. 809–824.
9. Borji, A.; Cheng, M.M.; Jiang, H.; Li, J. Salient object detection: A benchmark. *IEEE Trans. Image Process.* **2015**, *24*, 5706–5722. [CrossRef] [PubMed]
10. Avidan, S.; Shamir, A. Seam Carving for Content-aware Image Resizing. *ACM Trans. Graph.* **2007**, *26*. [CrossRef]
11. Corchs, S.; Ciocca, G.; Schettini, R. Video summarization using a neurodynamical model of visual attention. In Proceedings of the IEEE 6th Workshop on Multimedia Signal Processing, Siena, Italy, 29 September–1 October 2004; pp. 71–74.
12. Margolin, R.; Zelnik-Manor, L.; Tal, A. Saliency for image manipulation. *Vis. Comput.* **2013**, *29*, 381–392. [CrossRef]
13. Gao, D.; Han, S.; Vasconcelos, N. Discriminant saliency, the detection of suspicious coincidences, and applications to visual recognition. *IEEE Trans. Pattern Anal. Mach. Intell.* **2009**, *31*, 989–1005.
14. Ren, Z.; Gao, S.; Chia, L.T.; Tsang, I.W.H. Region-based saliency detection and its application in object recognition. *IEEE Trans. Circuits Syst. Video Tech.* **2014**, *24*, 769–779. [CrossRef]
15. Li, Q.; Zhou, Y.; Yang, J. Saliency based image segmentation. In Proceedings of the 2011 International Conference on Multimedia Technology, Hangzhou, China, 26–28 July 2011; pp. 5068–5071.
16. Hu, J.; Shen, L.; Sun, G. Squeeze-and-excitation networks. In Proceedings of the IEEE Conference on Computer Vision and Pattern Recognition, Salt Lake City, UT, USA, 18–22 June 2018; pp. 7132–7141.
17. Redmon, J.; Farhadi, A. YOLO9000: Better, faster, stronger. In Proceedings of the IEEE Conference on Computer Vision and Pattern Recognition, Honolulu, HI, USA, 21–26 July 2017; pp. 7263–7271.
18. Chen, L.C.; Papandreou, G.; Kokkinos, I.; Murphy, K.; Yuille, A.L. Deeplab: Semantic image segmentation with deep convolutional nets, atrous convolution, and fully connected crfs. *IEEE Trans. Pattern Anal. Mach. Intell.* **2017**, *40*, 834–848. [CrossRef] [PubMed]
19. Ma, Z.; Qing, L.; Miao, J.; Chen, X. Advertisement evaluation using visual saliency based on foveated image. In Proceedings of the 2009 IEEE International Conference on Multimedia and Expo, Cancun, Mexico, 28 June–3 July 2009; pp. 914–917.
20. Borji, A.; Sihite, D.N.; Itti, L. Quantitative analysis of human-model agreement in visual saliency modeling: A comparative study. *IEEE Trans. Image Process.* **2012**, *22*, 55–69. [CrossRef]
21. Wang, W.; Lai, Q.; Fu, H.; Shen, J.; Ling, H. Salient object detection in the deep learning era: An in-depth survey. *arXiv* **2019**, arXiv:1904.09146.
22. Cong, R.; Lei, J.; Fu, H.; Cheng, M.M.; Lin, W.; Huang, Q. Review of visual saliency detection with comprehensive information. *IEEE Trans. Circuits Syst. Video Technol.* **2019**, *29*, 2941–2959. [CrossRef]

23. Zhang, D.; Fu, H.; Han, J.; Borji, A.; Li, X. A review of co-saliency detection algorithms: Fundamentals, applications, and challenges. *ACM Trans. Intell. Syst. Technol.* **2018**, *9*, 1–31. [CrossRef]
24. Riche, N.; Mancas, M. Bottom-up saliency models for videos: A practical review. In *From Human Attention to Computational Attention*; Springer: Berlin/Heidelberg, Germany, 2016; pp. 177–190.
25. Wang, T.; Piao, Y.; Li, X.; Zhang, L.; Lu, H. Deep learning for light field saliency detection. In Proceedings of the IEEE International Conference on Computer Vision, Seoul, Korea, 27 October–3 November 2019; pp. 8838–8848.
26. Hong, D.; Yokoya, N.; Chanussot, J.; Zhu, X.X. An augmented linear mixing model to address spectral variability for hyperspectral unmixing. *IEEE Trans. Image Process.* **2018**, *28*, 1923–1938. [CrossRef]
27. Wang, Q.; Lin, J.; Yuan, Y. Salient band selection for hyperspectral image classification via manifold ranking. *IEEE Trans. Neural Netw. Learn. Syst.* **2016**, *27*, 1279–1289. [CrossRef]
28. Bianco, S.; Buzzelli, M.; Schettini, R. A unifying representation for pixel-precise distance estimation. *Multimed. Tools Appl.* **2019**, *78*, 13767–13786. [CrossRef]
29. Battisti, F.; Carli, M. Depth-based saliency estimation for omnidirectional images. *Electron. Imaging* **2019**, *2019*, 271. [CrossRef]
30. Fu, H.; Xu, D.; Lin, S.; Liu, J. Object-based RGBD image co-segmentation with mutex constraint. In Proceedings of the IEEE Conference on Computer Vision and Pattern Recognition, Boston, MA, USA, 7–12 June 2015; pp. 4428–4436.
31. Fu, H.; Xu, D.; Lin, S. Object-based multiple foreground segmentation in RGBD video. *IEEE Trans. Image Process.* **2017**, *26*, 1418–1427. [CrossRef]
32. Gutierrez-Maldonado, J.; Gutierrez-Martinez, O.; Cabas-Hoyos, K. Interactive and passive virtual reality distraction: Effects on presence and pain intensity. *Stud. Health Technol. Inform.* **2011**, *167*, 69–73.
33. Sitzmann, V.; Serrano, A.; Pavel, A.; Agrawala, M.; Gutierrez, D.; Masia, B.; Wetzstein, G. Saliency in VR: How do people explore virtual environments? *IEEE Trans. Vis. Comput. Graph.* **2018**, *24*, 1633–1642. [CrossRef] [PubMed]
34. De Abreu, A.; Ozcinar, C.; Smolic, A. Look around you: Saliency maps for omnidirectional images in VR applications. In Proceedings of the 2017 Ninth International Conference on Quality of Multimedia Experience (QoMEX), Erfurt, Germany, 29 May–2 June 2017; pp. 1–6.
35. Assens Reina, M.; Giro-i Nieto, X.; McGuinness, K.; O'Connor, N.E. Saltinet: Scan-path prediction on 360 degree images using saliency volumes. In Proceedings of the IEEE International Conference on Computer Vision, Venice, Italy, 22–29 October 2017; pp. 2331–2338.
36. Cheng, H.T.; Chao, C.H.; Dong, J.D.; Wen, H.K.; Liu, T.L.; Sun, M. Cube padding for weakly-supervised saliency prediction in 360 videos. In Proceedings of the IEEE Conference on Computer Vision and Pattern Recognition, Salt Lake City, UT, USA, 18–22 June 2018; pp. 1420–1429.
37. Wang, W.; Shen, J.; Sun, H.; Shao, L. Video co-saliency guided co-segmentation. *IEEE Trans. Circuits Syst. Video Technol.* **2017**, *28*, 1727–1736. [CrossRef]
38. Liversedge, S.P.; Findlay, J.M. Saccadic eye movements and cognition. *Trends Cogn. Sci.* **2000**, *4*, 6–14. [CrossRef]
39. Assens, M.; Giro-i Nieto, X.; McGuinness, K.; O'Connor, N.E. PathGAN: Visual scanpath prediction with generative adversarial networks. In Proceedings of the European Conference on Computer Vision (ECCV), Munich, Germany, 8–14 September 2018.
40. Rai, Y.; Gutiérrez, J.; Le Callet, P. A dataset of head and eye movements for 360 degree images. In Proceedings of the 8th ACM on Multimedia Systems Conference, Taipei, Taiwan, 20–23 June 2017; ACM: New York, NY, USA, 2017; pp. 205–210.
41. Gutiérrez, J.; David, E.; Rai, Y.; Le Callet, P. Toolbox and dataset for the development of saliency and scanpath models for omnidirectional/360 still images. *Signal Process. Image Commun.* **2018**, *69*, 35–42. [CrossRef]
42. Rosenblatt, M. Remarks on Some Nonparametric Estimates of a Density Function. *Ann. Math. Stat.* **1956**, 832–837. [CrossRef]
43. Li, J.; Xia, C.; Chen, X. A benchmark dataset and saliency-guided stacked autoencoders for video-based salient object detection. *IEEE Trans. Image Process.* **2017**, *27*, 349–364. [CrossRef]
44. Li, Y.; Hou, X.; Koch, C.; Rehg, J.M.; Yuille, A.L. The secrets of salient object segmentation. In Proceedings of the IEEE Conference on Computer Vision and Pattern Recognition, Columbus, OH, USA, 23–28 June 2014; pp. 280–287.

45. Alpert, S.; Galun, M.; Brandt, A.; Basri, R. Image segmentation by probabilistic bottom-up aggregation and cue integration. *IEEE Trans. Pattern Anal. Mach. Intell.* **2011**, *34*, 315–327. [CrossRef]
46. Su, Y.C.; Jayaraman, D.; Grauman, K. Pano2Vid: Automatic Cinematography for Watching 360° Videos. In Proceedings of the Asian Conference on Computer Vision, Taipei, Taiwan, 20–24 November 2016; Springer: Berlin/Heidelberg, Germany, 2016; pp. 154–171.
47. Gupta, R.; Khanna, M.T.; Chaudhury, S. Visual saliency guided video compression algorithm. *Signal Process. Image Commun.* **2013**, *28*, 1006–1022. [CrossRef]
48. Mechrez, R.; Shechtman, E.; Zelnik-Manor, L. Saliency driven image manipulation. *Mach. Vis. Appl.* **2019**, *30*, 189–202. [CrossRef]
49. Fan, D.P.; Wang, W.; Cheng, M.M.; Shen, J. Shifting more attention to video salient object detection. In Proceedings of the IEEE Conference on Computer Vision and Pattern Recognition, Long Beach, CA, USA, 16–20 June 2019; pp. 8554–8564.
50. Riche, N.; Duvinage, M.; Mancas, M.; Gosselin, B.; Dutoit, T. Saliency and human fixations: State-of-the-art and study of comparison metrics. In Proceedings of the IEEE International Conference on Computer Vision, Sydney, Australia, 1–8 December 2013; pp. 1153–1160.
51. Bravais, A. *Analyse Mathématique sur les Probabilités des Erreurs de Situation D'un Point*; Impr. Royale: Paris, France, 1844.
52. Peters, R.J.; Iyer, A.; Itti, L.; Koch, C. Components of bottom-up gaze allocation in natural images. *Vis. Res.* **2005**, *45*, 2397–2416. [CrossRef] [PubMed]
53. Kullback, S.; Leibler, R.A. On information and sufficiency. *Ann. Math. Stat.* **1951**, *22*, 79–86. [CrossRef]
54. Judd, T.; Durand, F.; Torralba, A. A benchmark of computational models of saliency to predict human fixations. In *CSAIL Technical Reports (1 July 2003—Present)*; CSAIL: Cambridge, MA, USA, 2012.
55. Rubner, Y.; Tomasi, C.; Guibas, L.J. A metric for distributions with applications to image databases. In Proceedings of the Sixth International Conference on Computer Vision (IEEE Cat. No. 98CH36271), Bombay, India, 4–7 January 1998; pp. 59–66.
56. Chinchor, N. MUC-4 evaluation metrics. In Proceedings of the 4th Message Understanding Conference, McLean, Virginia, 16–18 June 1992; pp. 22–29.
57. Fan, D.P.; Cheng, M.M.; Liu, Y.; Li, T.; Borji, A. Structure-measure: A new way to evaluate foreground maps. In Proceedings of the IEEE International Conference on Computer Vision, Venice, Italy, 22–29 October 2017; pp. 4548–4557.
58. Fan, D.P.; Gong, C.; Cao, Y.; Ren, B.; Cheng, M.M.; Borji, A. Enhanced-alignment measure for binary foreground map evaluation. *arXiv* **2018**, arXiv:1805.10421.
59. Zhao, Q.; Koch, C. Learning a saliency map using fixated locations in natural scenes. *J. Vis.* **2011**, *11*, 9. [CrossRef] [PubMed]
60. Li, J.; Levine, M.D.; An, X.; Xu, X.; He, H. Visual saliency based on scale-space analysis in the frequency domain. *IEEE Trans. Pattern Anal. Mach. Intell.* **2012**, *35*, 996–1010. [CrossRef] [PubMed]
61. Monroy, R.; Lutz, S.; Chalasani, T.; Smolic, A. Salnet360: Saliency maps for omni-directional images with cnn. *Signal Process. Image Commun.* **2018**, *69*, 26–34. [CrossRef]
62. Ling, J.; Zhang, K.; Zhang, Y.; Yang, D.; Chen, Z. A saliency prediction model on 360 degree images using color dictionary based sparse representation. *Signal Process. Image Commun.* **2018**, *69*, 60–68. [CrossRef]
63. Lebreton, P.; Raake, A. GBVS360, BMS360, ProSal: Extending existing saliency prediction models from 2D to omnidirectional images. *Signal Process. Image Commun.* **2018**, *69*, 69–78. [CrossRef]
64. Fang, Y.; Zhang, X.; Imamoglu, N. A novel superpixel-based saliency detection model for 360-degree images. *Signal Process. Image Commun.* **2018**, *69*, 1–7. [CrossRef]
65. Min, K.; Corso, J.J. TASED-Net: Temporally-Aggregating Spatial Encoder-Decoder Network for Video Saliency Detection. In Proceedings of the IEEE International Conference on Computer Vision, Seoul, Korea, 27 October–2 November 2019; pp. 2394–2403.
66. Gorji, S.; Clark, J.J. Going from image to video saliency: Augmenting image salience with dynamic attentional push. In Proceedings of the IEEE Conference on Computer Vision and Pattern Recognition, Salt Lake City, UT, USA, 18–22 June 2018; pp. 7501–7511.
67. Zhang, K.; Li, T.; Liu, B.; Liu, Q. Co-Saliency Detection via Mask-Guided Fully Convolutional Networks With Multi-Scale Label Smoothing. In Proceedings of the IEEE Conference on Computer Vision and Pattern Recognition, Long Beach, CA, USA, 16–20 June 2019; pp. 3095–3104.

68. Tsai, C.C.; Li, W.; Hsu, K.J.; Qian, X.; Lin, Y.Y. Image co-saliency detection and co-segmentation via progressive joint optimization. *IEEE Trans. Image Process.* **2018**, *28*, 56–71. [CrossRef]
69. Jeong, D.j.; Hwang, I.; Cho, N.I. Co-salient object detection based on deep saliency networks and seed propagation over an integrated graph. *IEEE Trans. Image Process.* **2018**, *27*, 5866–5879. [CrossRef]
70. Wei, L.; Zhao, S.; Bourahla, O.E.F.; Li, X.; Wu, F. Group-wise deep co-saliency detection. *arXiv* **2017**, arXiv:1707.07381.
71. Gauss, C.F. *Disquisitiones Generales Circa Superficies Curvas*; ITypis Dieterichianis: 1828; Volume 1. Available online: https://www.sophiararebooks.com/pages/books/4602/carl-friedrich-gauss/disquisitiones-generales-circa-superficies-curvas (accessed on 3 February 2020).
72. Pressley, A. Gauss' Theorema Egregium. In *Elementary Differential Geometry*; Springer: Berlin/Heidelberg, Germany, 2010; pp. 247–268.
73. Maugey, T.; Le Meur, O.; Liu, Z. Saliency-based navigation in omnidirectional image. In Proceedings of the 2017 IEEE 19th International Workshop on Multimedia Signal Processing (MMSP), Luton, UK, 16–18 October 2017; pp. 1–6.
74. Chatfield, K.; Simonyan, K.; Vedaldi, A.; Zisserman, A. Return of the devil in the details: Delving deep into convolutional nets. *arXiv* **2014**, arXiv:1405.3531.
75. Simonyan, K.; Zisserman, A. Very deep convolutional networks for large-scale image recognition. *arXiv* **2014**, arXiv:1409.1556g.
76. He, K.; Zhang, X.; Ren, S.; Sun, J. Deep residual learning for image recognition. In Proceedings of the IEEE Conference on Computer Vision and Pattern Recognition, Las Vegas, NV, USA, 26 June–1 July 2016; pp. 770–778.
77. Xu, P.; Ehinger, K.A.; Zhang, Y.; Finkelstein, A.; Kulkarni, S.R.; Xiao, J. Turkergaze: Crowdsourcing saliency with webcam based eye tracking. *arXiv* **2015**, arXiv:1504.06755.
78. Xiao, J.; Ehinger, K.A.; Oliva, A.; Torralba, A. Recognizing scene viewpoint using panoramic place representation. In Proceedings of the 2012 IEEE Conference on Computer Vision and Pattern Recognition, Providence, RI, USA, 16–21 June 2012; pp. 2695–2702.
79. Training Dataset | Salient360!—Visual Attention Modeling for 360° Content. Available online: https://salient360.ls2n.fr/datasets/training-dataset/ (accessed on 3 February 2020).
80. Saliency in VR: How Do People Explore Virtual Environments? Available online: https://vsitzmann.github.io/vr-saliency/ (accessed on 3 February 2020).
81. Cube Padding for Weakly-Supervised Saliency Prediction in 360° Videos. Available online: http://aliensunmin.github.io/project/360saliency/ (accessed on 3 February 2020).
82. anadeabreu/Testbed_Database. Available online: https://github.com/anadeabreu/Testbed_Database (accessed on 3 February 2020).
83. Jarodzka, H.; Holmqvist, K.; Nyström, M. A vector-based, multidimensional scanpath similarity measure. In Proceedings of the 2010 Symposium on Eye-Tracking Research & Applications, Austin, TX, USA, 22–24 March 2010; pp. 211–218.
84. Russakovsky, O.; Deng, J.; Su, H.; Krause, J.; Satheesh, S.; Ma, S.; Huang, Z.; Karpathy, A.; Khosla, A.; Bernstein, M.; et al. Imagenet large scale visual recognition challenge. *Int. J. Comput. Vis.* **2015**, *115*, 211–252.
85. Toshev, A.; Shi, J.; Daniilidis, K. Image matching via saliency region correspondences. In Proceedings of the 2007 IEEE Conference on Computer Vision and Pattern Recognition. IEEE, Minneapolis, MN, USA, 17–22 June 2007; pp. 1–8. [CrossRef]
86. Winn, J.; Criminisi, A.; Minka, T. Object categorization by learned universal visual dictionary. In *Tenth IEEE International Conference on Computer Vision (ICCV'05) Volume 1*; IEEE: Piscataway, NJ, USA, 2005; Volume 2, pp. 1800–1807.
87. Zhang, D.; Han, J.; Li, C.; Wang, J.; Li, X. Detection of co-salient objects by looking deep and wide. *Int. J. Comput. Vis.* **2016**, *120*, 215–232.
88. Dai, J.; Nian Wu, Y.; Zhou, J.; Zhu, S.C. Cosegmentation and cosketch by unsupervised learning. In Proceedings of the IEEE International Conference on Computer Vision, Sydney, Australia, 1–8 December 2013; pp. 1305–1312. [CrossRef]
89. Yao, X.; Han, J.; Zhang, D.; Nie, F. Revisiting co-saliency detection: A novel approach based on two-stage multi-view spectral rotation co-clustering. *IEEE Trans. Image Process.* **2017**, *26*, 3196–3209.

90. Rother, C.; Minka, T.; Blake, A.; Kolmogorov, V. Cosegmentation of image pairs by histogram matching-incorporating a global constraint into mrfs. In Proceedings of the 2006 IEEE Computer Society Conference on Computer Vision and Pattern Recognition (CVPR'06), New York, NY, USA, 17–22 June 2006; Volume 1, pp. 993–1000. [CrossRef]
91. Tang, K.; Joulin, A.; Li, L.J.; Fei-Fei, L. Co-localization in real-world images. In Proceedings of the IEEE Conference on Computer Vision and Pattern Recognition, Columbus, OH, USA, 24–27 June 2014; pp. 1464–1471.
92. Jerripothula, K.R.; Cai, J.; Yuan, J. Quality-guided fusion-based co-saliency estimation for image co-segmentation and colocalization. *IEEE Trans. Multimed.* **2018**, *20*, 2466–2477.
93. Cong, R.; Lei, J.; Fu, H.; Huang, Q.; Cao, X.; Ling, N. HSCS: Hierarchical sparsity based co-saliency detection for RGBD images. *IEEE Trans. Multimed.* **2018**, *21*, 1660–1671. [CrossRef]
94. Hsu, K.J.; Tsai, C.C.; Lin, Y.Y.; Qian, X.; Chuang, Y.Y. Unsupervised CNN-based co-saliency detection with graphical optimization. In Proceedings of the European Conference on Computer Vision (ECCV), Munich, Germany, 8–14 September 2018; pp. 485–501. [CrossRef]
95. Zheng, X.; Zha, Z.J.; Zhuang, L. A feature-adaptive semi-supervised framework for co-saliency detection. In Proceedings of the 2018 ACM Multimedia Conference on Multimedia Conference, Seoul, Korea, 22–26 October 2018; ACM: New York, NY, USA, 2018; pp. 959–966.
96. Cong, R.; Lei, J.; Fu, H.; Huang, Q.; Cao, X.; Hou, C. Co-saliency detection for RGBD images based on multi-constraint feature matching and cross label propagation. *IEEE Trans. Image Process.* **2018**, *27*, 568–579.
97. Long, J.; Shelhamer, E.; Darrell, T. Fully convolutional networks for semantic segmentation. In Proceedings of the IEEE Conference on Computer Vision and Pattern Recognition, Boston, MA, USA, 7–12 June 2015; pp. 3431–3440. [CrossRef] [PubMed]
98. Batra, D.; Kowdle, A.; Parikh, D.; Luo, J.; Chen, T. icoseg: Interactive co-segmentation with intelligent scribble guidance. In Proceedings of the 2010 IEEE Computer Society Conference on Computer Vision and Pattern Recognition, San Francisco, CA, USA, 13–18 June 2010; pp. 3169–3176.
99. Lai, K.; Bo, L.; Ren, X.; Fox, D. A large-scale hierarchical multi-view rgb-d object dataset. In Proceedings of the 2011 IEEE international conference on robotics and automation, Shanghai, China, 9–13 May 2011; pp. 1817–1824.
100. Ju, R.; Liu, Y.; Ren, T.; Ge, L.; Wu, G. Depth-aware salient object detection using anisotropic center-surround difference. *Signal Process. Image Commun.* **2015**, *38*, 115–126.
101. Sun, D.; Roth, S.; Black, M.J. Secrets of optical flow estimation and their principles. In Proceedings of the 2010 IEEE Computer Society Conference on Computer Vision and Pattern Recognition, San Francisco, CA, USA, 13–18 June 2010; pp. 2432–2439. [CrossRef]
102. Rubinstein, M.; Joulin, A.; Kopf, J.; Liu, C. Unsupervised joint object discovery and segmentation in internet images. In Proceedings of the IEEE conference on computer vision and pattern recognition, Portland, OR, USA, 23–28 June 2013; pp. 1939–1946.
103. Li, H.; Ngan, K.N. A co-saliency model of image pairs. *IEEE Trans. Image Process.* **2011**, *20*, 3365–3375.
104. Zhang, D.; Javed, O.; Shah, M. Video object co-segmentation by regulated maximum weight cliques. In Proceedings of the European Conference on Computer Vision, Zurich, Switzerland, 6–12 September 2014; Springer: Berlin/Heidelberg, Germany, 2014; pp. 551–566. [CrossRef] [PubMed]
105. RGBD Segmentation. Available online: http://hzfu.github.io/proj_rgbdseg.html (accessed on 3 February 2020).
106. Runmin Cong. Available online: https://rmcong.github.io/proj_RGBD_cosal.html (accessed on 3 February 2020).
107. Advanced Multimedia Processing (AMP) Lab, Cornell University. Available online: http://chenlab.ece.cornell.edu/projects/touch-coseg/ (accessed on 3 February 2020).
108. Image Understanding-Microsoft Research. Available online: https://www.microsoft.com/en-us/research/project/image-understanding/#!downloads (accessed on 3 February 2020).
109. Co-Saliency Database: Cosal2015-Junwei Han. Available online: http://www.escience.cn/people/JunweiHan/Co-saliency.html (accessed on 3 February 2020).
110. Cosegmentation and Cosketch by Unsupervised Learning. Available online: http://www.stat.ucla.edu/~jifengdai/research/CosegmentationCosketch.html (accessed on 18 November 2020).
111. Unsupervised Joint Object Discovery and Segmentation in Internet Images. Available online: http://people.csail.mit.edu/mrub/ObjectDiscovery/ (accessed on 3 February 2020).
112. Image-Pair. Available online: http://ivipc.uestc.edu.cn/project/cosaliency/ (accessed on 19 May 2011).

113. CRCV | Center for Research in Computer Vision at the University of Central Florida. Available online: https://www.crcv.ucf.edu/projects/video_object_cosegmentation/#Safari (accessed on 3 February 2020).
114. shenjianbing/vicosegment: Dataset for 'Video Co-saliency Guided Co-segmentation' (T-CSVT18). Available online: https://github.com/shenjianbing/vicosegment (accessed on 3 February 2020).
115. Prest, A.; Leistner, C.; Civera, J.; Schmid, C.; Ferrari, V. Learning object class detectors from weakly annotated video. In Proceedings of the 2012 IEEE Conference on Computer Vision and Pattern Recognition, Providence, RI, USA, 16–21 June 2012; pp. 3282–3289.
116. Russell, B.C.; Torralba, A.; Murphy, K.P.; Freeman, W.T. LabelMe: A database and web-based tool for image annotation. *Int. J. Comput. Vis.* **2008**, *77*, 157–173.
117. Chiu, W.C.; Fritz, M. Multi-class video co-segmentation with a generative multi-video model. In Proceedings of the IEEE Conference on Computer Vision and Pattern Recognition, Portland, OR, USA, 23–28 June 2013; pp. 321–328. [CrossRef]
118. Hochbaum, D.S.; Singh, V. An efficient algorithm for co-segmentation. In Proceedings of the 2009 IEEE 12th International Conference on Computer Vision, Kyoto, Japan, 29 September–2 October 2009; pp. 269–276.
119. Griffin, G.; Holub, A.; Perona, P. *Caltech-256 Object Category Dataset*; CalTech Report; CalTech: Pasadena, CA, USA, 2007.
120. Everingham, M.; Van Gool, L.; Williams, C.K.; Winn, J.; Zisserman, A. The pascal visual object classes (voc) challenge. *Int. J. Comput. Vis.* **2010**, *88*, 303–338.
121. Ullah, J.; Khan, A.; Jaffar, M.A. Motion cues and saliency based unconstrained video segmentation. *Multimed. Tools Appl.* **2018**, *77*, 7429–7446. [CrossRef]
122. Ochs, P.; Malik, J.; Brox, T. Segmentation of moving objects by long term video analysis. *IEEE Trans. Pattern Anal. Mach. Intell.* **2013**, *36*, 1187–1200. [CrossRef]
123. Li, F.; Kim, T.; Humayun, A.; Tsai, D.; Rehg, J.M. Video segmentation by tracking many figure-ground segments. In Proceedings of the IEEE International Conference on Computer Vision, Sydney, Australia, 1–8 December 2013; pp. 2192–2199. [CrossRef] [PubMed]
124. Perazzi, F.; Pont-Tuset, J.; McWilliams, B.; Van Gool, L.; Gross, M.; Sorkine-Hornung, A. A benchmark dataset and evaluation methodology for video object segmentation. In Proceedings of the IEEE Conference on Computer Vision and Pattern Recognition, Las Vegas, NV, USA, 26 June–1 July 2016; pp. 724–732.
125. Hu, Y.T.; Huang, J.B.; Schwing, A.G. Unsupervised video object segmentation using motion saliency-guided spatio-temporal propagation. In Proceedings of the European Conference on Computer Vision (ECCV), Munich, Germany, 8–14 September 2018; pp. 786–802.
126. Wang, W.; Shen, J.; Yang, R.; Porikli, F. Saliency-aware video object segmentation. *IEEE Trans. Pattern Anal. Mach. Intell.* **2017**, *40*, 20–33.
127. Li, H.; Chen, G.; Li, G.; Yu, Y. Motion Guided Attention for Video Salient Object Detection. In Proceedings of the IEEE International Conference on Computer Vision, Seoul, Korea, 27 October–2 November 2019; pp. 7274–7283. [CrossRef]
128. Yan, P.; Li, G.; Xie, Y.; Li, Z.; Wang, C.; Chen, T.; Lin, L. Semi-Supervised Video Salient Object Detection Using Pseudo-Labels. In Proceedings of the IEEE International Conference on Computer Vision, Seoul, Korea, 27 October–2 November 2019; pp. 7284–7293.
129. Cong, R.; Lei, J.; Fu, H.; Porikli, F.; Huang, Q.; Hou, C. Video saliency detection via sparsity-based reconstruction and propagation. *IEEE Trans. Image Process.* **2019**, *28*, 4819–4831.
130. Zhou, X.; Liu, Z.; Gong, C.; Liu, W. Improving video saliency detection via localized estimation and spatiotemporal refinement. *IEEE Trans. Multimed.* **2018**, *20*, 2993–3007. [CrossRef] [PubMed]
131. Chen, C.; Li, S.; Wang, Y.; Qin, H.; Hao, A. Video saliency detection via spatial-temporal fusion and low-rank coherency diffusion. *IEEE Trans. Image Process.* **2017**, *26*, 3156–3170. [CrossRef]
132. Xie, S.; Sun, C.; Huang, J.; Tu, Z.; Murphy, K. Rethinking spatiotemporal feature learning: Speed-accuracy trade-offs in video classification. In Proceedings of the European Conference on Computer Vision (ECCV), Munich, Germany, 8–14 September 2018; pp. 305–321. [CrossRef] [PubMed]
133. Szegedy, C.; Liu, W.; Jia, Y.; Sermanet, P.; Reed, S.; Anguelov, D.; Erhan, D.; Vanhoucke, V.; Rabinovich, A. Going deeper with convolutions. In Proceedings of the IEEE Conference on Computer Vision and Pattern Recognition, Boston, MA, USA, 7–12 June 2015; pp. 1–9.
134. Mathe, S.; Sminchisescu, C. Actions in the eye: Dynamic gaze datasets and learnt saliency models for visual recognition. *IEEE Trans. Pattern Anal. Mach. Intell.* **2014**, *37*, 1408–1424.

135. Wang, W.; Shen, J.; Guo, F.; Cheng, M.M.; Borji, A. Revisiting video saliency: A large-scale benchmark and a new model. In Proceedings of the IEEE Conference on Computer Vision and Pattern Recognition, Salt Lake City, UT, USA, 18–22 June 2018; pp. 4894–4903. [CrossRef] [PubMed]
136. Brox, T.; Malik, J. Object segmentation by long term analysis of point trajectories. In Proceedings of the European conference on Computer Vision, Hersonissos, Crete, Greece, 5–11 September 2010; Springer: Berlin/Heidelberg, Germany, 2010; pp. 282–295.
137. Tron, R.; Vidal, R. A benchmark for the comparison of 3-d motion segmentation algorithms. In Proceedings of the 2007 IEEE Conference on Computer Vision and Pattern Recognition, Minneapolis, MN, USA, 17–22 June 2007; pp. 1–8.
138. Tsai, D.; Flagg, M.; Nakazawa, A.; Rehg, J.M. Motion coherent tracking using multi-label MRF optimization. *Int. J. Comput. Vis.* **2012**, *100*, 190–202.
139. Wang, W.; Shen, J.; Shao, L. Consistent video saliency using local gradient flow optimization and global refinement. *IEEE Trans. Image Process.* **2015**, *24*, 4185–4196. [CrossRef]
140. Kim, H.; Kim, Y.; Sim, J.Y.; Kim, C.S. Spatiotemporal saliency detection for video sequences based on random walk with restart. *IEEE Trans. Image Process.* **2015**, *24*, 2552–2564. [CrossRef] [PubMed]
141. Liu, Z.; Li, J.; Ye, L.; Sun, G.; Shen, L. Saliency detection for unconstrained videos using superpixel-level graph and spatiotemporal propagation. *IEEE Trans. Circuits Syst. Video Technol.* **2016**, *27*, 2527–2542. [CrossRef]
142. Grundmann, M.; Kwatra, V.; Han, M.; Essa, I. Efficient hierarchical graph-based video segmentation. In Proceedings of the 2010 IEEE Computer Society Conference on Computer Vision and Pattern Recognition, San Francisco, CA, USA, 13–18 June 2010; pp. 2141–2148. [CrossRef]
143. Toyama, K.; Krumm, J.; Brumitt, B.; Meyers, B. Wallflower: Principles and practice of background maintenance. In Proceedings of the Seventh IEEE International Conference on Computer Vision, Kerkyra, Greece, 20–27 September 1999; Volume 1, pp. 255–261.
144. Fukuchi, K.; Miyazato, K.; Kimura, A.; Takagi, S.; Yamato, J. Saliency-based video segmentation with graph cuts and sequentially updated priors. In Proceedings of the 2009 IEEE International Conference on Multimedia and Expo, Cancun, Mexico, 28 June–3 July 2009; pp. 638–641.
145. Rodriguez, M.D.; Ahmed, J.; Shah, M. Action mach a spatio-temporal maximum average correlation height filter for action recognition. In Proceedings of the 2008 IEEE Conference on Computer Vision and Pattern Recognition, Anchorage, Alaska, 24–26 June 2008; pp. 1–8.
146. Marszalek, M.; Laptev, I.; Schmid, C. Actions in context. In Proceedings of the 2009 IEEE Conference on Computer Vision and Pattern Recognition, Miami, FL, USA, 20–26 June 2009; pp. 2929–2936.
147. Mital, P.K.; Smith, T.J.; Hill, R.L.; Henderson, J.M. Clustering of gaze during dynamic scene viewing is predicted by motion. *Cogn. Comput.* **2011**, *3*, 5–24.
148. Shifting More Attention to Video Salient Object Detection—Media Computing Lab. Available online: http://mmcheng.net/DAVSOD/ (accessed on 11 February 2020). [CrossRef]
149. Computer Vision Group, Freiburg. Available online: https://lmb.informatik.uni-freiburg.de/resources/datasets/ (accessed on 11 February 2020).
150. SegTrack v2 Dataset. Available online: https://web.engr.oregonstate.edu/~lif/SegTrack2/dataset.html (accessed on 11 February 2020).
151. Project VOS (IEEE TIP 2018). Available online: http://cvteam.net/projects/TIP18-VOS/VOS.html (accessed on 11 February 2020).
152. fperazzi/davis: Package Containing Helper Functions for Loading and Evaluating DAVIS. Available online: https://github.com/fperazzi/davis (accessed on 11 February 2020).
153. Actions in the Eye: Human Eye Movement Datasets. Available online: http://vision.imar.ro/eyetracking/description.php (accessed on 11 February 2020).

154. wenguanwang/DHF1K: Revisiting Video Saliency: A Large-scale Benchmark and a New Model (CVPR18, PAMI19). Available online: https://github.com/wenguanwang/DHF1K (accessed on 11 February 2020).
155. Belongie, S.; Malik, J.; Puzicha, J. Shape matching and object recognition using shape contexts. *IEEE Trans. Pattern Anal. Mach. Intell.* **2002**, *24*, 509–522.

 © 2020 by the author. Licensee MDPI, Basel, Switzerland. This article is an open access article distributed under the terms and conditions of the Creative Commons Attribution (CC BY) license (http://creativecommons.org/licenses/by/4.0/).

MDPI
St. Alban-Anlage 66
4052 Basel
Switzerland
Tel. +41 61 683 77 34
Fax +41 61 302 89 18
www.mdpi.com

Applied Sciences Editorial Office
E-mail: applsci@mdpi.com
www.mdpi.com/journal/applsci

www.ingramcontent.com/pod-product-compliance
Lightning Source LLC
LaVergne TN
LVHW070137100526
838202LV00015B/1841